Understanding Radar

Understanding Radar

Second Edition

Henry W. Cole

OXFORD
BLACKWELL SCIENTIFIC PUBLICATIONS
LONDON EDINBURGH BOSTON
MELBOURNE PARIS BERLIN VIENNA

Blackwell Scientific Publications
Editorial Offices:
Osney Mead, Oxford OX2 0EL
25 John Street, London WC1N 2BL
23 Ainslie Place, Edinburgh EH3 6AJ
238 Main Street, Cambridge,
Massachusetts 02142, USA
54 University Street, Carlton
Victoria 3053, Australia

Other Editorial Offices:
Librairie Arnette SA
2, rue Casimir-Delavigne
75006 Paris
France

Blackwell Wissenschafts-Verlag
Meinekestrasse 4
D-1000 Berlin 15
Germany

Blackwell MZV
Feldgasse 13
A-1238 Wien
Austria

First edition published by
Collins Professional and Technical Books 1985
Reprinted by BSP Professional Books 1988
Second edition published by Blackwell
Scientific Publications 1992

Set by DP Photosetting, Aylesbury, Bucks
Printed and bound in Great Britain by
Hartnolls Ltd, Bodmin, Cornwall

DISTRIBUTORS

Marston Book Services Ltd
PO Box 87
Oxford OX2 0DT
(*Orders:* Tel: 0865 791155
Fax: 0865 791927
Telex: 837515)

USA
Blackwell Scientific Publications, Inc.
238 Main Street
Cambridge, MA 02142
(*Orders:* Tel: 800 759-6102
617 225-0401)

Canada
Oxford University Press
70 Wynford Drive
Don Mills
Ontario M3C 1J9
(*Orders:* Tel: 416 441-2941)

Australia
Blackwell Scientific Publications
(Australia) Pty Ltd
54 University Street
Carlton, Victoria 3053
(*Orders:* Tel: 03 347-0300)

British Library
Cataloguing in Publication Data
A Catalogue record for this book is
available from the British Library

ISBN 0–632–03124–7

Library of Congress
Cataloguing in Publication Data
Cole, Henry W.
Understanding radar/Henry W. Cole.—
2nd ed. p. cm.
Includes bibliographical references and index.
ISBN 0–632–03124–7
1. Radar. I. Title.
TK6575.C62 1992
621.3848—dc20 92-10157
CIP

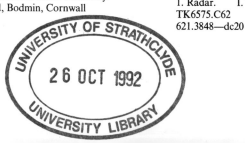

Contents

Preface to the Second Edition

This second edition of *Understanding Radar* follows the same format and style as the First Edition. It now incorporates descriptions of new and important techniques emergent since the book was first written. Among these are, synthetic aperture radars of mind-blowing resolution, phased array antennas, over-the-horizon radars of fantastic range, and new display systems, together with the new International Aviation communications concept – the Aeronautical Telecommunications Network (ATN). Most of these new systems have come into being largely because of the enormous power and speed of modern computers and digital processors.

Not all radars are modern and they have an extremely long service life. The fundamental principles of primary and secondary radar also remain the same. Thus, much of the original text will be relevant for many decades. For this reason it has been retained and the opportunity for its revision has been taken. The new topics have been selected from a very extensive candidate list as embodying the essence of new technologies which will be found in a wide variety of specific modern systems. To have included all on the list would have been outside the compass of my mind and the available writing time. However, I entertain the hope that this new edition will go further down the road to its aim – the better understanding of radar by its users.

At the time of publication of the First Edition I called attention to some of the reasons why radar users were so seldom consulted or listened to when equipment and systems for their use were being planned or purchased. A prime reason was the presumption on the part of planning or procurement teams that radar users were too innocent of technical matters to understand the issues at stake. Thus, the prime motivation and aim of the book was to give users of radar a better insight into its techniques, and the confidence and encouragement to enquire further. This objective has been retained in this new edition.

Judging from my own experience of reactions to the First Edition, the work has gone some way down the road towards achieving its aim; many radar users – people not given to sycophancy, or politeness for its own sake – have commented favourably to me along these lines. I have also noted an increase in the number of occasions where users have been co-opted into radar planning and procurement teams, remaining important voices throughout the long process of specifying, designing, assessing and procuring radar systems for their use. Future radars will almost certainly employ one or another of the new techniques described in this Second Edition and I hope it continues its useful purpose of making clearer the seeming mysteries of modern radar technology.

Henry W. Cole

Acknowledgements

I wrote the First Edition of this book with the great encouragement of Mr Roy Simons – at that time Technical Director of Marconi Radar, and once again gratefully acknowledge his faith in my ability to accomplish the work. Equally, am I grateful to our current Technical Director, Mr John Mark FIEE, who has echoed the encouragement of his predecessor in this new work. As before, all the new material in this Second Edition has been read by specialist colleagues to ensure that it will not mislead readers – again my grateful thanks are due to them for their comments and suggestions. I also express grateful thanks to Mr J. Colston, Director of GEC-Marconi Research Centre, and to other manufacturers for kind permission to reproduce their photographic material.

A work such as this demands much revision and changes of text. Despite the miracles of word processors, someone has to decipher my scrawl, to type it into the machine and see it properly printed. My grateful thanks for her speed, diligence and above all, patience, go to Sylvia Simpson for attending these needs. Equally is it true of modern-day computer-generated drawings. For these my thanks go to Andy Emberton for their preparation and layout – its nice to get the boss working for you!

Finally, with the promise that it will not happen again, I thank my family who so often have taken second place during the book's preparation.

List of Abbreviations and Terms

a (A)	amperes
ac	alternating current
a/c	aircraft
ACP	azimuth change pulse
ACR	airfield control radar
ADLP	airborne data link processor
Adsel*	'address selective' SSR system (UK term)
a.f.c.	automatic frequency control
a.m.	amplitude modulation
Anoprop	anomalous propagation
ASDE	airfield surface detection equipment
ASMI	airfield surface movement indicator
ASP	adaptive signal processor
ASR	airfield surveillance radar
a.t.c.	air traffic control
ATM	air traffic movement
ATN	aeronautical telecommunications network
ATR	anti-transmit/receive
az	azimuth
CFAR	constant false alarm rate
CFO	cross field oscillator
CH	'chain home' (the UK's WW2 defence radar)
coho	coherent oscillator
CRT	cathode ray tube
CW	continuous wave
DABS*	direct addressed beacon system (US term)
dB	decibel
DME	distance measuring equipment
D.P.S.K.	differential phase shift keying (modulation method)

* Both these have been rationalised in the new Mode 'S' system

D.S.P.	digital signal processor
e.h.t.	extra high tension (voltage)
E–M	electro-magnetic
e.m.f.	electromotive force
E.R.R.	en-route radar
ESA	electronically scanned array
f	frequency
FFT	fast Fourier transform
FIR	flight information region
f.m.	frequency modulation
FMCW	frequency modulated carrier wave
FTC	fast time constant (see also STC)
GDLP	ground data link processor
GTC	gain-time control (see also STC)
HPD	horizontal polar diagram
Ht	height
Hz	Hertz
i.f.	intermediate frequency
IFF	identification, friend or foe (synonymous with SSR)
IISLS	improved ISLS
ISAR	inverse synthetic aperture radar
I^2SLS	another common notation for IISLS
$I_{(n)}$	a sub-factor of $I_{(total)}$
ISLS	interrogation sidelobe suppression
ISO	International Standards Organisation
$I_{(total)}$	improvement factor
K	constant
K	1000 times (e.g. Kft = 1000s of feet)
Km	Kilometres
Kts	knots
LO	local oscillator
LTA	local tracking average
LVA	large vertical aperture
m	metres
M	Mega (10^6)
MHT	multi-head tracking
MMI	man machine interface
m.sec	milli-seconds
mtd	moving target detector
mti	moving target indicator
n.mls	nautical miles (6080 ft)
n. sec	nano-second (10^{-9} seconds)
OSI	open systems interconnection (ISO-defined)
OTH	over the horizon (radar)

P_d	probability of detection
P_{fa}	probability of false alarm
ppi	plan position indicator
prf	pulse repetition frequency
PSD	phase sensitivie detector
r.f.	radiated frequency
RSLS	receiver sidelobe suppression
Rx	receiver
SAR	synthetic aperture radar
S.C.V.	sub-clutter visibility
SPI	special position identification
SSR	secondary surveillance radar (see also IFF)
stalo	stable local oscillator
STC	sensitivity-time control (see also GTC)
S.T.C.	short time constant (see FTC)
Super CV	super-clutter visibility
Tacan	tactical area navigation
TAR	terminal area radar
Target	'The object of interest' from the radar point of view. The interest is usually benign.
TMA	terminal manoeuvring area
TMHT	true multi-head tracking
TR	transmit/receive
TVW	time varying weights
TWT	travelling wave tube
TV	television
Tx	transmitter
U.H.F.	ultra high frequency
VPD	vertical polar diagram
v (V)	volts
μ	micro (10^{-6})

* Both these have been rationalised in the new Mode 'S' system

Part 1

Matters Common to Primary and Secondary Radar

1
Fundamentals

1.1 Introduction

This work has two major components: theoretical and practical. Both are subject to change. That the practical scene changes is easy to appreciate because technology is changing – ever faster. It is not so easy to see why theory should change. The realisation rests in remembering the scientific method which is:

(a) Curiosity about effects and their causes.
(b) Construction of an hypothesis to explain their relationship.
(c) Generation of theory supporting the hypothesis.
(d) Testing the theory by controlled experiment.
(e) Modifying the theory if experiment does not confirm it.

Theories are tools by which the engineer produces machines to answer people's needs. The better the theories and the engineer's understanding of them, the better do they serve people.

But theories remain theories and nobody yet has produced any that are in themselves complete and truly universal. Physicists tell us that Newton's theories are very serviceable until masses move close to the speed of light. Atomic scientists seeking the truth in physics have the need to invoke as yet unmeasurable qualities of particles such as 'charm' and 'colour', 'weak' and 'strong' forces.

In the radar field also, theory is incomplete. Those theories concerning propagation, the analysis of targets and clutter, weather detection, human appreciation of data are being continually refined.

It might seem then that any work such as this will be outdated before it is printed. Fortunately conditions are much more stable because both modern theory and practice produce ideas and equipment which are serviceable for decades. The users will therefore, it is hoped, benefit by what is presented here. If their equipment is ten years old and destined for another ten years' use or for replacement by a more up-to-date system, then in these pages I hope to present matter of value to them.

The use of mathematical expression has been limited, favouring pictures and words where they serve best. But there are occasions where the user will be put on better terms with his technical supporters by understanding the mathematics of certain aspects. Here, the equations receive more explanation than generally found. Those readers for whom this is unnecessary will be wise enough to recognise that there are others for whom it is helpful.

1.1.1 An important basic concept

Many elements of radar system depend upon signals having the property of *coherence*. It is important enough for me to risk the accusation of being too simplistic in making sure the idea is understood clearly. Coherence rests in two other related properties – frequency and phase.

1.2 Frequency and phase

Imagine your watch (an old-fashioned one with hands) is absolutely accurate. Once every hour the minute hand will execute one revolution at a very regular rate: its frequency of rotation is one cycle per hour. By international agreement 'frequency' is expressed as the number of times per second a regular event occurs. So one cycle per hour is expressed as a frequency of 1/3600 Hz. 'Hz' is a contraction of the name of Heinrich Hertz, the famous German phyicist of the nineteenth century. Imagine now that you have two such watches, one on each wrist, and both telling the same time. The minute hands will rotate at the same rate, i.e. at the same frequency. Because their angle relative to 12 o'clock will always be equal, they are said to be 'in phase'. This is illustrated in fig. 1.1(a) and (b).

Now set one of the watches half an hour ahead (180° angle difference between the two minute hands relative to 12 o'clock). The minute hands of the two accurate watches will still rotate at the same rate, i.e. their frequency of rotation will be the same, *but* the angle between each hand will always be 180° different, i.e. they are 180° 'out of phase', the circumstance shown in figs. 1.1(c) and (d).

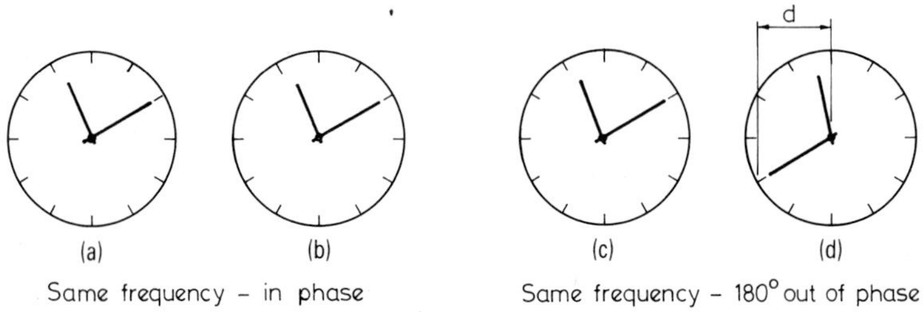

Fig. 1.1. Frequency and phase relationships

1.2.1 Translation into signal terms

In fig. 1.1(d) the distance between 12 o'clock position and the tip of the minute hand is represented as 'd'. As time passes, the history of d will be as shown in fig. 1.2. This is a graph of the familiar sine function and characterises the most simple form of signals in a continuous wave (CW) system. The amplitude of such a signal is analogous to the values of d; the frequency is analogous to the time to execute one cycle of values of d – i.e. one cycle per hour or 1/3600 Hz in this case.

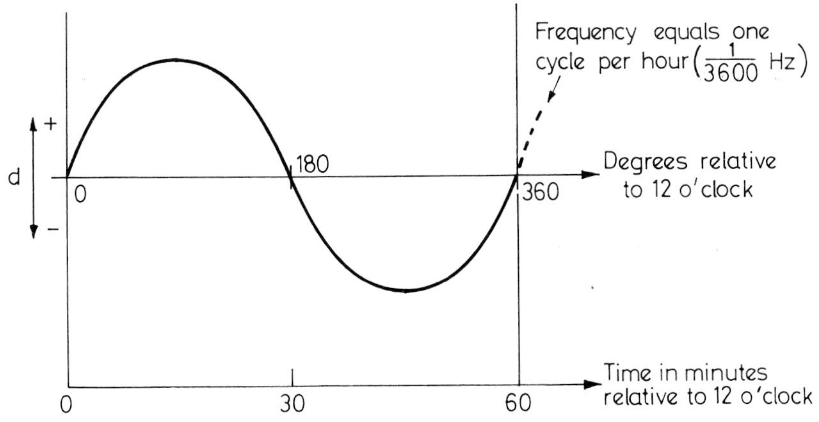

Fig. 1.2. Horizontal movement (d) of the end of the minute hand in fig. 1.1(d) relative to 12 o'clock.

Of course radar engineering involves much higher frequencies – many millions of cycles per second – but there is no difference in principle, only difference of numbers.

1.2.2 Coherence

Most readers will be clear about the term 'synchronised': a series of events accurately and continuously related in time. An example would be the dipping of oars of a boat-race crew or, better still, the high-kicking legs of dancers in a crack chorus line. Coherence is a special case of synchrony. Imagine the chorus line stretching far away, out of sight. If all the dancers immaculately performed the same routine, knowing the steps of the first (each movement of foot, leg and body in the continuum of time) would enable one to say that they applied to the last in line. The common knowledge of the routine and its execution, shared by all the dancers, confers 'coherence' upon them as a system; they are all moving 'in phase' with each other.

Another model would be a ruler of infinite length whose bench mark or zero was fixed on the ground upon which the ruler was laid and whose gradations were at 1 m intervals. Imagine further that somewhere along its length an irregular section was removed. Any distance which was an integral number of metres apart would still be indicated as such by the ruler, even if the distance included its removed section. We know this because the ruler with its fixed 1 m intervals and fixed reference of zero on the ground forms a 'coherent' system. In these two models we see that their coherence has to do with knowledge of behaviour in a continuum by use of a reference – the first is the continuum of time, the second, that of space. The first is self-referenced (i.e. each of the dancers knows the routine and acts 'in phase') and the system can be called 'self-coherent'. The second uses a reference outside the system (a bench mark on the ground) and it would cease to be coherent if one section of the ruler was moved relative to the other – unless the movement was an exact integral of 1 m.

The two models will be found analogous to certain radar systems and will be referred to later in the text.

1.3 Electro-magnetic radiation

Since the whole phenomenon of radar is rooted in electro-magnetic radiation it is as well to attempt an understanding of its deep mystery. The first hints of its existence and nature came as early as 1680 when Isaac Newton showed it possible, using a prism of glass, to split white light into rays of differing colour; thus giving a clue to its propagation by wave motion, each colour having a different wavelength. Faraday showed in 1831 that a beam of light passed through a glass block in a magnetic field would have its plane of polarisation rotated when the field was present: the stronger the field, the greater was the rotation. Thirty-two years later James Clark Maxwell developed the theory of electro-magnetic waves and their propagation which we use today.

Most readers will be familiar with the notion that an electric current flowing in a conductor will generate an accompanying magnetic field (electro-magnets). Equally, a conductor in a circuit moved across a magnetic field will generate a current in the conductor (electric power generators: dynamos).

In both cases, movement is implicit. In the first case, the current is moving; in the second, the conductor. If there is no movement, there is no induced effect. For example an electric storage cell (such as a car battery) after being disconnected would create a static electric field between its terminals but it would not change, so no magnetic field would be associated with it. Equally, a permanent magnet would produce a static magnetic field between and around its poles, but no electric field would result. The two fields only become inextricably linked when a *change* of the electric or magnetic fields takes place.

In what is termed 'free space' it is possible to generate changing magnetic or electric fields. An instance would be a communications antenna where the electric potential between earth and the antenna is varied at the radio frequency to be broadcast. The fast-changing potential creates an identically changing electric field. There is thus a concomitant magnetic field established and it is at right angles to the electric field. But empty space is not a conductor; how can currents exist in it? Maxwell argued that a changing electric field involves *electric displacement* and that this in effect is an electric current. His equations were developed on this idea.

Imagine a point in space whose potential is made to vary. Suppose this change – for simplicity's sake – is sinusoidal. At any instant the electric field around the point will have strength diminishing as distance along a straight line away from the point increases. Thus any change in the field strength at source (i.e. the point in question) will produce changes in field strength at distances away from the point. Because the potential changes have no mass, the rate of travel of their effect from the reference point is the velocity of light. The changing electric field produces a companion changing magnetic field orthogonal to it. The effects of its changes travel outwards in the same way at the same speed. The two travelling fields represent a wave of energy whose power is the product of the two vectors representing the electric and magnetic field changes. Their companionship is mutually supportive and can be likened to the situation of a castaway on the open sea. If he satisfies his thirst by drinking sea water he becomes more thirsty and needs more to drink. The changing electric field is the 'thirst', the sea water the magnetic field. The situation is represented diagramatically in fig. 1.3.

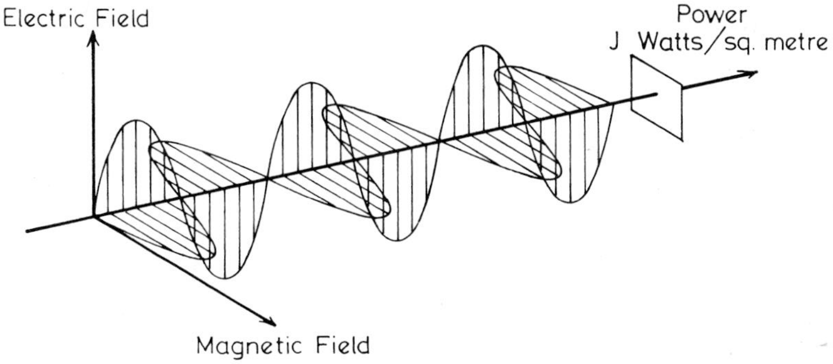

Fig. 1.3. Representation of an electro-magnetic wave.

1.3.1 Polarisation

By convention the wave has its plane of polarisation (vertical, horizontal or at intermediate angles) referred to the electric field vector. The plane of polarisation contains both the electric vector and the direction of propagation. Thus if the electric vector is vertical, the wave is said to be vertically polarised. Figure 1.3 illustrates linear vertical polarisation. In a practical system this may be translated to horizontal polarisation by simply rotating the source through 90°. It is possible to generate non-linear polarisations, e.g. circular or elliptical, both of which have great use. A description of how and why it is produced is given below.

Non-linear polarisation

In an air traffic control surveillance radar system, signals returned from cloud, rain, hail or snow can obscure aircraft signals received at the same time. The air traffic controllers thus need means of rejecting the weather signals and simultaneously preserving the aircraft signals. Equally they need to be able to see where the weather is and to have some knowledge of its severity, to be able to advise pilots. To this end the radar designer uses non-linear polarisation techniques. How do they work?

Clouds of consequence to pilots usually contain water drops which are kept aloft by internal circulating up-draughts of air. The drops are precipitated when their weight is too heavy to be upheld by the cloud's air circuit. Ice crystals, water drops, hail stones, etc., have predominantly spherical or circular shape and thus their reflecting characteristics are markedly different from aircraft in that their overall symmetry is greater. If the energy incident upon them was to be organised into two orthogonal vectors, those reflectors which were spherical would return as much energy in the vertical as the horizontal planes; aircraft would return energy in which either horizontal or vertical components predominated. This circumstance is the first step to realising how it is possible to discriminate weather from aircraft signals. The designers go a step further and produce these two vectors in such a manner as to give either circular or elliptical polarisation to the radiated energy.

Circular polarization

Having generated linear polarisation, the designer now does something which seems quite mad. He takes the electric vector of the transmitter's output, splits it into halves, rotates one half through 90° and delays it relative to the other half. To save our sanity, the resultant situation is represented in fig. 1.4(a). For clarity's sake only the electric vectors are represented. The effect of the phase delay is to produce correspondence between peaks and zeroes of the two orthogonal electric vectors. Now consider the effect of this along the line of plane wave propagation. The plane wave at any instant will be at right angles to the direction of propagation and will contain both vectors. In this plane the electric field will be the vector sum of the two orthogonal components. At point 1 it will be vertical, at 2 horizontal and so

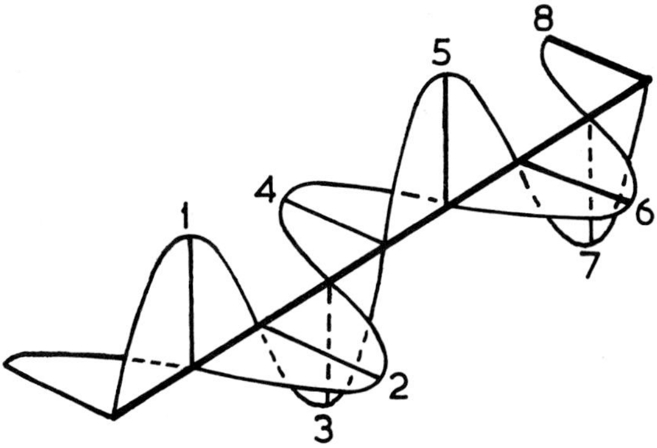

(a)

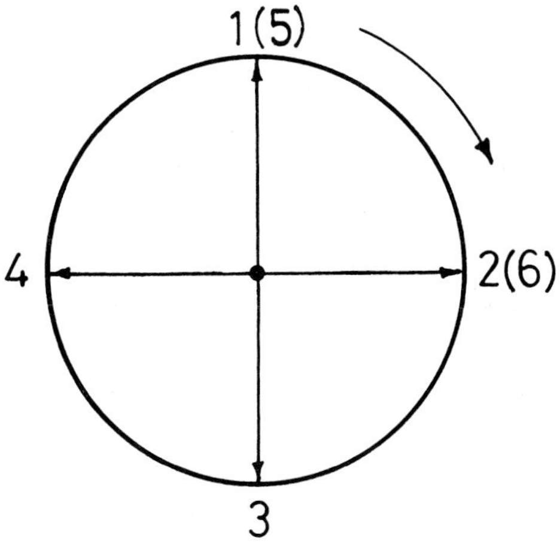

(b)

Fig. 1.4(a) and (b). Circular polarisation: The electric vector is split into equal horizontal and vertical components. Quarter wavelength delay of one produces rotation of resultant.

on. Referring to fig. 1.4(b) we see something else has happened. The resultant vector rotates as it progresses down the line of propagation.

The direction of rotation depends upon the original phase relationship between the two halves. Altering this can produce what is called 'left' or 'right' handed circular polarisations. A circularly polarised wave has equal horizontal and vertical components. If it impinges upon a sphere then the reflected energy will have this same equality (but less magnitude). However, a phase reversal takes place upon reflection and the reflected energy finally reaches the device which generated it. When right-hand polarisation is reflected it becomes left-hand polarisation. The device can only accept waves of the same 'handed' rotation. Thus the reflected signals are rejected. This only occurs however if the signal has equal horizontal and vertical components. This will be very much the case for rain and hail and hardly ever the case for other targets such as aircraft or ground clutter. Circular polarising elements which reject rain and preserve wanted signals can take different forms. The most common is mounted in the waveguide run between antenna and transmitter/receiver. Sometimes the element is rotatable so that the splitting of energy into horizontal and vertical components is not equal in magnitude. This produces elliptical polarisation which can optimise rain rejection under given weather conditions by varying the ellipticity.

1.4 Decibel notation

The engineer's use of decibel notation to express equipment performance is sometimes confusing to many. I hope the following explanation clarifies the situation.

Why is the system used? Simply as a matter of convenience and to prevent errors. In engineering, and particularly electronics, the quantities to be expressed range from millions to millionths; sometimes much greater and smaller quantities need expression. For example in primary radar, received signal powers are in the order of micro microwatts, i.e. ten to the power of minus twelve watts (10^{-12} W). If this were written in full in a calculation it would be 0.000000000001 W, rather tiresome to write and with a great chance of error in counting the zeroes. By using the logarithm (to base ten) of the ratio between two powers, the notation is much shorter. To prevent errors and make the numbers more tractable a device is used whereby the logarithmic number is multiplied by ten, again purely for convenience and to express 'deci'bels. In the example given above the decibel quantity, one micro microwatt expressed as a ratio (relative to one watt) would be $10 \times \log_{10} 0.000000000001$, which equals 10×-12 or -120 dB. In much larger quantities, the device works equally well.

Suppose we wish to express the power gain of a circuit which amplifies an input by a factor of four times as shown in fig. 1.5. The ratio of output power to input power is 4:1. Expressed in decibels according to the above, the gain (G) would be:

$$\begin{aligned} G &= 10 \times \log_{10} 4 \\ &= 10 \times 0.6021 \\ &= 6.021 \text{ dB} \end{aligned}$$

It is common, in fact usual, to neglect the last fractions of the decibel value, again for convenience, especially since they represent such a small real value and so we have the gain of four times expressed as 'a gain of 6 dB'.

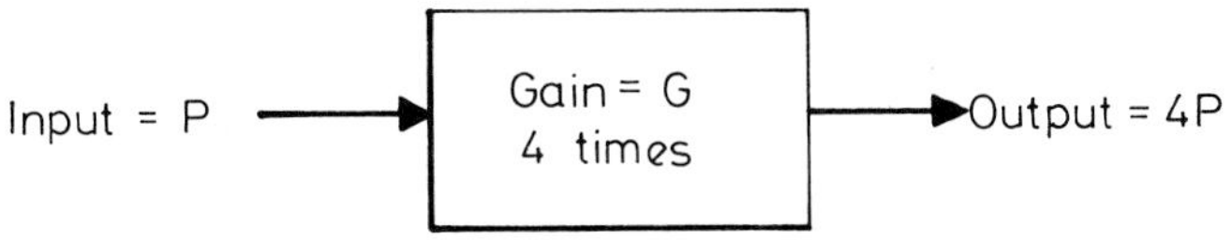

Fig. 1.5. Simple amplifier.

As in normal arithmetic we have the advantage of simply adding and subtracting the logarithmic values to effect multipying and dividing operations – the equivalent of gain and loss respectively. Thus a complete chain of amplifiers and attenuators, commonly found in the electronics world, can have the relationships between input and output power very conveniently calculated with reduced chance of error as shown in fig. 1.6.

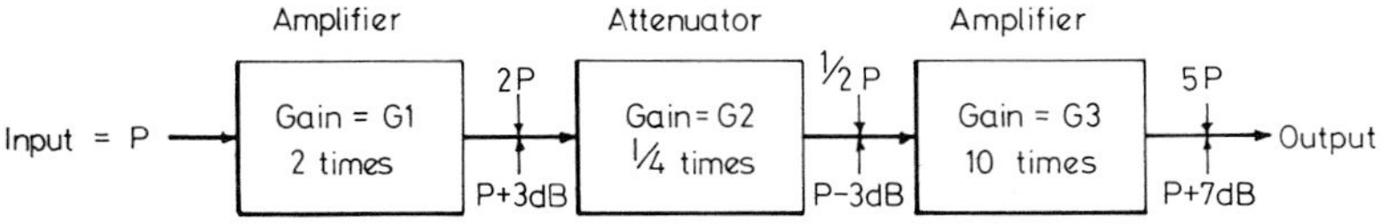

Fig. 1.6. Amplifiers and attenuator in series showing relationship between magnification factors and their decibel equivalents.

Another source of confusion is caused by forgetting that the system is based upon power ratios. It is common to find amplifiers and receivers specified in terms of their ability to magnify an input *voltage* from one value to another. Because the power dissipated in a given resistive load is proportional to the square of the voltage across it ($P = E^2/R$) an amplifier of gain = 6 dB will magnify an input voltage by a factor of two and not four. Expressed in decibel notation we have then:

$$G \text{ dB} = 20 \log_{10} \frac{V_{\text{out}}}{V_{\text{in}}} = 10 \log_{10} \frac{P_{\text{out}}}{P_{\text{in}}}$$

When relationships have to be expressed in graphical form the decibel notation is again often convenient, but no more than convenient. Quite often the parameters to be plotted are in logarithmic progression. When converted to decibel form, the scales become linear.

Definition: The ratio of two powers expressed in decibels is ten times the logarithm (to base ten) of the ratio.

To put things in perspective a set of decibel values is given in table 1.1.

Table 1.1

Ratio P_1/P_2	Ratio in dB
1.0	0
2.0	+3 (10 × 0.3010)
0.5	−3
4	+6 (10 × .6021)
0.25	−6
5	+7 (10 × 0.6999)
10	+10
100	+20
1000	+30
10^n	$+(n \times 10)$

1.5 Spectrum and bandwidth

The model taken to describe frequency in section 1.2 consisted of a pure sine wave signal. One way of describing its character would be to express its amplitude changes with time as:

$$A = P \sin wt \tag{1.1}$$

where P = amplitude (peak to peak)
$w = 2\pi \times$ frequency
t = time (s) that the frequency is allowed to run

so wt expresses an angle in radian measure.

Another way would be to represent its amplitude and frequency in graphical form, as in fig. 1.7. This describes the *spectrum* of the signal; in the example presented the signal has an amplitude of 5 V at a frequency of 3.4 MHz. If it were received from a transmitter, this signal would convey only

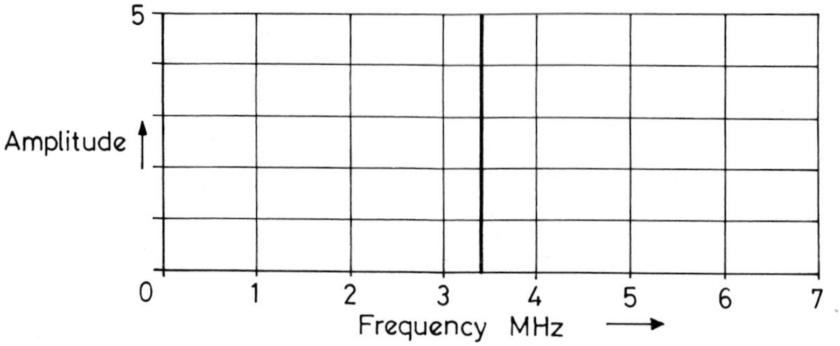

Fig. 1.7. Spectrum of a pure sine wave at a frequency of 3.4 MHz with amplitude = 5.

one piece of information: that the signal was present. To convey more information it is necessary to alter this pure signal in some way. The alteration is brought about by *modulation* of the signal.

Should the pure sine wave (call it the carrier frequency) have another sine wave of lower frequency (as modulation) added to it in such a way that the carrier's amplitude varied in proportion to the modulation, the spectrum of the combined output would be as shown in fig. 1.8. Three spectral lines appear: the central line is that of the carrier frequency (f_0) and the other two (f_1) are those of the modulation. There are two because the electronic mixing of the two frequencies creates $f_0 \pm f_1$. In some communications systems one of these so called 'side bands' is suppressed.

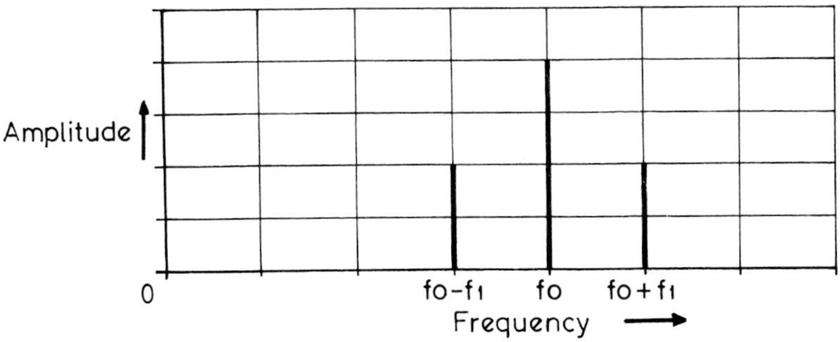

Fig. 1.8. Spectrum of a pure sine wave of frequency f_0, modulated by another of f_1.

1.5.1 Pulse Modulation

In radar systems it is usual to form an output which is a carrier frequency (in the microwave band) which is turned on for very short periods and off for much longer times. It is also common for the 'on' period to produce a single amplitude of the carrier. Its character in time is as shown in fig. 1.9. This 'pulse'

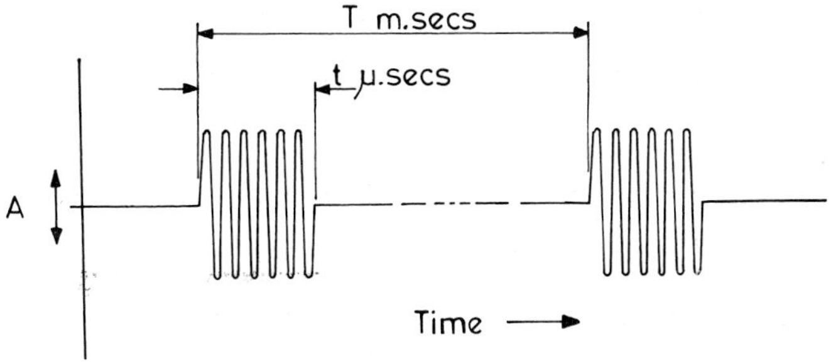

Fig. 1.9. Common form of pulsed radar output, which is present for duration t and repeated at intervals separated by T.

amplitude modulation produces an output spectrum of special form. The frequencies within the spectrum are:

(a) The carrier;
(b) The pulse repetition frequency ($1/T$), where T is the inter-pulse period;
(c) The compound of frequencies making up the square shape of the pulse.

Fourier analysis shows that a square wave shape can be constructed from a number of pure sine waves of differing frequency added in specific amounts. When this magic mix is performed the spectrum of the pulsed carrier output takes the form shown in fig. 1.10. The envelope of the spectral lines follows the expression:

$$A = \frac{\sin x}{x} \tag{1.2}$$

where x = angle in radians.

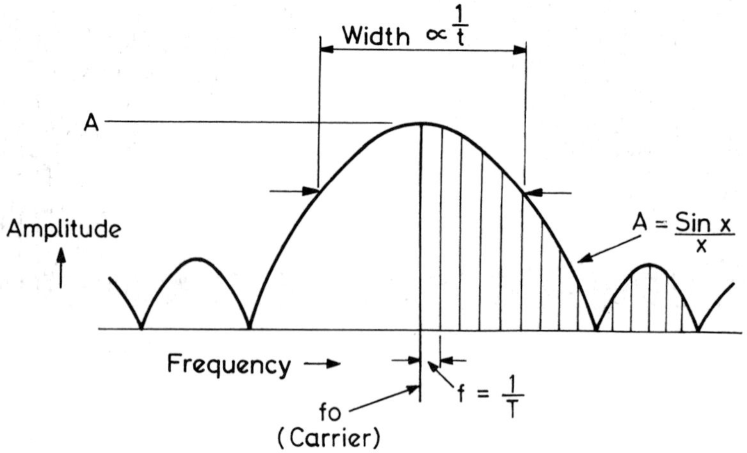

Fig. 1.10. Idealised spectrum of pulsed radar output of the form in fig. 1.9.

Points of note are:

(a) In an ideal system the overall spectrum is symmetrical about f_0.
(b) The narrower the modulating pulse, the higher are the frequencies going into its construction and the wider becomes the main lobe of the pattern.
(c) The higher the pulse repetition frequency (p.r.f.) the closer are the p.r.f. lines. ($f = 1/T$)
(d) The envelope has a main beam and sidelobes, as does an antenna horizontal polar diagram (see Chapter 3).

A question arises: why is the spectrum a set of lines spaced at the p.r.f. with no components between? The answer is that there *are* lines in between representing the Fourier analysis components and their intermodulations, but their magnitude is extremely low relative to the amplitude of the successive sidebands created by the predominant frequencies.

To give an idea of practical values, the photograph in fig. 1.11 is included. It is the spectral picture of an L band radar having a p.r.f. of 770 Hz and a pulse duration of 1.6 μs.

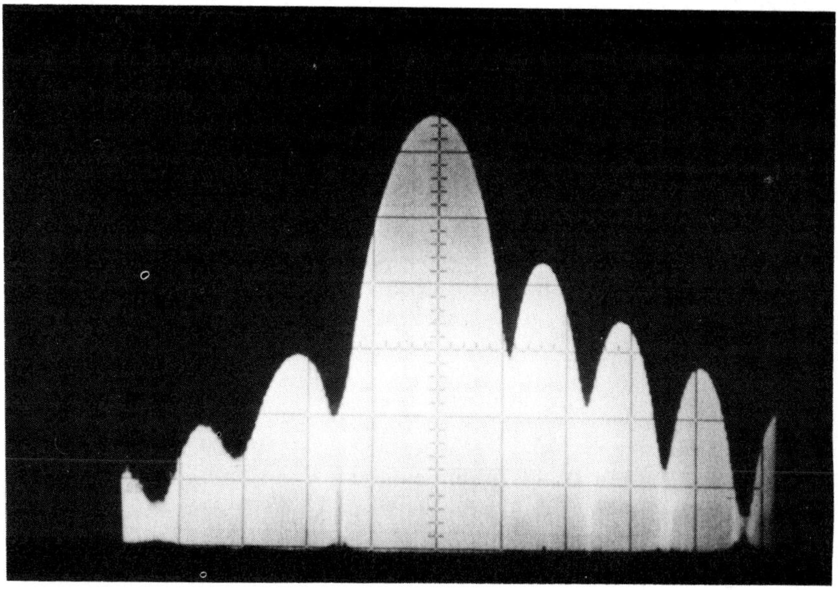

Fig. 1.11. Actual spectrum of an L band radar output f_0 = 1300 MHz, t = 1.6 μs, T = 1.3 ms. Vertical scale: 10 dB/cm. Horizontal scale: 500 KHz/cm.

1.5.2 Bandwidth

In a communications system the optimum condition for passing information is to have a receiver capable of receiving all (but only) those spectral lines transmitted. If it was of infinite bandwidth the information would be lost in the jumble of signals from many transmissions and noise so the bandwidth has to have a finite value.

Designers have thus to make the receiver selective of the carrier frequency used and to tailor the band of frequencies accepted so that it embraces the spectral lines containing the information transmitted. For the case of radar, the way the designer goes about solving this problem is given in chapter 2.

2
Signal detectability

2.1 Noise

Veteran and most practising radar operators will be familiar with the effects of 'noise' on their radar screens. There are however an increasing number of modern radar systems which present pictures which are 'noise-free' and so an increasing number of new radar users to whom the idea of 'noise' may be foreign. It is fitting then to review the subject.

Why is it important? Noise is the ultimate factor which limits the radar's ability to detect signals. What is it? Taking the practical radar situation, we can say that all signals enter the receiver as minute voltages across the receiver's two input terminals. They are magnified by the receiver's gain. This gain is designed to be frequency-selective, i.e. the gain exists across a controlled band of frequencies; then the receiver has a bandwidth (call it B).

In an ideal receiver, intuitively one would expect zero signal input to result in no output. Imagine that a signal was deliberately prevented from entering the system by connecting its two input terminals together. If the gain of the receiver chain was sufficient (and it usually is) and its output examined, we would find it has one – even though the input of signals had been prevented. This output is the result of electrical noise at the receiver's input. It is caused by the random jostling of molecules in the conductor connecting the two input terminals. This random molecular movement generates minute random voltages across the terminals and thus they are magnified by the receiver gain and produce random voltages at the receiver output. The amount of noise is proportional to the receiver input temperature and it would only cease to be present at absolute zero (−273 degrees Centigrade). In the ideal receiver the minimum noise level is mathematically expressed as:

Input noise power $= KTB$

where T = temperature of the input's noise generator in degrees absolute
K = Boltzman's constant
B = Bandwidth through which the noise passes.

Boltzman's constant expresses the energy which translates into the noise voltages and is 1.372×10^{-23} joules per degree Kelvin.

It is usual to express the performance of a practical receiver in terms of the degree to which it falls short of the ideal by use of the term 'noise factor' or more usually its decibel equivalent 'noise figure', thus:

$$\text{Input noise power (actual)} = FKTB$$

where F is the noise factor. Thus for a noise factor of four the receiver would have an input noise power which was four times worse than the ideal. We have seen from table 1.1 in chapter 1 that this is equivalent to noise figure of 6 dB.

2.2 The effects of noise

How does noise affect the radar system? The radar is designed to create recognisable events, for example a transmitted pulse of energy leads to reflection recognised as small pulses of the same history. The user sees these as either discrete 'blips' of light on a radar display or symbols stimulated to be at the same positions.

The wanted pulse signals can be described as events characterised by voltages across the radar receiver's terminals, growing at a measurable rate, remaining for a known period and then collapsing again. The radar system has to recognise these events and process them for display; the logic used by the designer is 'we know what to expect as a result of our transmission and will emphasise the characteristics we know the received wanted signal should have'.

Let us take a simple case of a radar designed to detect only the amplitude of a signal returned from an aircraft unobscured by any other returned signals. Let this be represented by the voltage history across the receiver terminals as in fig. 2.1. As described above, this will be added to the noise history which it will find when it arrives. Suppose this to be as in fig. 2.2. The combined effect will be as in fig. 2.3. It is obvious that the noise history of the amplitude blurs that of the signal. The larger the signal, for given noise, the less the signals events will be blurred. Suppose only one criterion was to be used for the presence of a signal, does its amplitude exceed a certain preset value?

In the case of threshold 1 in fig. 2.3, there is one event which meets this criterion, in the case of threshold 2 there are two and for threshold 3 there are three. It is seen by using this simple criterion that the random nature of noise can create more events, the closer the threshold is brought to mean noise level. With a low threshold the wanted signals will be among a larger group of events mostly caused by noise voltages crossing the threshold.

These noise voltage crossings are 'false alarms' and it follows that for a given threshold level the number of false alarms grows as noise approaches it. It also follows that the less noise in the system, the lower the threshold can be

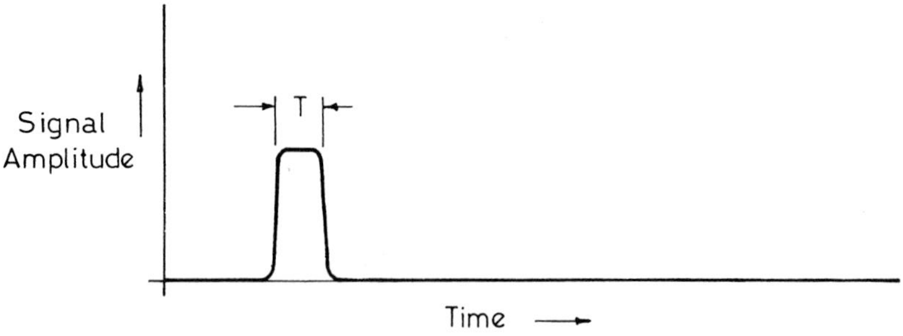

Fig. 2.1. Signal history.

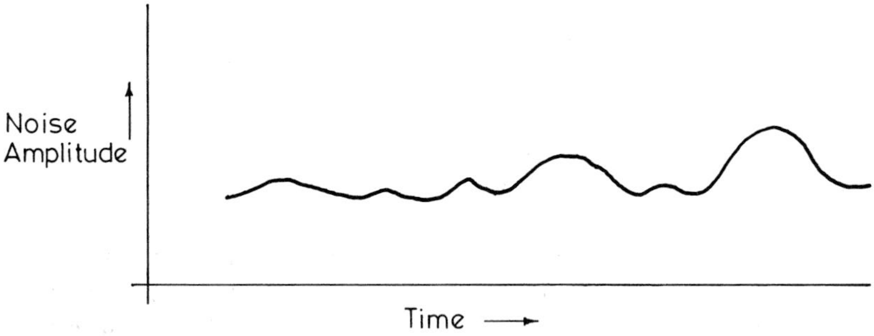

Fig. 2.2. Noise history.

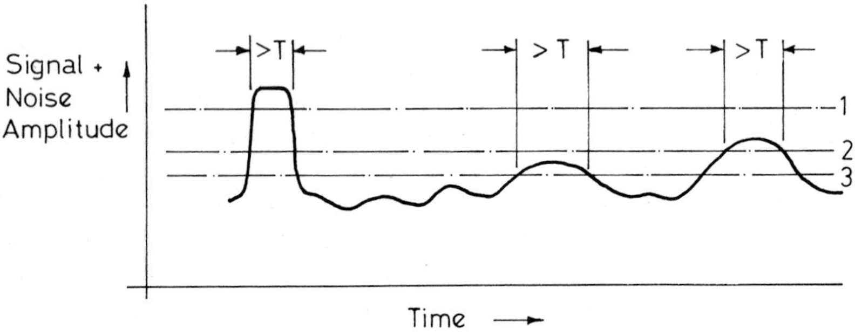

Fig. 2.3. Signal plus noise history. As detection threshold is lowered, detection sensitivity increases – and so does false alarm rate.

for a given acceptable false alarm rate. For this reason design engineers have striven hard and long to reduce noise figures of receiver systems. The alternative means of improving signal to noise ratio is to enhance the output transmission by increasing power or antenna gain or by changing modulation techniques – all very expensive by comparison. An idea of this struggle to reduce noise figure can be seen from the graph in fig. 2.4.

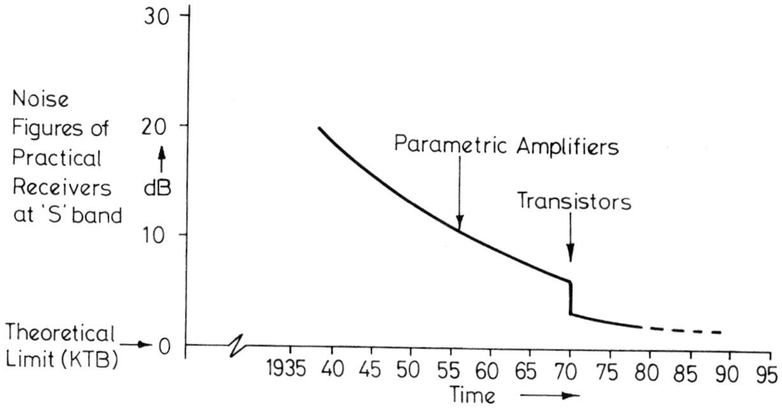

Fig. 2.4. Outline of progress to perfect receivers.

2.3 Noise through a bandwidth

If the nature of noise is truly random, then given an infinite bandwidth, all the differing rates of change of noise amplitude (representing all frequencies) are equiprobable. The radar receiver however has limited bandwidth and only those components of noise made from frequencies passed by this bandwidth appear at the receiver output. It is readily seen that the narrower this bandwidth, the less is the noise power at the output. Narrowing the pass band cannot be continued beyond a certain point however, because it still has to have sufficient width to reproduce the signal history faithfully, i.e. its spectral lines. The classical relationship is represented in simple form in fig. 2.5(a).

A photograph of a ppi (plan position indicator) is shown in fig. 2.5(b) to illustrate the nature of noise. It was exposed for one antenna revolution. The display is of a high power L-band radar. Range is 120 nautical miles radius and signals are displayed prior to any processing to show the 'rubbish' among which wanted signals can be found. The centre of the display has been deliberately blanked out to save the display tube's phosphor layer being burned by the continual registration of ground clutter signals, some of which can be seen at the blanked areas' edges. To the south-west round to about 270° can be seen small areas of weather clutter. Aircraft are registered as small bright arcs whose extent depends upon the number of times in the beam's passage across the target the resultant signals exceed detection criteria. Radial lines of discrete bright dots appear. These are interference signals received from a neighbouring radar (see chapter 9). The other random 'speckles' are caused by noise. The reason for their apparent radial line structure and decrease in brightness from display centre to edge is explained in Part 4.

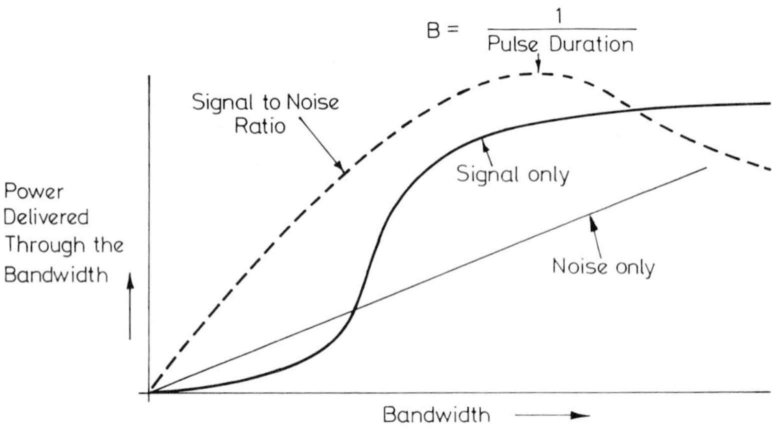

Fig. 2.5(a). Once the bandwidth is wide enough to give signal fidelity, greater width introduces noise at a rate which degrades the signal to noise ratio.

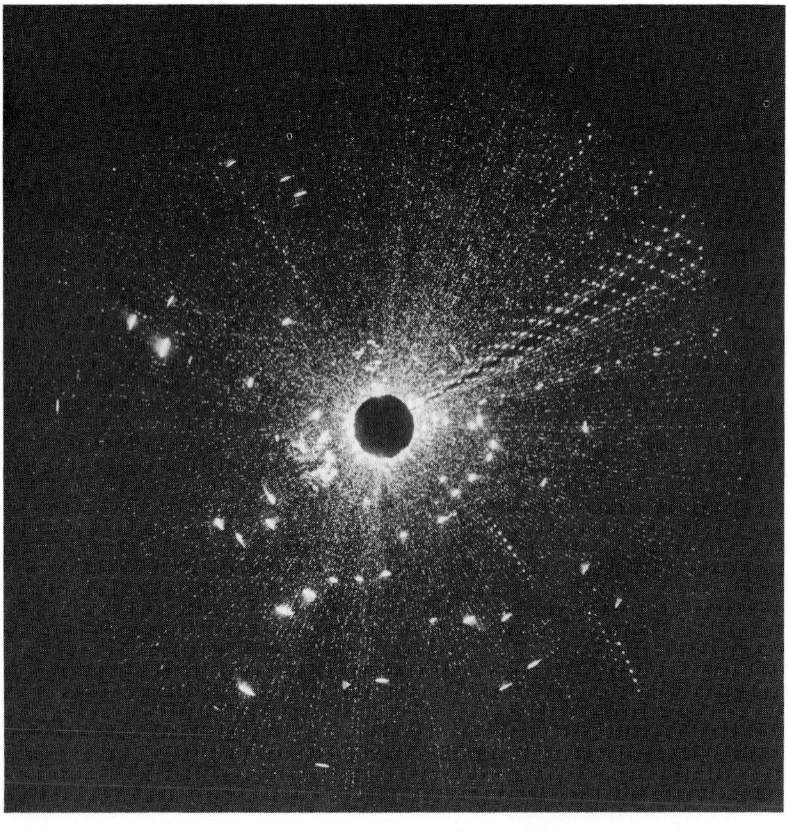

Fig. 2.5(b). Real-time ppi display of noise and radar signals prior to signal processing. Description of content given in text. (*Photo courtesy of Marconi Radar.*)

2.3.1 Probability of detection and false alarms

In pulse radar systems the signal is made up of varying amounts of different frequency components but not in a random fashion as is noise. The fast edges of the signal are represented by high frequencies, e.g. a pulse edge rising in 0.1 μs has a large component of a frequency of 10 MHz. If the signal duration is 0.5μs a major component will be one of 2 MHz: there is a reciprocal relationship between time and frequency.

What would be a practical value to set for false alarms? To a large degree this depends upon the radar's operational role. Take two extreme cases: a military defence radar operator whose role is to detect intruders at great range would be tolerant of a fairly high rate of false alarms since this condition gives the best chance of seeing a target at the earliest time. An operator using a radar to monitor aircraft on final approach would be very intolerant of false alarms since these might wrongly indicate conflicting traffic or aircraft turns.

An illustration of the inter-relationship between probability of false alarm and probability of detection is given in fig. 2.6. It is based on Skolnik's exposition of the radar equation (1). These two curves represent conditions found after detection of the signals and noise and are unipolar, i.e. all values above zero. Their character is related. The signal plus noise curve follows a Gaussian distribution law and that of the noise alone has Rayleigh distribution. The greater the signal, the further to the right the peak of signal plus noise curve goes and similarly with the noise alone. Both curves are the result of integration and so the area under them is of concern, representing as it does probability of detection. Detections of *noise only* above threshold represent false alarms among the group of true detection of *signals mixed with noise.*

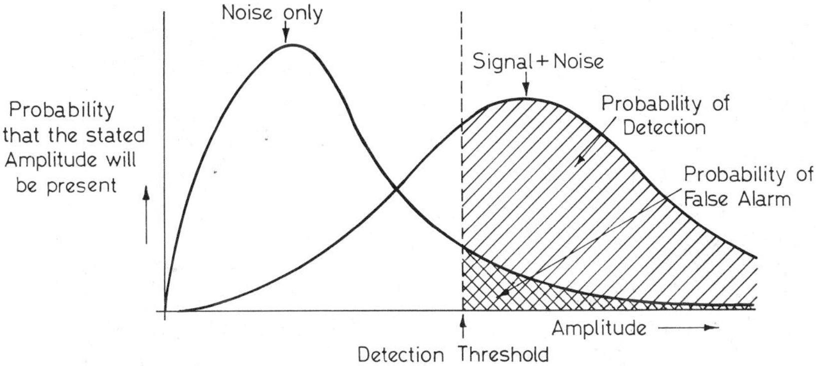

Fig. 2.6. Showing the relationships between probability of detection (P_d) of a non-fluctuating signal and probability of false alarm (P_{fa}). If the threshold is lowered, P_d increases but so does P_{fa}. The exchange is highly non-linear.

From these two related curves it can be seen that as the threshold is lowered, the probability of detection (represented by the *area* under the curve) increases from threshold and above, but so does the probability of false alarms. The converse is obviously true for an increase in threshold level. It is also clear that the inter-relationships are complex.

False alarms in a pulsed radar system should be considered in two related regimes. The first is concerned with the probability that a single event will simulate falsely, by accident, the wanted signal's behaviour. The second is the probability that a number of these false alarms will simulate the wanted signal's behaviour *repeatedly* at the same ranges and bearings expected of a real target.

Designers will refer to these events related to signal detection or false alarms in a single interpulse period (i.e. as a result of a single transmission and resultant reception) as 'first threshold crossings'. The 'second threshold' crossings relate to detection signals over a number of interpulse periods to see if they match an expected pattern (i.e. the beamwidth history is examined). For instance, a real target would generate a series of first threshold crossings at virtually the same range across the antenna's beamwidth. A second threshold criterion could be (and often is) 'do the first threshold crossings occur at a given range for any three (or n) successive interpulse periods'. This criterion might be accompanied by others such as 'do the succession of first threshold crossings continue with no more than one miss' and finally 'if two or more successive misses are present the target ceases to be detected'. These logical processes are common to both the human observer of a radar ppi (plan position indicator) and to automatic plot detectors or extractors.

2.3.2 Minimum detectable signal (S_{min})

For a false alarm to be created it is necessary for it to look like a wanted signal. That is, it not only has to cross an amplitude threshold but to stay above it for a duration very like that of the wanted signal. Remembering the reciprocal relationship between pulse duration and the bandwidth necessary to pass it without distortion, we see that for example a 1 MHz bandwidth requires the false alarm to exist for at least 1 μs. Since it is possible for 10^6 pulses to exist in one second through the 1 MHz bandwidth then the system must be set up, by means of adjusting the threshold, so that a false alarm probability of less than one in a million exists if false alarm *times* are to be greater than one second. This value of 10^{-6} probability of false alarm for a single pulse is commonly used in radar specifications.

However, a radar operator is hardly likely to construe a single pulse as a signal from a wanted target. The single events must occur repeatedly at a constant range. The probability of this happening by chance is extremely remote. It has been shown that it is possible to operate a radar successfully with a much higher single pulse false alarm probability of 10^{-4}.

Much experimental work has been done to establish the minimum signal level (S_{min}) needed for an operator to achieve given probability of detection when using the traditional ppi displaying unprocessed radar outputs in real time. An experimentally measured figure within Marconi Radar was between 5 and 6 decibels signal to noise ratio for a 50% probability of detection at 10^{-6} probability of false alarm for a non-fluctuating target having four pulses per beamwidth. This is in accordance with modern accepted theory invoked in this work. To raise this probability of detection to higher percentages than fifty requires better signal to noise ratio; precisely how much depends upon the nature of the target, as will be seen in chapter 7.

3
Antenna beams

3.1 Beam formation

It is common for users to be given an idealised picture of antenna beams in radar systems. The assumption is made that this will content them. It is however rewarding to take a closer look at the matter, for in practice the short-fall from the ideal brings about certain effects which the user usually sees to his disadvantage.

The deeper question 'why do we need beams' would seem not to need answering because most users realise this as fundamental to the radar principle. That is, by concentrating radiation into knowable directions gives target bearing information – the narrower the beam, the more accurate the target bearing data. This concentration, in both azimuth and elevation planes, is brought about by a variety of engineering techniques, ranging from batteries of radiating elements in a line (linear arrays) to beam-forming reflectors illuminated by various sources (sometimes microwave lens elements are used).

I do not intend to explain the workings of all types of antennae but to show by simple principles, common to most of them, how the ideal shapes are attempted and to what degree the ideal is achieved. Let us take the most simple form of antenna, an array of radiating elements in line, with reflectors to each so placed that their radiation is essentially in one hemisphere. Allow them to be equally spaced in a straight line and to be of equal gain. Allow further that the power received or radiated by each is equal and of the same phase. Such an array of six elements is shown in fig. 3.1.

Suppose we have a signal source radiating omni-directionally at a range much greater than the width of the array shown in fig. 3.1 – great enough for the assumption that the signal arriving at the array is a plane wave (i.e. the phase of energy from the source across the array is the same). This occurs

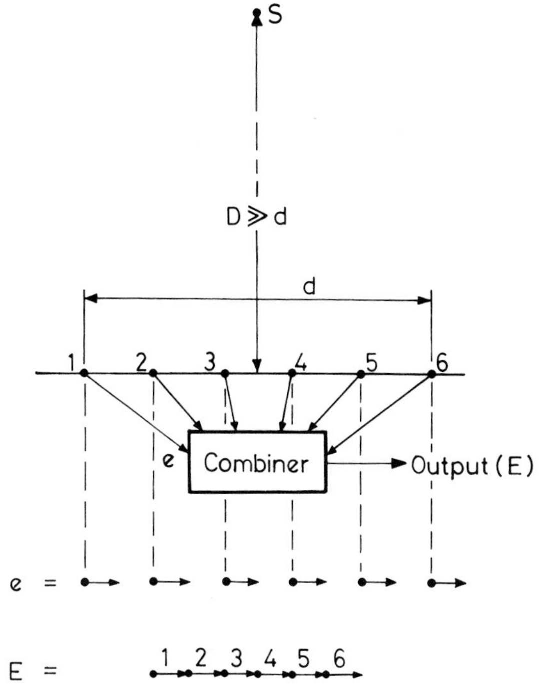

Fig. 3.1. Receiving elements 1 to 6 form a linear array. S is a signal source radiating to the array. Contributions from each element (e) sum vectorially to E by a combining network.

beyond what, by convention, is called the Rayleigh distance (D), after its propounder Lord Rayleigh, the famous physicist:

$$D = \frac{d^2}{\lambda} \tag{3.1}$$

where d = effective aperture width in metres
λ = wavelength in metres.

In fact the divergence from a plane wave across the array for a source at D is 45° and most designers use a value of $2D$ in their work.

Arrangements are made to collect energy from each of the elements and combine them with no phase shift; the output from the combining network will be the vector sum of contributions from each of the six elements. Allow the combining process to be loss-less and examine what happens to the combined signal when the signal source moves across the array in the plane containing the array.

Obviously when the signal source is at right angles to the centre point of the array, because we have posited a plane wave, all contributions will be equal in phase (it does not matter what the phase is, only that it is equal for all) and equal in amplitude. The combined output by vector addition, will have six times that of the output of one element.

Now suppose the source is moved laterally to a point away from the array's boresight. Because the energy reaching the array is in a plane wave, i.e. equiphase across its front, the energy from each of the six elements will be phase shifted relative to each other by the same amount since they are equispaced. The phase difference between signals from each element is proportional to the element spacing and the frequency of the radiation.

The total output from a combining network will be the vector sum of six equal values, each shifted in phase by the same amount. The principle is illustrated in fig. 3.2 and the effects of moving the signal source successively away from the array's boresight are shown in fig. 3.3. The eight examples given are all to the same scale. The gain of the antenna array as a function of angle off boresight will be the resultant of the vector addition.

It is seen that at a certain point off-axis the sum of the signals from three of the elements is exactly equal and opposite to the remaining three and a null is formed. This, it is obvious, will occur at two points of symmetry either side of the boresight and thus a beam shape is formed. In the example given the null occurs when there is 60° phase shift between contributions from each element. The vectors form a closed loop – a hexagon.

3.1.1 General cases

The main beam formation centres around two circumstances: one is where the wavefront is parallel to the aperture, the other is when the wavefront is at an angle relative to the aperture such that there is 360° phase shift across the array from end to end. Thus the nulls forming the main beam occur over an angle $\theta = 2 \sin^{-1} 1/n$ where n is the number of wavelengths in the aperture. We see from this that the wider the physical size of the aperture the narrower is the beam. But the actual width is not intrinsically the determining factor. The important parameter is how many wavelengths there are in the width.

Another important parameter is the beamwidth containing most of the energy. By convention this is taken to be that between the points where the gain falls to half the peak value (i.e. the '–3dB points'). A good approximation of this is $\theta = 51/n$ where n is the number of wavelengths.

3.2 Sidelobes and their reduction

Consider now what happens when the signal source is moved yet further away from boresight. As shown in fig. 3.3 the amplitude of the combined signals from the six elements gets bigger again (see $d\phi = 65°$), reaches a maximum (see $d\phi = 90°$) of much lower amplitude than the main beam but nevertheless forms a coherent beam. Subsidiary beams will go on being discovered as the

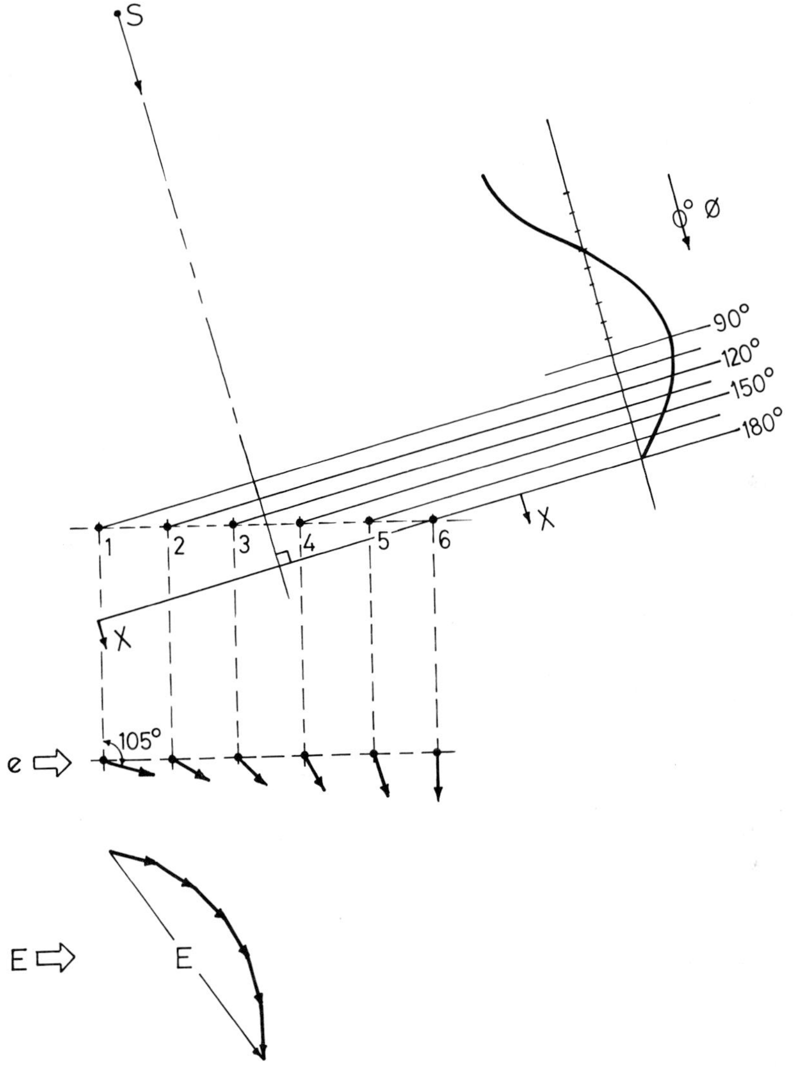

Fig. 3.2. S = signal source, X–X = plane wavefront across the 6-element array, Ø = phase of energy from wavefront to S, e = energy incident on each element, E = vector summation of element outputs.

angle of the plane wave of the input energy increases relative to the boresight of the array. These subsidiary beams are well known as *sidelobes* and are an inseparable characteristic of any antenna of practical form. The beam shape formed by the six-element array is shown in fig. 3.4. It is drawn in terms of power gain relative to peak. Since $P = A^2$, in decibel notation the power gain $P = 20 \log_{10} A$.

In the example taken the major sidelobe is at a level of −12.4 dB relative to the peak of the main beam. This is another way of saying that if the signal source were to be in line with the sidelobe it would produce an output from the array of about one seventeenth of that at the main beam peak (pedantically 0.058 of maximum gain).

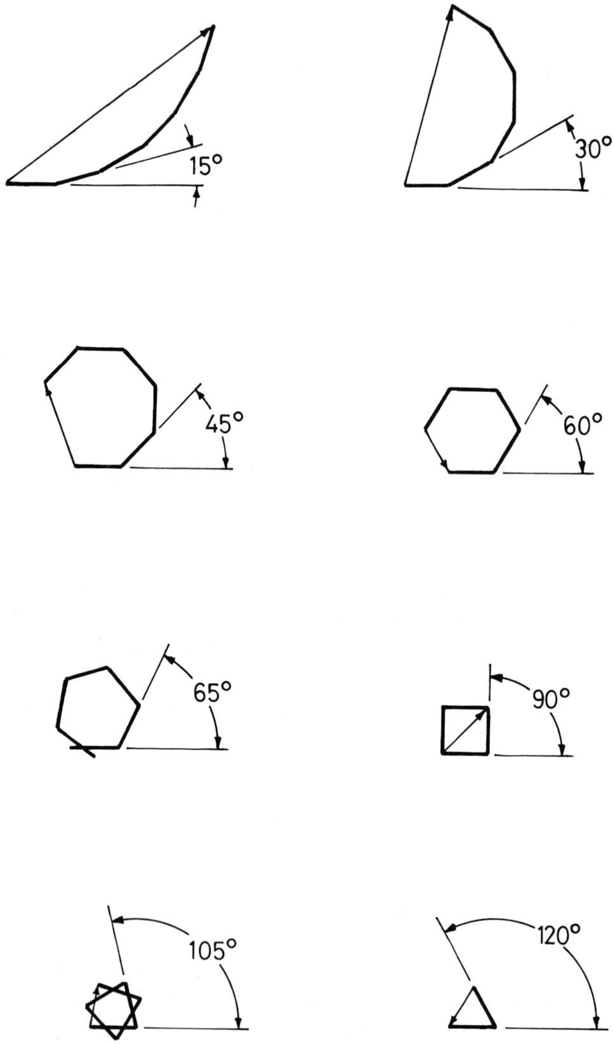

Fig. 3.3. Phase angles between vectors increase as source moves away from array boresight. When phase angle is 60° in the 6-element array, the resultant is zero. As angle off boresight grows again, so does the resultant, forming sidelobes.

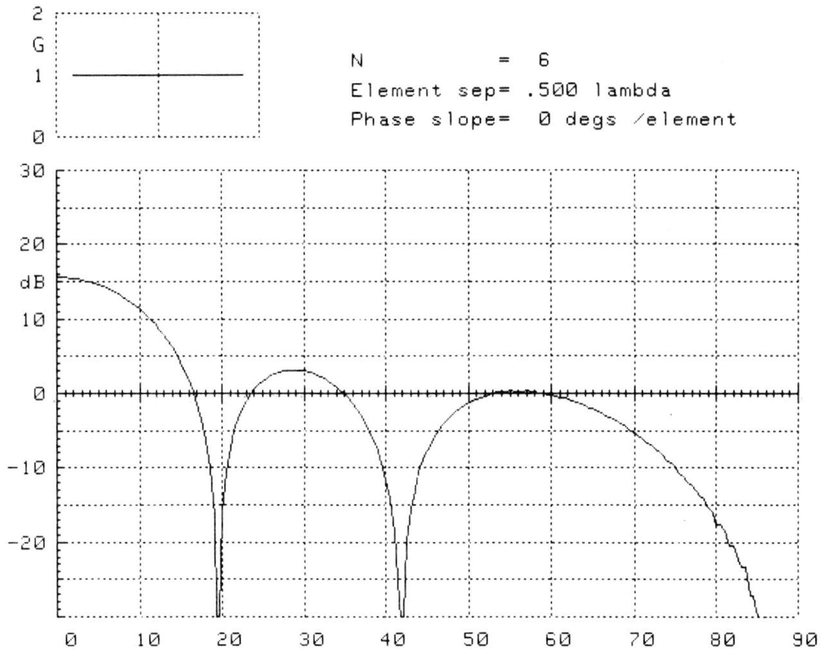

Fig. 3.4. Beamshape of a 6-element array with uniform phase and amplitude distribution. (The latter is indicated in the small box above the graph.)

In an array wherein each element contributes the same output for equal input (the so-called 'uniform distribution' array), the amplitude of the sidelobes varies slightly with the number of elements. When the number of elements is infinite the array can be considered as a continuum. Under these conditions the beam shape is characterised by the expression:

$$A = \frac{\text{Sin } x}{x} \tag{3.2}$$

where A is the resultant amplitude of the vector addition of all contributions across the array and x is the angle away from boresight in radians. The first major sidelobe then reduces to –13.3 dB.

Examples of this effect are shown in figs 3.5(a) and (b). It should be noticed that the vector addition, if continued for increasing phase difference between a small number of elements, produces a *series* of main beams and subsidiary sidelobes. Why do we not see these multiple main beams in practice? It is because the spacing between elements is only half a wavelength or slightly greater.

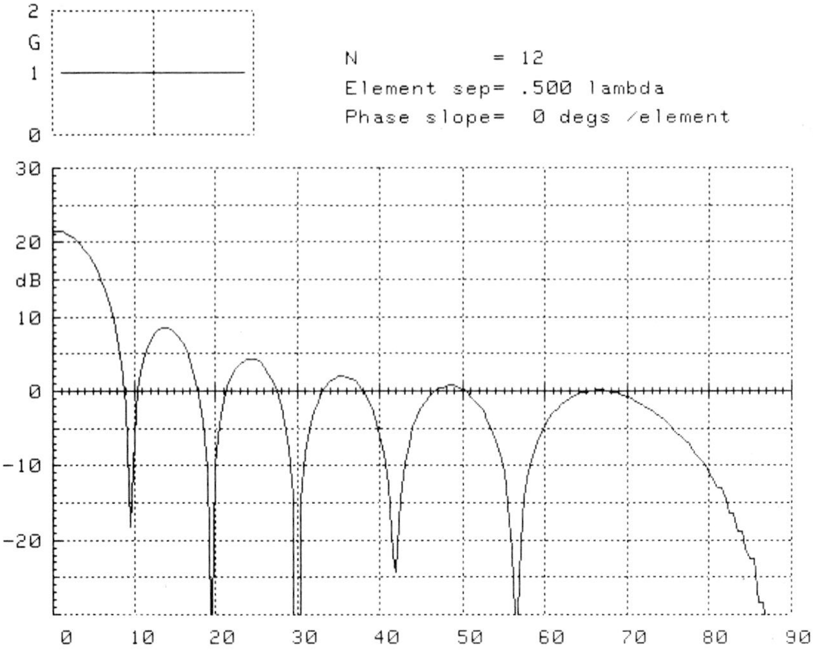

Fig. 3.5(a). Beamshape of a 12-element array with uniform amplitude distribution. Note increase in gain over a 6-element array.

3.3 The illuminated reflector antenna

The example of a linear array was taken as being the easiest means to explain beam formation and its consequent sidelobes. But by far the most common antenna is the shaped reflector illuminated by a 'feed' element. Most people will be familiar with the idea that a concentrated beam is formed when a parabolic reflector is illuminated at the parabola's focus point. That is because any ray from the focus is returned from the reflector in such a way that the path length of the ray from source to a straight line parallel to the reflector's aperture is the same across the whole aperture. That is, at a large distance away from the assembly, any radiation is in the form of a plane wave, i.e. of equal phase across a line normal to the aperture boresight. This is illustrated in fig. 3.6.

Analysing the situation in the same way as for the linear array it can be seen that because the reflector has resulted in plane wave radiation (and therefore reception), the same analysis holds good: any plane wave entering the antenna aperture will result in the vector addition of all the contributions across the reflector's aperture (now a continuum instead of discrete elements). As the plane wave entering the antenna tilts away from the normal to its aperture, so the contributions will be phase shifted relative to each other.

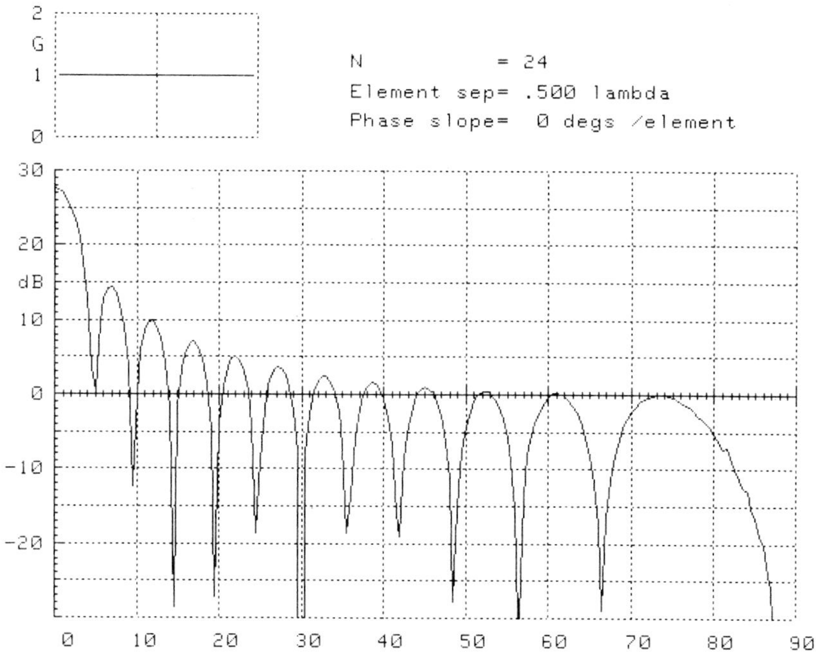

Fig. 3.5(b). Beamshape of a 24-element array with uniform amplitude distribution. Gain = 12 dB greater than for a 6-element array.

Their vector addition will therefore follow the same history as shown in fig. 3.5(a) and (b).

The explanation above suggests that there is a unique focal omni-directional feeder point through which this all happens. In practice this is not the case, for as in the linear array, the feed illuminating the reflector can have amplitude taper; trade-off between beam shape and sidelobe levels can be, and is, done.

In forming the beam for transmission, attempts are made to concentrate the output energy in the main area of the reflector, tapering it down at the edges. If this is not done carefully, one of two consequences follows. If the energy is tapered too much the effective area of the antenna is reduced, producing a wider beam and less gain than desired. If it is not tapered enough there is a risk that the illuminating element (commonly a horn feed) will 'look past' the reflector's edges and gather energy reflected from mountains and other large objects at azimuths far away from the main beam's azimuth. Perhaps you have seen wide arcs of clutter about 120° either side of that gathered by the main beam? This unwanted effect is, not unnaturally, called 'spill-over'. The arcs are generally wide because the effective beamwidth of the spill-over beam is much wider than the antenna's main beam. These effects contribute to the antenna's inability to be 100% efficient.

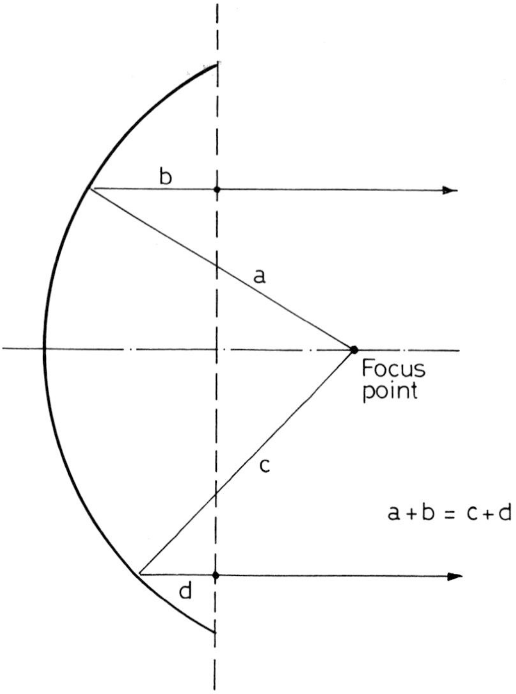

Fig. 3.6. Simple parabolic reflector illuminated at its focal point. Path lengths from this via the reflector are equal, resulting in plane waves propagating across reflector aperture.

A radar system's efficiency in discriminating between wanted and unwanted targets to as great a range as possible depends largely upon its ability to concentrate radiated energy into desired directions. If the antenna has sidelobes, two related unwanted effects result. One is that energy goes into undesired directions (albeit at reduced power). As a result of this, targets (particularly unwanted clutter targets) in primary radar are illuminated in directions other than the main beam and these clutter signals received by the sidelobes can be present with wanted signals from the main beam simultaneously, causing corruption of the wanted signals. The other effect is to reduce the maximum gain of the antenna; the sidelobes contain power which should otherwise be in the main beam.

The example used to explain the forming of a beam assumed that all elements of the array contributed equally to the performance. It is possible, and indeed usual, to arrange matters so that contributions from the elements near the ends of the array are reduced relative to those near the centre. This considerably reduces sidelobes. Various 'tapers' are used and the antenna engineers are forever seeking new distributions – for in reducing sidelobe level, the main beam shape is altered, sometimes to the disadvantage of the system. An example of different distribution is given in fig. 3.7 to illustrate these tantalising problems in antenna design.

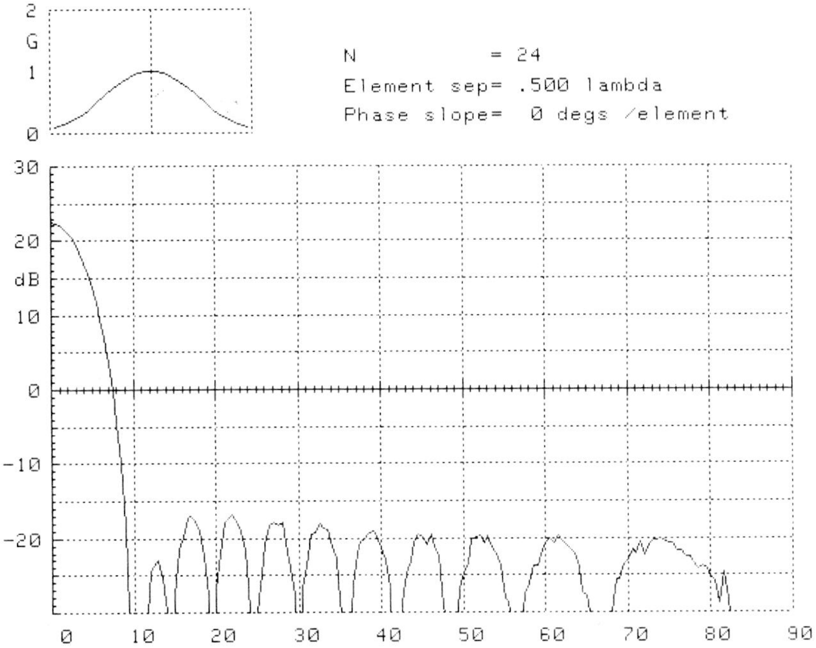

Fig. 3.7. Beamshape of a 24-element array with uniform phase but amplitude tapered distribution. Taper is indicated in the small box above the graph. Note reduction in sidelobe levels relative to non-tapered distribution (fig. 3.5(b)).

3.4 Beams are three-dimensional

So far we have considered beam shape in the azimuth plane only. Antennae are required to radiate also in the elevation plane. Beam shapes in this plane can be analysed in just the same way as described above, but it is more difficult because amplitude distributions are generally more complicated.

We usually see antenna beam shapes represented in graphical form, their gain at angles away from peak expressed in decibels relative to peak gain as in fig. 3.7. All antennae radiate in all directions simultaneously; their differences rest in *how much* they radiate *in what* directions. It is difficult but important to realise that the antenna's beam is a three-dimensional volume. For instance the diagrams of figs. 3.8(a) and (b) represent the horizontal and vertical free space patterns of one antenna. Taking a look at the azimuth or horizontal

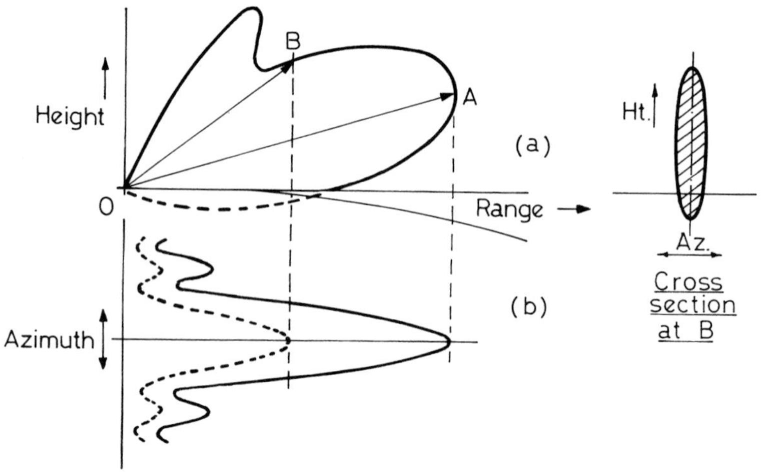

Fig. 3.8. Typical horizontal and vertical beamshapes. Azimuth scale is deliberately expanded for clarity. Cross-section shows the 3-dimensional nature of the beam. Dotted part of the elevation pattern represents ground-incident energy.

polar diagrams (HPD) at two different elevation angles as illustrated shows this three-dimensional property clearly.

3.4.1 The 'cosecant squared' pattern

To make the best use of radiated power the antenna must be designed to distribute it in space in the most economical manner which in turn is determined by the user's operational requirement.

In an air traffic control radar, for instance, its targets lie in a thin flat cylindrical box whose radius is many times its height. Typically no targets will fly at heights above 60 000 ft (approximately 10 nautical miles) but they need to be seen in an area whose radius is as much as twenty times this (200 nautical miles). An idealised form of this shape is shown in fig. 3.9(a). An airborne radar designed for ground mapping would lead to an idealised pattern which was the converse of this, as shown in fig. 3.9(b). The vertical polar diagram indicates equal received signal to noise ratios as a function of slant range and height.

Equations of chapter 7 in part 2 are written in terms of the maximum range achievable for given probability of detection. This value is obtained at the absolute peak of the radar beam. The ranges achieved at angles away from this singular 'line of shoot' are determined by what happens to the antenna gain at these different angles. Thus it is possible to express this idea as:

$$R^4\ (\phi,\theta) = kG^2\ (\phi,\theta) \qquad (3.3)$$

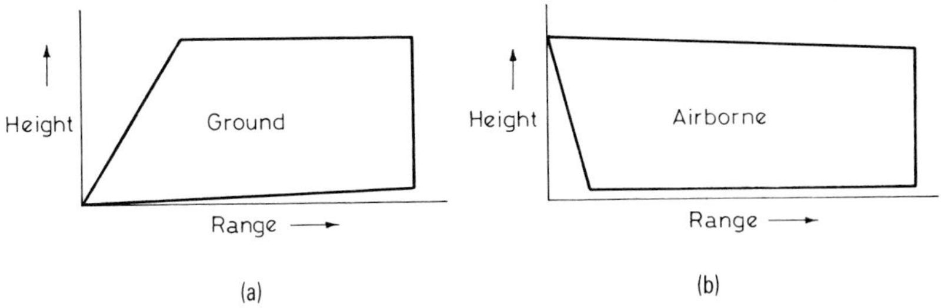

Fig. 3.9(a) and (b). Idealised vertical polar diagrams of 'cosecant squared' shape.

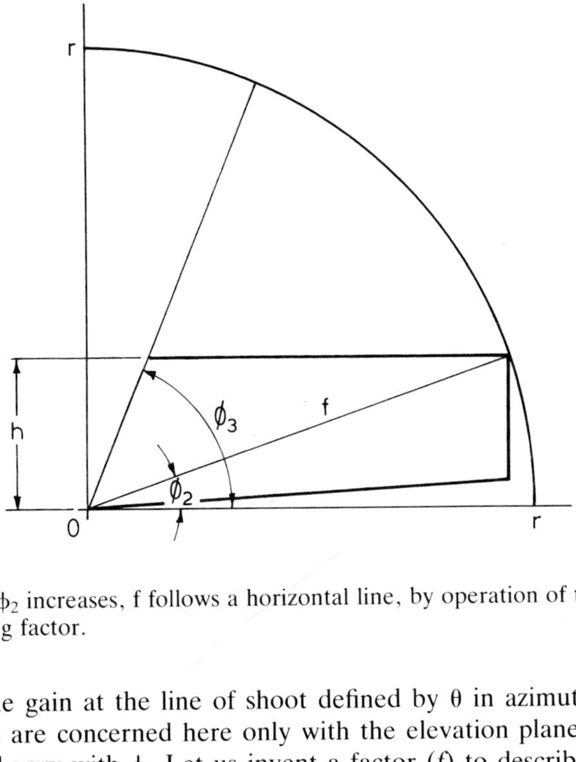

Fig. 3.9(c). As ϕ_2 increases, f follows a horizontal line, by operation of the 'cosecant squared' shaping factor.

where G is the gain at the line of shoot defined by θ in azimuth and ϕ in elevation. We are concerned here only with the elevation plane and, thus, how G should vary with ϕ. Let us invent a factor (f) to describe this. This leads to:

$$G(\theta,\phi) = G_{max.} f(\theta,\phi) \tag{3.4}$$

Translating fig. 3.9(a) into terms of the shaping factor, f, we can redraw it as fig. 3.9(c). Between ϕ_2 and ϕ_3, h must be held constant otherwise it would grow as ϕ increased, to the same value as r at $\phi_3 = 90°$. This would generate much unnecessary high cover which would be wasted. From fig. 3.9(c) we have:

$$\frac{h}{f} = \sin \phi = \frac{1}{\text{Cosec } \phi}$$

therefore $\frac{f}{h} = \text{Cosec } \phi$

and $f = h \text{ Cosec } \phi$ (3.5)

Thus if h is required to be a constant f must vary as Cosec ϕ. The shaping factor, f, expresses the change of magnitude of the antenna gain so the power gain varies as f^2, leading to:

$$G(\phi) = G_{\text{max.}} \text{ Cosec}^2\phi \quad (3.6)$$

The above description is what one might call the 'classical' explanation of the sometimes puzzling expression 'cosecant squared' pattern. Rarely are such patterns actually achieved in practice because nature abhors such sudden changes as indicated by the idealised curves.

3.5 Reflection effects

The usual explanation is very simple but it is worthwhile to look a little deeper. The classical case is described as follows.

Consider a radiating source 0 at a height h, above a reflecting surface, as in fig. 3.10. There will be two paths by which energy can reach, and be returned by a distant illuminated target reflector; the direct path OA and the reflected path OPA. Suppose all the energy reflected at P was to be directed to the target, i.e. the surface had a reflection coefficient (ρ) of -1.0. Thus there is no loss on reflection, but a phase reversal. This energy would combine at A with that following the direct path OA. The incident power will be the vector addition of the two. Thus if they were equal in magnitude and in phase at the target the incident power would be twice that of OA. If they were in antiphase they would cancel to zero.

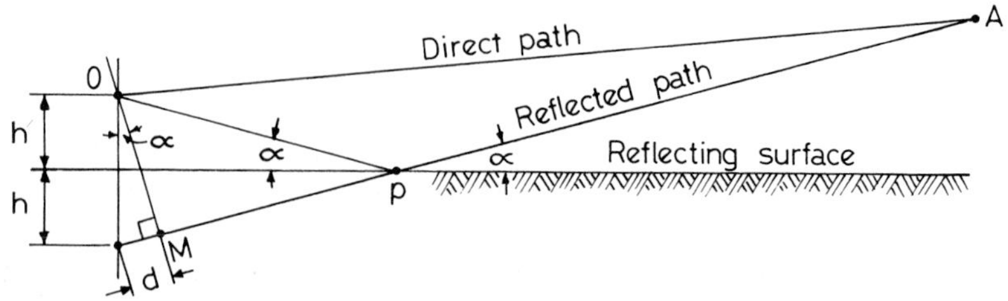

Fig. 3.10. Reflection geometry.

This circumstance would apply in the elevation domain first when the path length differences were such that:

$$OA = OPA - \frac{\lambda}{2}$$

That is OPA is one half wavelength greater than OA. In practical fact, a pulse of 2 μs duration and 1000 MHz in frequency would contain 2000 cycles of r.f. energy. The pulse energy from OPA would reach the target half a cycle late (also suffering a phase reversal on reflection). The net effect is that the two pulses of energy would virtually overlap except for the two half cycles at the beginning and end of the pulse. These conditions occur repeatedly as the elevation range is explored. The elevation angles at which the constructive and destructive effects occur are a function of the height of the radiating source, and the wavelength of radiation.

The elevation angles may be calculated from the following argument. In fig. 3.10 the lengths OA and OPA are assumed to be very much greater than h, the height of the antenna's phase centre above a plane reflecting surface. The geometry allows consideration of an image of the real phase centre of the antenna below the reflecting surface at depth h. We have posited that the path length differences (OPA−OA) must be integers of half a wavelength.

The assumption above allows the arc 0M to be considered as forming a right-angle with MA. Thus the path length difference we seek is d:

$$d = 2h \sin \alpha \tag{3.7}$$

We also need d to be integral numbers of half wavelengths, thus

$$\frac{n\lambda}{2} = 2h \sin \alpha \tag{3.8}$$

which, when transposed, gives

$$\sin \alpha = \frac{n\lambda}{4h} \tag{3.9}$$

For angles up to 4°, sin θ = Tan θ = θ radians to a very high accuracy (to about 1 part in 7000). Thus we may write

$$\theta_n = \frac{n\lambda}{4h} \tag{3.10}$$

where θ (in radian measure) is the elevation angle at which maxima and minima occur in sequence, n representing integers. Odd integers are lobes, even integers are gaps. The effect is illustrated in fig. 3.11.

The model given above is a very simple one. The explanation has been in terms of geometrical ray theory. It is the limiting case of the more complex practical situation. In practice the mechanism is as follows. The antenna will

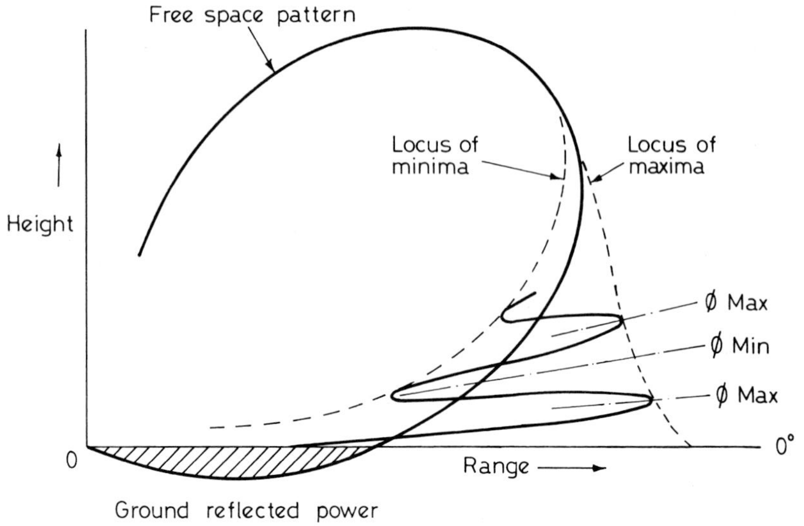

Fig. 3.11. Ground reflected energy modulates the free space pattern.

illuminate an area of the reflecting surface. This surface will reflect energy in two ways:

(a) By diffuse and disorganised omni-directional scattering.
(b) By directional specular reflection.

In the former, no phase relationships between reflected energy from adjacent points are maintained. In specular reflection the phase relationships *are* held. If the surface is smooth enough to give rise to specular reflection, the myriad points in the illuminated area will each direct energy with coherent phase relationships and some of it at angles such that it reaches the point target. The reflecting area will have differing power densities incident upon it since the antenna pattern has usually been deliberately designed to reduce reflection effects and also to concentrate it in a narrow azimuth beam. Thus we can consider the area as having a power density 'footprint' upon it. The resultant power density directed at the target is the sum of all the contributions in that line from all the specularly reflecting points, each having their own small intensity. As can be imagined the full calculation is extremely complex; the simple model given is the limiting case of a perfect and infinite reflecting surface. It is made practical by allowing for reflecting surface imperfections and assuming the reflecting area is large enough (and it usually is) to support the theory.

A number of points should be noted. First, the reflection coefficient for most surfaces is not unity. It also has different characteristics for horizontally, vertically and circularly polarised waves. The actual values are themselves a function not only of the dielectric property of the reflecting surface but also of how electrically 'smooth' the surface is. This is usually referred to as the 'roughness criterion' and expresses undulations of the surface in terms of their magnitude relative to a wavelength. For instance, a rough-ploughed field seen at shallow or 'grazing' angles of up to about 10° would appear 'smoother' at 50 cm wavelength than at 10 cm. Typical values would be as shown in fig. 3.12, curves (a) and (b). It is usual to take the mean of such values for circularly polarised waves. A comprehensive set of curves giving reflection coefficients for differing terrain and wavelength is to be found in reference 3.

3.5.1 *Effect of beam shaping*

Beam shaping in the azimuth plane has already been described. The antenna gain function is also deliberately shaped in elevation to ensure power is

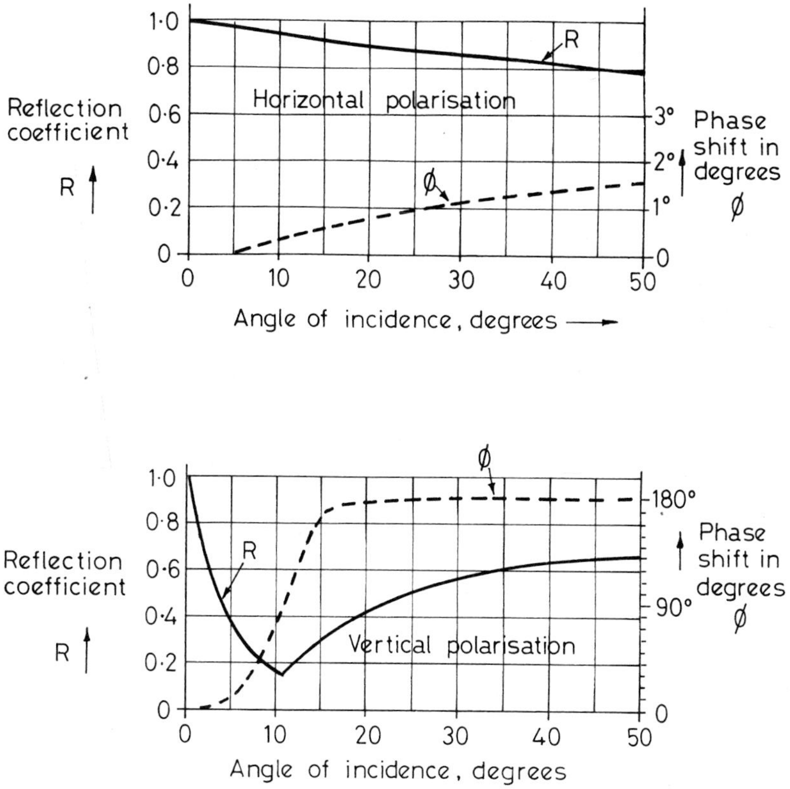

Fig. 3.12. Reflection co-efficient (R) and phase of reflected energy (ϕ) variations with angle of incidence.

directed where it is needed. One such shape expressed as a height/range diagram is shown in fig. 3.13. It is typical of the 'cosecant squared' pattern found in a.t.c. radars and is designed to reduce power at negative angles of elevation which creates unwanted vertical lobing. It also restricts radiation to altitudes in which wanted targets will be found.

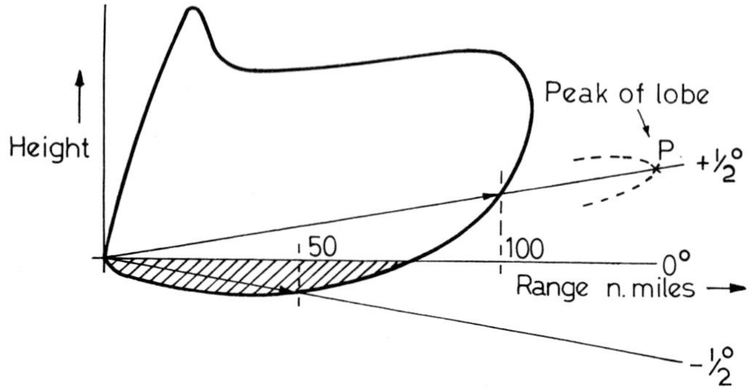

Fig. 3.13. An example of beam modulation.

Suppose the elevation angle at which the direct and ground reflected energy combined constructively (i.e. formed the first of a series of lobes) was at + 0.5°. At 23 cm wavelength this would occur if the antenna was about 8 m above flat ground.

From the free-space diagram we see that the energy reflected from the ground (directed from the antenna at negative angles of course) reduces as elevation decreases. Thus the power reflected into the + 0.5° elevation angle is far less than that by the direct path at + 0.5°. Remembering the reflection coefficient is less than unity (assumed to be 0.9, the two-way range effect is 0.9 squared), we see for the hypothetical range figures given that the actual range of the cover diagram at + 0.5° elevation will grow from 100 nautical miles to $R = 100 + (50 \times 0.8) = 140$ nautical miles.

From this it is obvious that the greater the rate of change of gain of the antenna at negative angles of elevation, the less becomes the magnitude of the lobes and gaps formed in space. This is another way of saying that the designer seeks a 'sharp bottom edge' to the vertical radiation pattern.

An interesting example of how this reflection effect, usually unwanted, can be put to good use is found in Marconi Radar's series of 50 cm wavelength radars. In these, the antenna is deliberately mounted close to flat ground (commonly the airfield itself). The net result is to create a cover diagram with a long forward lobe giving valuable increased low radar cover and a first gap that doesn't intrude too deeply into high elevation cover. A measured cover diagram of one such radar is shown in fig. 3.14.

All the above is predicated upon a reflecting surface which directs power to

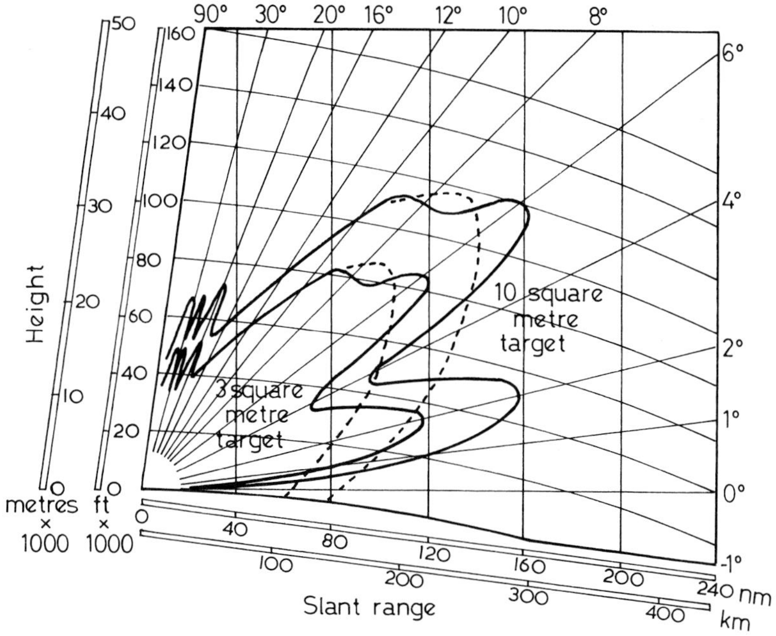

Fig. 3.14. Vertical polar diagrams of a 50 cm radar whose phase centre is 12 ft (3.69 m) above flat ground. (*Data courtesy of Marconi Radar.*)

arrive virtually simultaneously with that via the direct path. This means that the surface has to be a horizontal plane over an extended area and must have no lateral tilt (any tilt would cause the reflected energy to be directed away from the vertical plane containing the direct radiation). Airfields, water surfaces and natural planes are examples forming good reflecting surfaces. But in practice many sites have surface slopes which vary with azimuth and whose character changes from grass to concrete to 'urban area'. Thus as the antenna beam comes under the influence of these changing conditions the resultant vertical polar diagram also changes.

At very short wavelengths (23 cm and less) the net effect can be pictured as a beam in the shape of a hand with many long fingers rotating about the wrist. As the hand rotates, the fingers grow or lessen in length and change their spacing as the ground over which transmission is made changes its character, height, slope and extent.

4
Propagation

4.1 The atmosphere

Despite the fact that the Earth is flattened slightly at the poles and widened at its equator, for what follows it is in order to consider it truly a sphere. Overlaying the two shapes (a true sphere and a true Earth) gives correspondence at about 48° north and south of the Equator. Although the distortion from a true sphere creates trouble for navigators – the nautical mile is the length of an arc subtending an angle of one minute measured from the Earth's centre and will vary across the surface if it isn't a sphere – a mean value of 6080 ft (1.86 km) is taken. On these bases the radius of the Earth is 3437 nautical miles (6371 km). Most of the Earth's surface is covered by water and the rest by land whose surface varies considerably in terms of electric properties i.e. conductivity and dielectric constant. Surrounding the Earth is its atmosphere: a mixture of gases, notably water vapour, which exists in significant quantities up to about 30 000 ft (9000 m). Its density and quantity vary significantly, not only across the whole surface of the Earth but locally as well, particularly at land–sea interfaces. Its variations in the height domain are particularly important in considering propagation and its anomalies.

Atmospheric pressure decreases with height and so therefore will the atmospheric density. A pressure of 1000 mB at sea level will fall to about half this value at 18 000 ft (4000 m) and to about a tenth at 50 000 ft (15 000 m). At around 100 000 ft (28 000 m) the atmosphere can be considered not to exist.

The sun's effect upon the atmosphere can therefore be seen to be greatest near the Earth's surface and to diminish with height. Temperature changes occur at the rate of about 1° C to 3° C per 1000 ft (280 m) at the lowest levels and at lower rates as height increases. Such changes are seen up to heights of about 45 000 ft (12 000 m). It is in this environment that radar signals travel. Their speed 'in vacuo' (C_0) is given as 3×10^5 km/second. This is modified slightly by the atmosphere to values dependent upon its electrical characteristics. It can be expressed as:

$$C_{at} = \frac{C_0}{\sqrt{E}} \tag{4.1}$$

where C_{at} = velocity in the atmosphere
E = the dielectric constant
C_0 = velocity in vacuo (free space)

Thus a ratio is seen by transposition of (4.1) between the velocity of propagation in vacuo and in an atmosphere.

$$\frac{C_0}{C_{at}} = \sqrt{E} = n$$

This will be familiar, from physics, as a measure of the refractive index of a medium and we are constantly reminded of its effect when seeing (at an oblique angle) how a stick or straw apparently bends when it is put into water. The refractive index of water and air are different: the denser the medium, the slower is the propagation velocity. It has expression as Snell's Law.

In the fig. 4.1(a) a ray passes through a medium of refractive index, n_1, into another of refractive index n_2. The angles made by the ray in a given plane are θ_1 and θ_2 respectively. Snell's Law states:

$$n_1 \cos \theta_1 = n_2 \cos \theta_2 \tag{4.2}$$

It is easy to appreciate the effect of a ray passing through a medium whose refractive index is continually changing, as in fig. 4.1(b), where the ray gradually bends away from a straight line. This is precisely what happens to radar beams as they are formed in space because the density changes with height in a gradual fashion. Imagine a ray in the beam directed along an elevation angle of 3°. Taking the Earth's radius as 3437 nautical miles the ray would achieve height of about 50 000 ft at a distance of some 150 nautical miles. Thus it will transverse through that section of the atmosphere affecting propagation most. As height is increased the atmosphere's density reduces

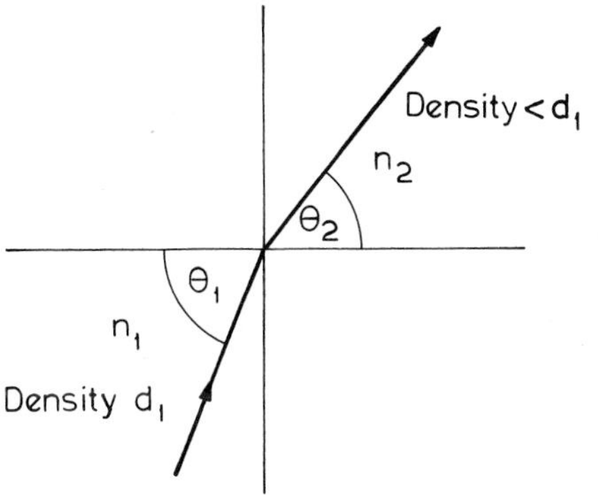

Fig. 4.1(a). Ray bending at the interface of two disparate media (Snell's Law). n is the refractive index.

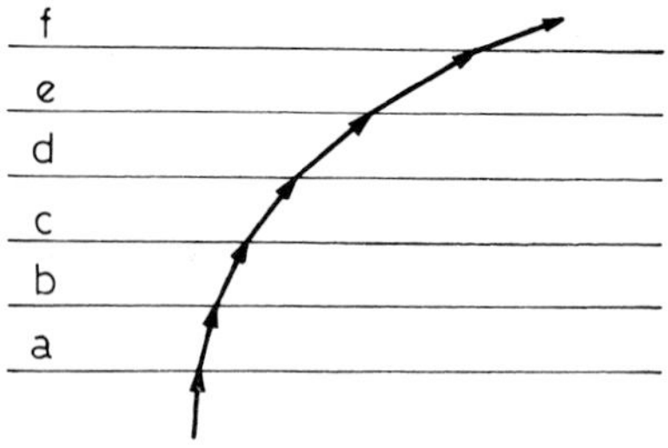

Fig. 4.1(b). Continual ray bending as the medium gradually changes density. The effect is greatly exaggerated in the illustration.

and so in turn does its refractive index. The effect is to bend the ray downwards and it becomes curved. The amount of curvature is dependent upon the way the atmosphere's refractive index changes and this in turn depends upon the temperature, humidity and pressure values it experiences.

4.2 The standard atmosphere

From the above it might seem that the many interdependent variable parameters prevent one from knowing what actually is going on. The way through the difficulty is found by introducing the notion of a standard atmosphere, and taking it as a reference point. An assumption (a good working one) is made that the atmosphere has water vapour content, temperature and pressure changes with increasing height, such that in the lower regions of most interest to radar users the rate of change of its refractive index, $dn/dh = -12$ parts per million per 1000 ft (280 m).

It must be said, however, that seldom is a standard atmosphere found in the same place for long because day and night, summer and winter climatic changes are present all over the surface of the globe.

4.3 The bigger Earth

An unfortunate and awkward consequence follows from the refraction effects described above. The real and apparent positions of a target are different, as illustrated in fig. 4.2. Suppose the radar was a height finder with a beam narrow in elevation. Knowledge of the angle of launch of the beam along the path OA and the range to the target would allow calculation, by geometry, of the target height. But because the beam is bent downwards and follows the

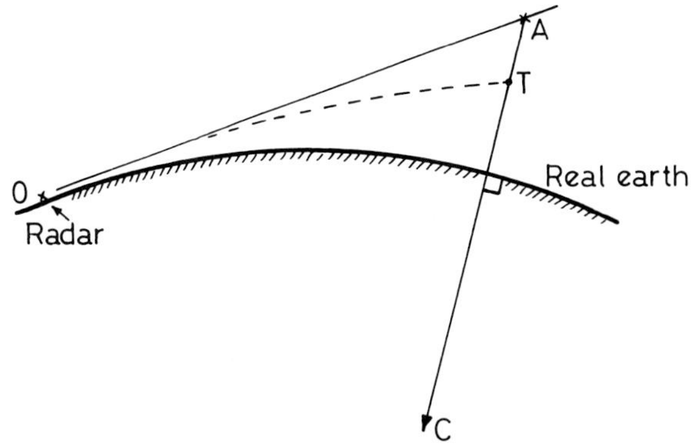

Fig. 4.2. Target at T appears to the radar to be at A. By assuming the earth's radius (AC) is larger, the points T and A are made to coincide.

path OT, the height calculation would obviously be in error. Such errors will be a function of range and the degree of beam bending. How to overcome this difficulty?

Propagation engineers have a device, using the idea of a standard atmosphere which does this. It works as follows. If a graticule representing a normal curved Earth were laid down with concentric circles about the Earth to represent height above ground and radial lines from its centre equally spaced to represent equal distances around the surface, rays plotted upon it would be curved for the reasons given above. But for a standard atmosphere a constant lapse rate of refractive index has been taken, so if the Earth's curvature was corrected, i.e. made less, the rays would straighten out.

The appropriate correction for the standard atmosphere is to make the Earth have a radius 1.33 (4⁄3) times normal. Most radar vertical cover diagrams are drawn on this basis and an example is given in fig. 4.3. The lines representing range are drawn parallel, even though properly they should pass through the Earth's centre. The divergence is small enough (for most radar purposes) to be neglected. By adopting this technique another awkwardness is literally 'straightened out'.

When carrying out radar performance calculations and plotting radar vertical coverage diagrams from known beams shapes it is highly inconvenient continually to account for ray bending. How much better if the rays were simple straight lines. The '4⁄3 Earth' technique does just this.

An example of the convenience and 'rightness' of using the device is seen in the effect on horizon distances. Consider a radar antenna some tens of metres above the true Earth's surface. The optical horizon of the antenna would be the tangent point of a straight line from the antenna to the surface. The distance of the point from the antenna would be a matter of the simple geometry of the straight line, the circle's radius and the antenna height. Now

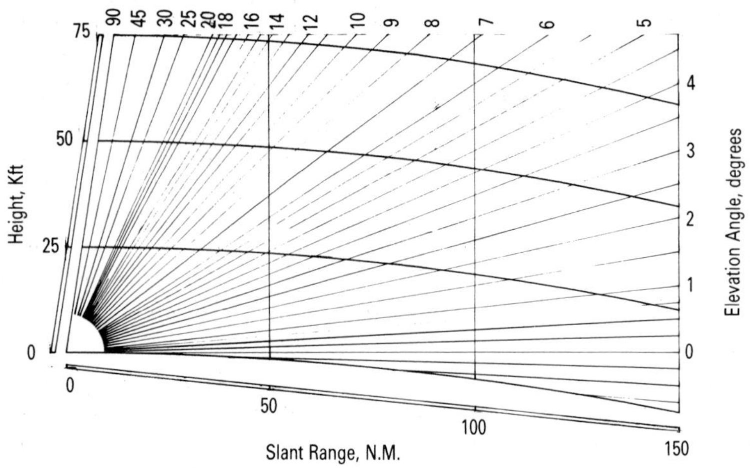

Fig. 4.3. The '4/3 Earth' model is satisfactory up tp 10 Kft. A more sophisticated model is illustrated to apply to greater heights (see ref. 21).

consider radar energy launched along this line. As it passed over and beyond the tangent point, its path would grow in height above the Earth's surface and be subject to the ray bending effect described earlier. Thus the radar horizon of the true Earth occurs at a greater distance from the antenna than does the optical horizon.

The effect is tolerable, even welcome, in surveillance radars because low coverage is increased. Height finding radars however are adversely affected by the assumption of a standard atmosphere when attempting to measure true height at low angles. Corrections to such measurements can be made if the physics of the atmosphere between radar and target are known by meteorological data such as can be gathered by radio-sonde systems. In these temperature, pressure and humidity changes with height are measured by sensors carried by a free balloon tracked either by theodolite or a tracking radar. The data is automatically signalled to ground by a small radio transmitter carried by the balloon. However, further assumptions have to be made that no atmospheric changes occur between meteorological measuring times. Relative heights between targets in the same area can still be relied upon since the energy returned from the targets travels the same route out and back to the radar. The errors introduced will normally be very small and it would be very unlikely that an atmospheric aberration would affect the measured height difference between two aircraft in similar positions.

4.4 Anomalous propagation (anoprop)

Anomalous propagation usually means 'any propagation condition different from that assumed in a standard atmosphere'. It will have been realised that by this definition, because of the vagaries of weather, very seldom are propagation conditions *other* than anomalous to some degree. In fact in some parts of the world, 'anomalies' are more nearly the normal condition.

Enough will have been said for it to be realised how such conditions can come about. Climatic changes which severely disturb the atmosphere's refractive index are the root cause. These disturbances take three main forms of consequence to radar operation. Each follows from a discontinuity in the lapse rate of coefficient of refraction. Layers of warm or cold air can be formed over the Earth's surface. The transition between warm and cold or cold and warm layers can be sharp enough to define two distinct lapse rates for the coefficient of refraction, one for each layer.

At the interface between layers the beam bending can be enhanced or diminished, producing disturbance to antenna beam shape. Sometimes this interface is so well defined and of such a height above the Earth that radar energy launched at low angles cannot penetrate it and is totally reflected downwards, following the Earth's curvature as does the layer of air between the interface and the Earth's surface. Such a condition is termed 'ducting' or 'super-refraction' and since the trapped energy follows the Earth's curvature, the radar has no horizon. The effects of this are discussed in part 2.

The third form of 'anoprop' is a combination of the other two layering effects, where an elevated duct is formed. This duct virtually traps the energy

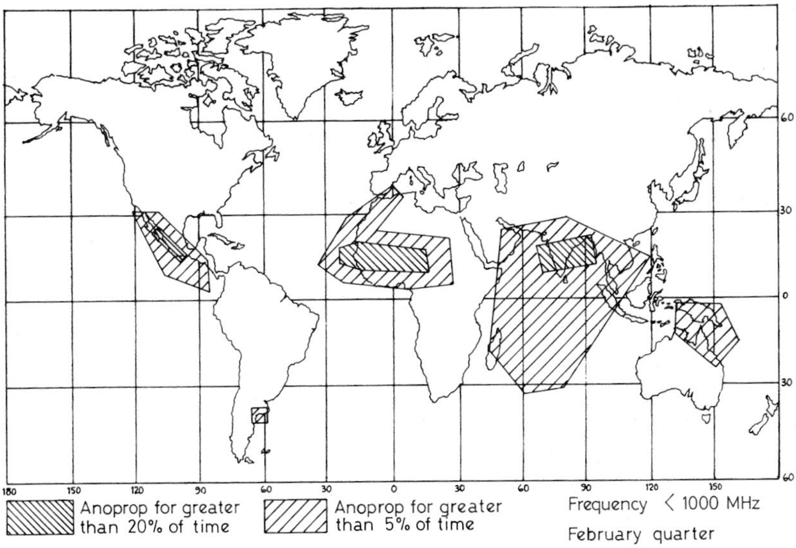

(a)

Fig. 4.4(a)–(h). A general guide to the incidence of anomalous propagation showing seasonal variations for two frequency bands.

in the layer between the two interfaces. The angles of attack within the duct have to be small enough to produce total internal reflection. In all these effects, angles are only of the order of about 1° but this is large enough to have profound impact for many radars. These effects can be present at all radar wavelengths. The onset of anoprop is, however, earlier in general for the shorter wavelengths, as is the effect's duration.

Figures 4.4(a) – (h) have been included as an indication of how anomalous

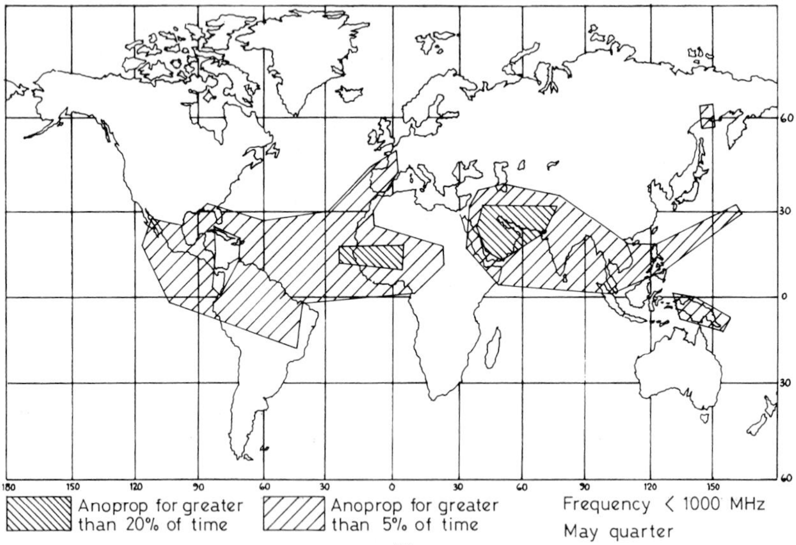

(b)

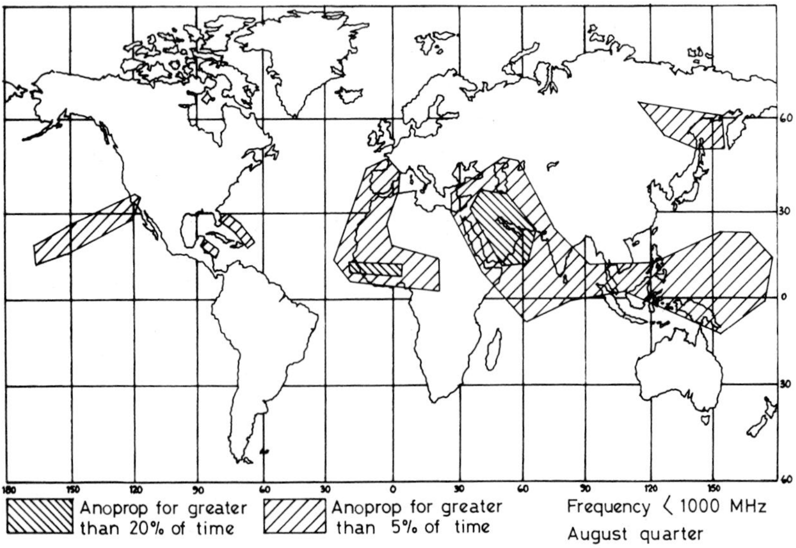

(c)

propagation occurs over the globe, its extent and average changes within the year. The frequency spectrum has been broken into two important microwave bands. The figures have been derived by R. Drage from data carefully compiled from reference 3.

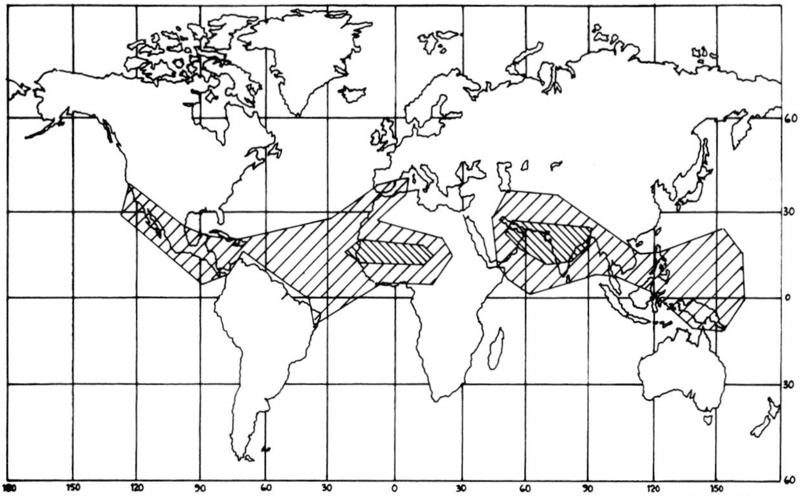

Anoprop for greater than 20% of time

Anoprop for greater than 5% of time

Frequency < 1000 MHz

November quarter

(d)

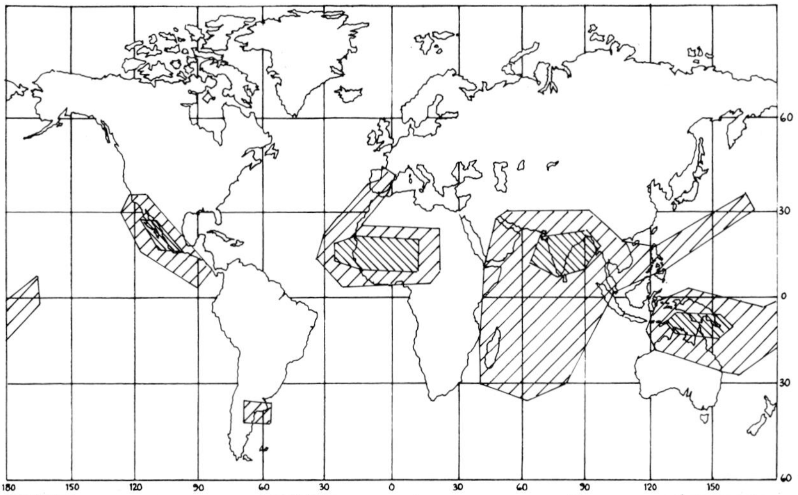

Anoprop for greater than 20% of time

Anoprop for greater than 5% of time

Frequency < 3000 MHz

February quarter

(e)

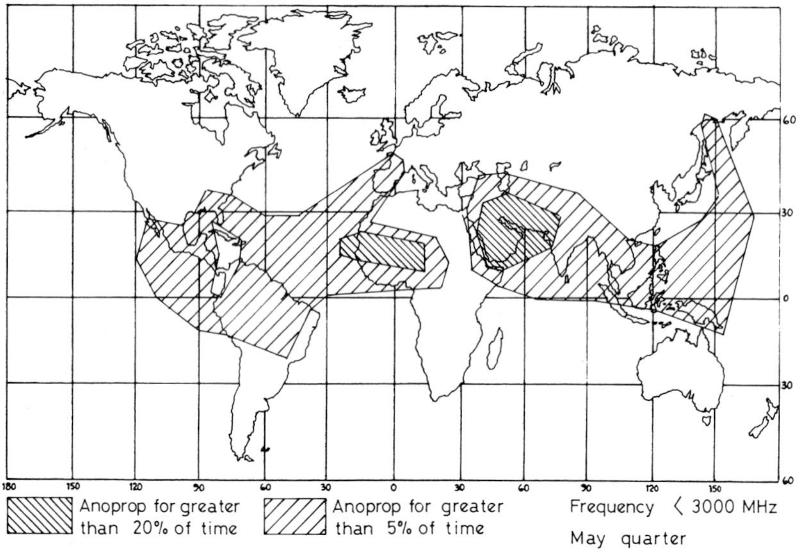
180
150
120
90
60
30
0
30
60
90
120
150
60
30
0
30
60
Anoprop for greater than 20% of time
Anoprop for greater than 5% of time
Frequency 〈 3000 MHz
May quarter

(f)

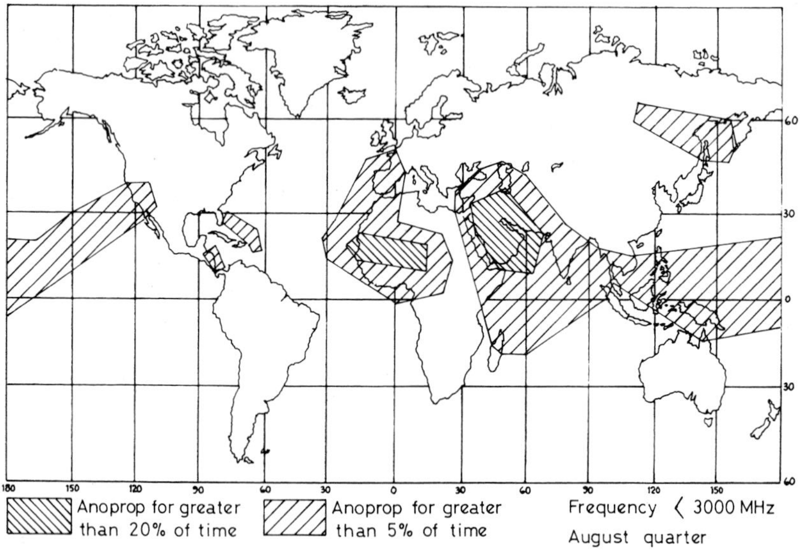
180
150
120
90
60
30
0
30
60
90
120
150
60
30
0
30
60
Anoprop for greater than 20% of time
Anoprop for greater than 5% of time
Frequency 〈 3000 MHz
August quarter

(g)

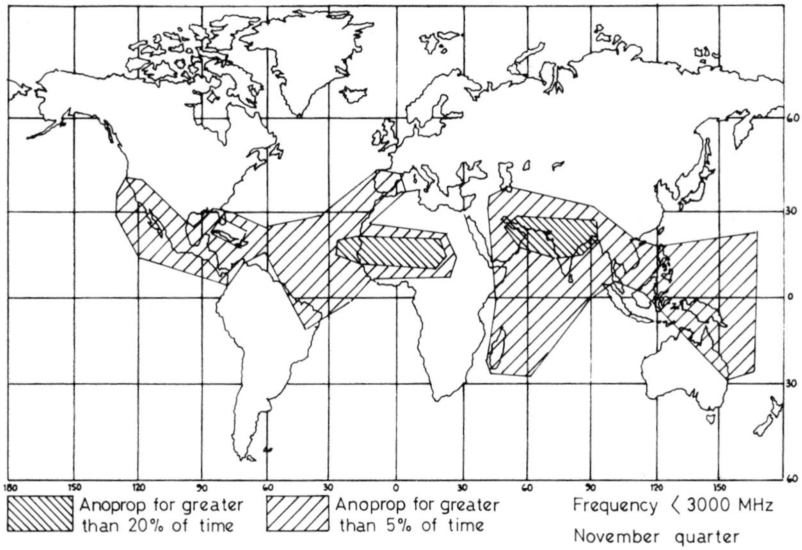

(h)

Wavelength

An electro-magnetic (E–M) wave's history in space is represented in fig. 1.3. Its speed of propagation in a vacuum is 2.997925×10^8 metres/s. This is usually rounded up to 300 000 km/s (186 000 statute miles/s). The speed of propagation in dry air is very slightly less (see equation (4.1)) because the dielectric constant of air is 1.00059 instead of unity *in vacuo*. The difference is, however, very small (about 0.1%) and for practical radar range measuring purposes it can be neglected.

As you sit reading this, a multitude of E–M waves are passing by (and sometimes through) you; their energy is low enough not to cause harm. Suppose one of these waves had a frequency of 100 MHz which was coming perhaps from a VHF radio station. One hundred cycles would pass you in a millionth of a second. The wavelength is characterised by the distance occupied in space by one complete cycle – that between neighbouring wave peaks for example. This distance is the speed of propagation divided by the time to execute one cycle, and for a frequency of 100 MHz the wavelength (λ) will be:

$$\frac{300\,000 \text{ km/s}}{100 \times 10^6 \text{ cycles/s}}$$

So, $\lambda = 3$ m.

The simple general equation showing the reciprocal relationship between wavelength (λ and frequency (f) is:

$$\lambda \text{ (m)} = \frac{3 \times 10^8}{\text{f (Hz)}}$$

5

Output power generation and distribution

5.1 Output power generation

In both primary and secondary radar it is necessary to generate microwave power and to distribute it as efficiently as possible. This chapter is concerned with these two matters at a fundamental level, from which particular engineering designs can be understood. There are two prime methods of microwave power generation used in radar, across its spectrum from about 8 mm to 50 cm wavelengths. These are:

(a) Self-oscillating devices.
(b) Power amplification devices.

Their relative advantages and disadvantages are discussed in part 2, chapter 8.

In the self-oscillating group is one of the most commonly used cross field oscillators (CFOs), the magnetron. In the power amplification group there are linear beam amplification tubes such as the klystron and travelling wave tube. High power transistor amplifiers are also now coming into use. All these devices can be characterised in simple form which illustrates fundamental differences between them. That difference is found in the way in which they can be made to work as coherent systems.

5.1.1 Self-oscillators

The fundamental mode of operation of self-oscillators is illustrated in fig. 5.1. The CFO, usually a magnetron, produces its output when a high energy pulse is applied across it. The pulse may be generated in a number of ways but all use the principle of taking power at a low level for a long time and releasing it at high level for a much shorter time. This transformation is governed by timing or trigger circuits. The action is shown in fig. 5.2.

Points of importance to note are:

(i) A proportion of the high energy voltage pulse has to be established across the magnetron before it starts oscillating.

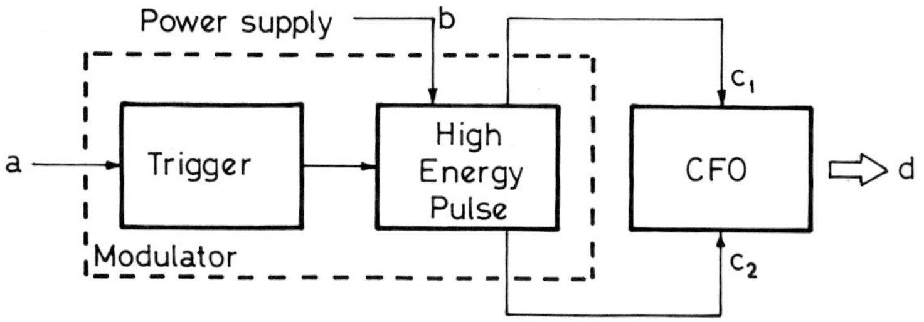

Fig. 5.1. The generation of output from a cross-field oscillator (e.g. magnetron). The letters (cross-refer to fig. 5.2) show the nature of waveforms in the system.

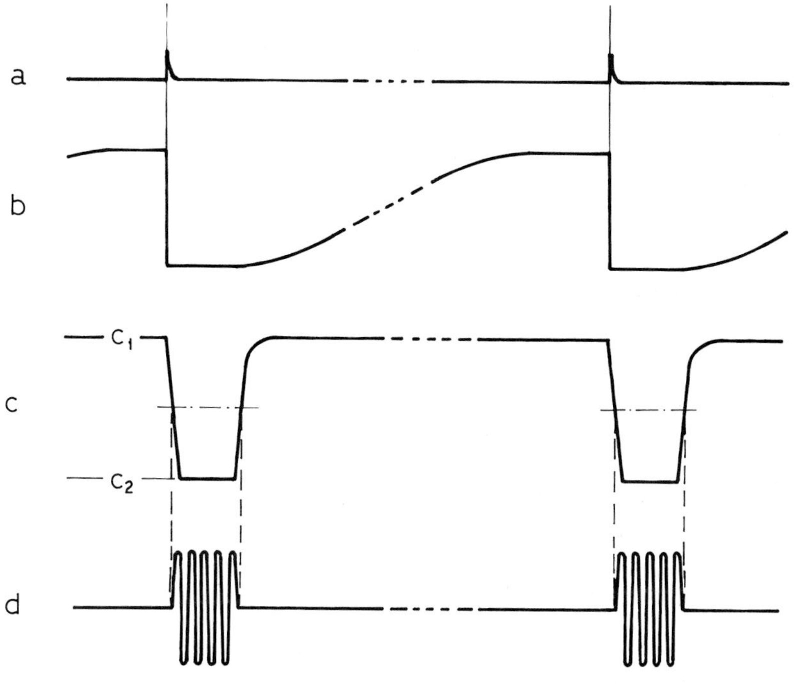

Fig. 5.2. The r.f. pulse (d) is generated when the high potential difference ($C_1 - C_2$) is applied across the CFO.

(ii) The number of cycles in the final output is:

$$n = ft \tag{5.1}$$

where f = frequency in Hz

t = time of pulse duration in seconds

So an L-band transmitter whose output frequency is 1300 MHz and pulse duration is 2.5 μs will release $1300 \times 10^6 \times 2.5 \times 10^{-6} = 3125$ cycles.

These will occupy 2.5/6.18 nautical miles in space or approximately 750 m.

(iii) The phase of the oscillations between each successive pulse varies.

5.1.2 Power amplifiers

The important difference between this class of microwave transmitter and self-oscillators is exemplified by the high power klystron. It is represented in simple form in fig. 5.3. The high energy pulse development is the same in principle as used in self-oscillators. The difference lies in the way the microwave output pulse is produced. The oscillator, usually of very high stability, is left running continuously and, neglecting engineering practicalities for the moment, can be considered as being 'gated' on and off by the pulse modulation. The action is shown in fig. 5.4. Because each pulse is virtually a 'bite' taken out of a continuous wave of high stability, the phase of successive output pulses is related and knowable by reference back to the oscillator producing the amplified pulse version. In the self-oscillator there is no such in-built source and the reference, when needed as in mti systems, has to be established by other means; it is commonly known as the 'Coho-Stalo' technique, wherein to produce the necessary coherence, a sample of

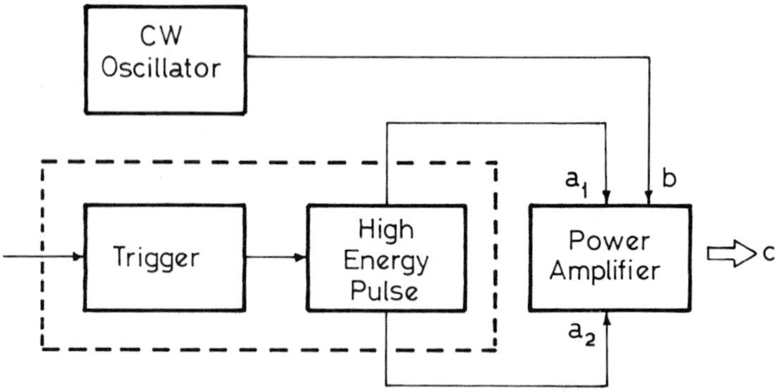

Fig. 5.3. High energy pulse (c) generated by power amplification in the output device. Waveforms in the system cross-refer to fig. 5.4.

the transmitted pulse is taken and converted to the i.f. by mixing with the receiver's very stable local oscillator (stalo). The resultant pulse is used to phase-lock another oscillator (the coherent oscillator – coho) at each new transmission.

5.2 Power distribution – the problem

There are modern radar systems, mainly airborne, (forward or sideways

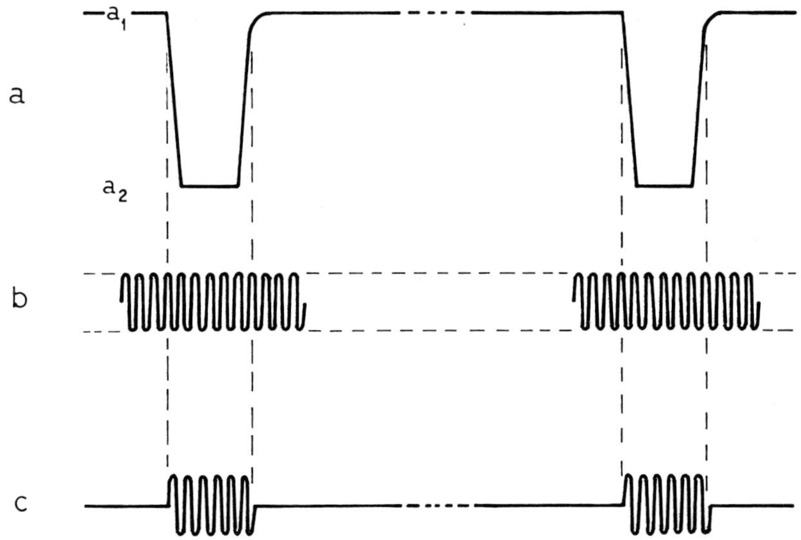

Fig. 5.4. Output pulse (c) is virtually a gated sample of a continuous wave input (b).

looking) wherein the radiator consists of clusters of very small transmitter/receiver elements arranged in a planar array. These elements, small both physically and electrically, form the system's antenna. Each element acts as an independent phase centre which, when combined with the rest in space, creates the necessary power density and directivity. The problem of transferring the transmitter's microwave power to the radiator is therefore avoided, because the transmitter/receiver *is* the antenna.

However, until such systems become cheap enough to use on a large scale for long range surveillance purposes the problem of power distribution has to be solved. How? There are numerous ways; the choice for use depends on frequency and power level.

There is a significant difference between the power distribution of direct and alternating currents. We are concerned with the latter, their frequency being very high indeed. Direct currents flowing in a conductor are distributed evenly throughout the conductor's cross-section with local variations of current density caused by any unevenness of resistance in the conductor. The flow of current is governed purely by the ohmic resistance. In the case of alternating currents, the conductor's inductance and capacitance have to be taken into account as well as the pure ohmic resistance. At very high frequencies the inductance becomes a significant factor and its effect causes the current density to be least in the conductor's centre and greatest at its surface. This is called the 'skin effect'. The higher the frequency the greater is this effect – so marked is it at high frequencies that conductors formed from tubes of metal can be used.

At the ultra-high frequencies used in radar, currents in conductors are determined much more by mutual and self inductance and capacitance parameters. Their analysis invokes transmission line theory which is not appropriate (or necessary) to present here. Suffice to say that it is the designer's aim to make the connection arrangement between the power source and the load (in our case the radar transmitter and antenna respectively) appear 'transparent'. In transmission line theory this 'transparency' is brought about by the process of *impedance matching.* It attempts to make the effective capacitance and inductance characteristics of source and connection the same over the band of frequencies of concern, with as little loss as possible. The aim is illustrated simply in fig. 5.5.

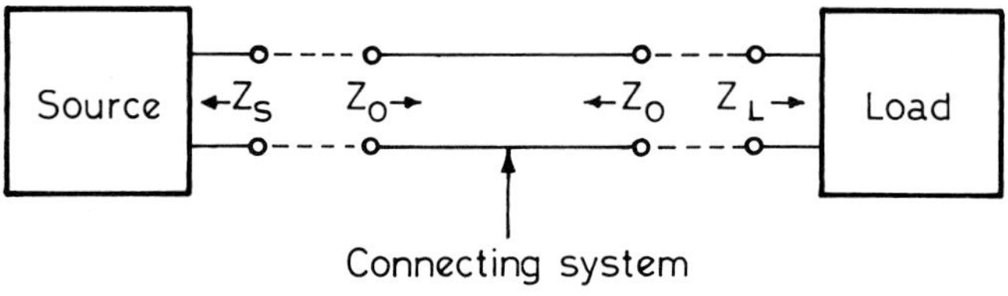
Source
Load
Connecting system
Ideal match is when $Z_S = Z_O = Z_L$
where Z = impedance of devices

Fig. 5.5. Maximum efficiency of power transfer from source to load is obtained by impedance matching.

Of course, the connecting system is never ideal. All systems lead to losses (resistive, imperfection in matching, radiation, dielectric, etc.). A simple analogy of matching may be drawn by considering a window in your house. Its purpose is to allow the viewer to see clearly through a wall and to prevent weather from entering. Plain, even, colourless, glass would be a good match. Its evenness and plainness would be a measure of the match; its clarity or lack of it a measure of loss.

5.3 Power distribution – solutions

5.3.1 Twin feeders

The first solution to the problem was found by using two parallel conductors a uniform distance apart: a twin feeder line. The electrical characteristics of this type of line can be governed by careful choice of spacing between the conductors and their size. These factors set calculable values to the all-important parameters of inductance, capacitance, resistance, etc., which determine the feeder line's impedance and loss. This type of feeder is unsatisfactory for long runs at radar frequencies. Losses are made worse by radiation from the feeder conductors themselves.

5.3.2 Coaxial feeders

A successful and much used method of overcoming the twin feeder's disadvantages (physical size, problems of support, radiation losses) is to make the pair of conductors into a coaxial arrangement – a single conductor in the centre of another in the form of a tube. Again, the parameters of inductance, capacitance and resistance can all be chosen to suit the matching requirements of the device to be interconected. Because the system is virtually closed (by its coaxial nature) there is very little radiation loss.

The space between inner and outer conductors of a coaxial feeder must be an insulator with very low dielectric loss. Commonly, plastic materials are used to fill the void. Sometimes air or inert gas is used as the filler, low loss spacers being inserted at regular intervals in the feeder's length to ensure the middle conductor is held concentrically with the outer shell. A disadvantage of coaxial feeders is their inability to support the very high voltages of a radar's output without making the feeder very large indeed. An example of one of the largest forms of coaxial feeder is shown in fig. 5.6. This is capable of distributing 500 kW peak power at 50 cm with very little loss.

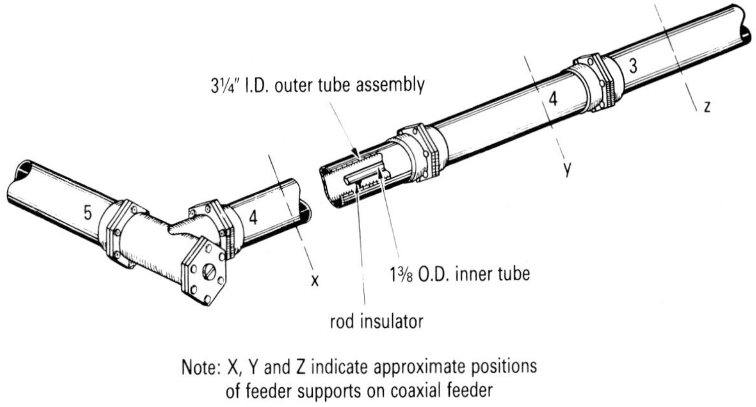

Fig. 5.6. Example of rigid co-axial feeder for high peak power use. (*Drawing courtesy of Marconi Radar.*)

5.3.3 Waveguides

The low power handling capacity of coaxial feeders is made good by the waveguide technique. Since waveguides are hollow and almost always air-filled, their size makes them very resistant to high voltage breakdown. Also, because the wavelengths in radar are small, the waveguide has manageable proportions up to wavelengths of about 30 cm.

Propagation of microwaves in a waveguide involves very little loss compared to free space since the power is confined within the guide itself. The proportions of the waveguide, whether circular or rectangular, are of importance since for a given size they determine the minimum frequency of waves they will propagate. Figure 5.7 shows a cross-section through a waveguide.

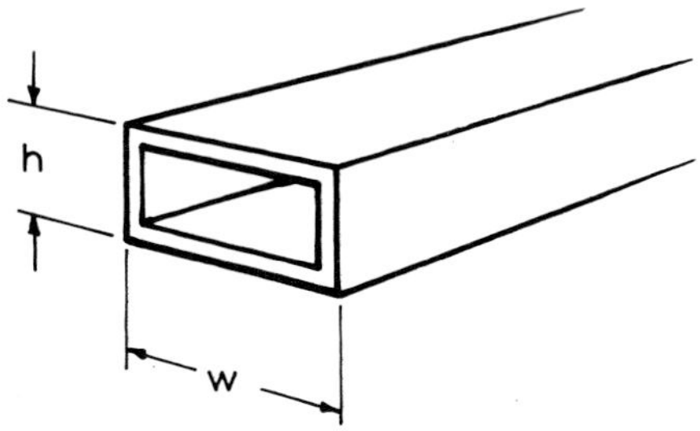

Fig. 5.7. Cross-section of waveguide.

A guide will propagate electro-magnetic energy of wavelengths shorter than a limiting value set by

$$\lambda c = 2w \tag{5.1}$$

where λc = propagation velocity in free space
w = internal width of guide.

The dimension h (fig. 5.7) is not critical and is usually about half w:wide enough to prevent voltage breakdown.

The relationships between the electric and magnetic fields within the guide are exactly as depicted in section 1.3, but the wavelength within the guide is slightly shorter than that in free space because the propagation velocity of electro-magnetic waves in the guide effectively increases above the free space value. Power is launched into the guide by probes of various shapes designed to provide a correct impedance match between the source and the connection system. A typical arrangement is shown in fig. 8.4(a) in Part 2. On the left is a glass-domed circular ring within which is a large coupling loop. The centre stalk of the loop gathers energy from the magnetron cavities which are launched into a specially shaped 'circular-to-rectangular' waveguide section (within the black body of the magnetron assembly shown on the right of fig. 8.4(a)). The rectangular waveguide coupling flange can be seen at the bottom of the assembly. The glass dome has a seal around it to allow the waveguide to be pressurised, if need be; pressurising the guide permits higher power to be handled without waveguide arcing.

An essential difference between waveguide transmission systems and the others is that waveguides do not need a direct current connection between their terminals. Sections of guide can be laid end to end without mechanical or electrical connection and still function. This is because the guide constrains the propagated waves and there is, for short distances of about a wavelength, very little divergence of the wavefront of energy at the guide's end.

However, the name of the game is to minimise transmission loss and waveguide sections are mechanically (and hence electrically) connected in order to present a smooth continuous path for the waves. It is easy to appreciate the difficulty and expense of designing sections of waveguide whose ends mate up precisely with their neighbours. Losses are bound to occur if there is misalignment or small changes of dimension at the transition from one section of guide to another. To reduce such losses and make assembly of waveguide runs easier, the ends or flanges of the waveguide sections incorporate a profile of special shape. This profile forms a choke of high impedance to leakage at the joints. The microwave energy is therefore encouraged to continue its journey down the next waveguide section rather than escape at the joints. A typical choke joint is shown in fig. 5.8.

Negotiating bends in either plane is not difficult, the corners of the bends being either radiused or incorporating flat plates at the right-angle to reflect energy along the correct path. Sometimes short sections of flexible waveguide are used where awkward alignment problems exist. These usually have higher loss (albeit small) and are avoided where possible.

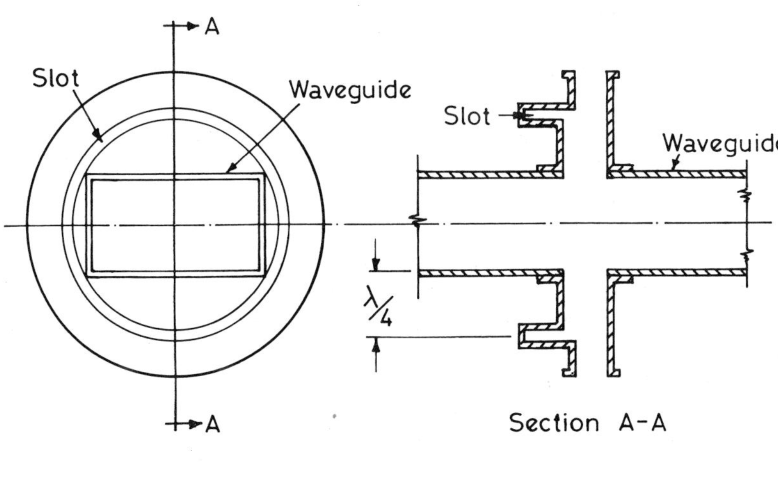

Fig. 5.8. A waveguide choke joint.

5.3.4 Rotating joints

Connection to a rotating antenna creates a problem. It is overcome by use of rotating joints. There are several techniques in design. One such is to make a transition from rectangular to cylindrical waveguide, as shown in fig. 5.9(a). The cylindrical part allows the top half to rotate relative to the bottom by provision of a suitable circular mechanical joint. Another form of rotating joint, for relatively low power transmission such as found in secondary surveillance radars, is the 'around-the-mast' type. The diagram of fig. 5.9(b) illustrates its appropriate name.

Coaxial feeders, by their concentric nature, lend themselves to rotating joints of relative simplicity. Groups of these may be formed by cunning use of the outer of one coaxial feeder as the inner of another.

5.4 Power separation – duplexing

Using a common antenna for both transmission and reception brings a severe problem to the designer. Transmission power can peak to millions of watts; the received power can be as little as micro microwatts. How to prevent the transmitter power from destroying the sensitive receiver? Duplexing (that is handling separate powers at one frequency) is performed by a variety of different devices. Whichever is used depends upon the power levels to be separated. The most commonly found in primary radar is the TR cell technique and its principle is illustrated in fig. 5.10.

The two cells are small gas discharge tubes strategically placed within the waveguide runs to transmitter and receiver. Their placement and form are designed to produce minimum disturbance to the matching between transmitter, receiver and antenna. As the transmitter pulse is developing, the voltage across the ATR cell begins to grow. At a certain point the voltage causes the gas in the tube to ionise and form a conductive path from transmitter to antenna. As the transmitter pulse progresses past the ATR cell some of the pulse power will travel towards the receiver terminals. Before reaching them it encounters the TR cell. When sufficient voltage is present across the TR cell, it too 'breaks down' and effectively short-circuits the receiver terminals. Thus the full power of the transmitter pulse is prevented from entering the receiver input stages. When the transmitter pulse is finished, the gas in both tubes deionises and restores to quiescence. In this condition the transmitter is prevented from robbing the receiver of any signals returned to the antenna and the short-circuit across the receiver is removed, thus allowing returned signals to enter the receiver.

The gas discharge tube technique is far from perfect because the reaction time is limited by the behaviour of the gas in the tubes. TR cell protection to the receiver is incomplete, because the cell requires a certain voltage across it

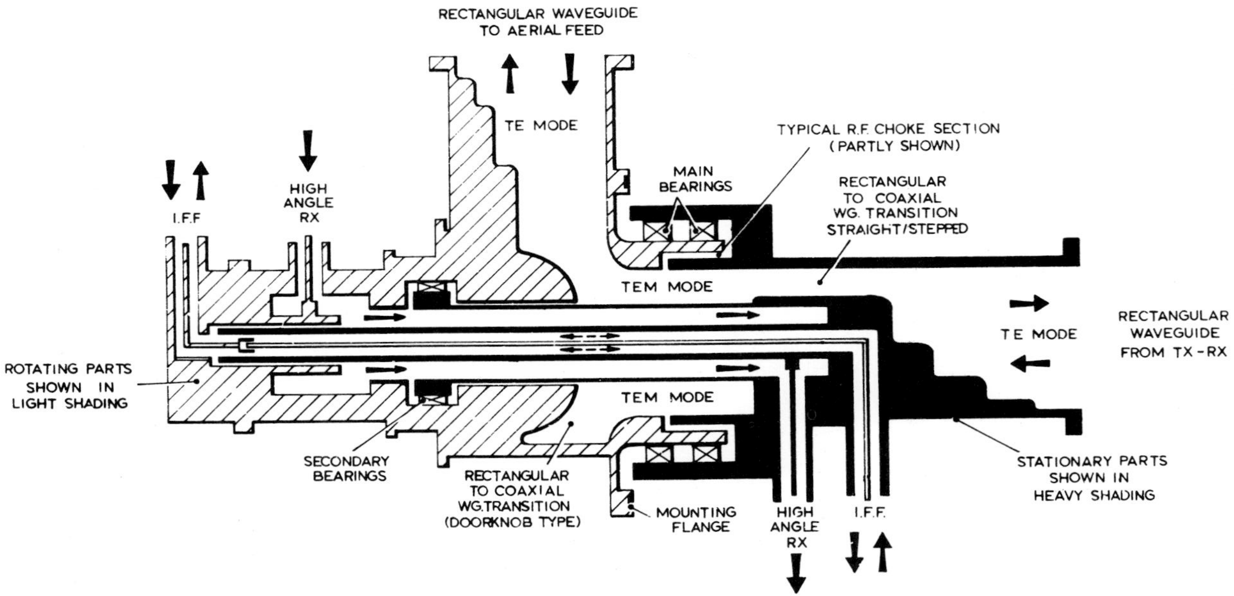

Fig. 5.9(a). A combined 3 channel rotating joint. High power primary radar output is carried via the waveguide joint having rectangular to coaxial transitions. Low power (receive only) signals in a dual-beam primary radar are passed by a co-axial joint (High Angle Rx). I.F.F. output and input signals are handled by a similar coaxial joint. I.F.F. is synonymous with S.S.R. The whole assembly is usually mounted with rotation axis vertical. (*Drawing courtesy of Marconi Radar.*)

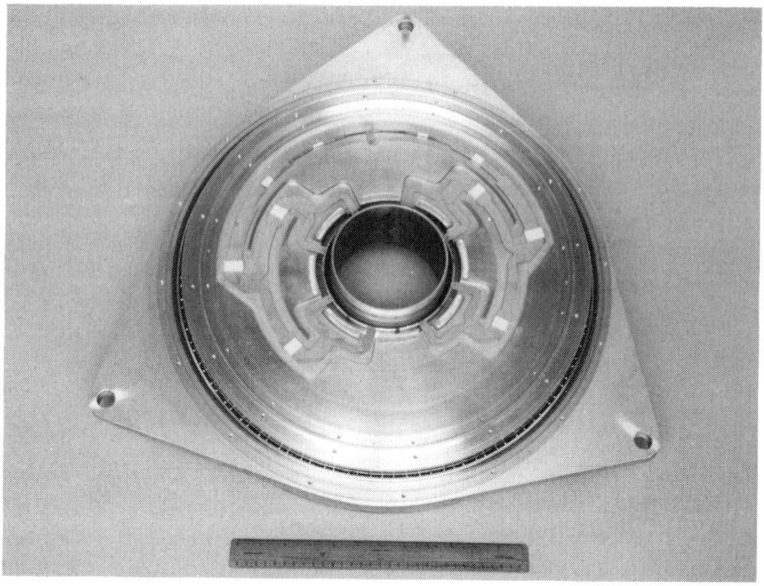

Fig. 5.9(b). Bottom half of an 'around-the-mast' rotating joint. Connection between top and bottom is by non-contacting strip-line feeder seen in this half as maze-like tracks. Top half strip-lines are circular tracks. The two halves rotate independently so one can be fixed. Connections from tracks to outer shells terminate in standard cable sockets. (*Photo courtesy of Marconi Research Centre.*)

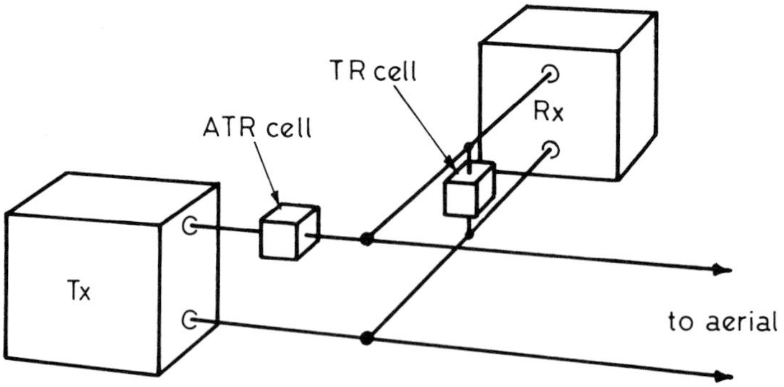

Fig. 5.10. TR (transmit/receive) and ATR (anti-TR) cell arrangement.

before it becomes effective. During the short time taken for the tube to ionise, power can leak into the receiver. This can be high enough to cause damage when the transmitter is of very high power. In such cases the TR cell is often pre-primed by a high voltage pulse before the transmitter pulse is generated. Sometimes a pre-TR cell is introduced ahead of the TR cell. Another difficulty is that the cells do not deionise quickly and so the full receiver sensitivity is not restored as soon as the transmission is over. The delay to quiescence can be as long as 70 μs (representing about 6 nautical miles) during which time the system only gradually regains its full performance.

In many modern radar systems the ATR cell is replaced by a solid state ferrite isolation device which allows power to pass from transmitter to antenna but presents attenuation of about 20 dB (1/100 of the power) from antenna to transmitter. Thus any signal destined for the receiver will only lose 1/100 of its little power by its passage to the transmitter. These devices do not suffer the disadvantages of the gas discharge ATR cells; their reaction time is much quicker. They also have much longer life. Similarly it is possible to replace TR cells with solid state devices in the form of clusters of PIN diodes. The PIN diode is a device which allows low-loss conduction of microwave energy when put in a conducting state and gives high isolation when non-conducting. However, the diodes must be put into a conducting state before the rf energy reaches them and so, to prevent burn-out, have to be pre-primed ahead of the transmitter pulse arrival time.

Ferrite isolators (circulators) can be used to prevent transmitted power from damaging an associated receiver provided the isolation is greater than the difference between peak transmitted power and that which would harm the receiver. The technique is particularly useful in secondary surveillance radar (SSR) systems, where modest transmitter power at one frequency is used outside the passband of that of the receiver. There are two isolating factors in the SSR case: one is the intrinsic isolation of the ferrite (or circulator) device, the other is the insensitivity (and hence protection) of the receiver's passband at the transmitted frequency. This is commonly more than one million to one, so the total isolation is of the order of one in 100 million (80 dB). Thus if the peak transmitted pulse was 1 kW (30 dB relative to 1 W) the receiver would have not more than 30−80 dBW = −50 dBW = 1×10^{-5} W of unwanted input. This would be entirely tolerable by the receiver.

5.5 Power combining – diversity

Many radar systems are installed with duplicated electronic equipment to increase the 'availability' of the system to the user. If each of the duplicated channels were to have a system failure rate probability of one in 2000 hours (i.e. 83 days or 11 weeks and 6 days) then if both channels were operated together, and either selectable for use, their combined failure rate would be 4000 hours – a very desirable increase. This can only be true, however, if time to utilise the second channel is very short. This in turn means running both channels continuously in their useable state and having a fast change-over system. If the mode of use is to have one channel connected to the antenna and the other connected to a dummy load with a switch to select either 'on the air', then one channel is dissipating its energy to little purpose. It adds nothing to the radar performance beyond increased system availability.

To gain full advantage of operating two transmitters 'on the air' at the same time the resultant signals from both must be simultaneously present for them to be added. The simple means of doing this (i.e. to make each transmitter radiate at precisely the same time) is not available because unless complicated cophasing arrangements are made, there is every likelihood of catastrophic failure of the transmitters. Even if cophased the peak voltage in the system will increase four times. So they must be operated one after the other, usually a few pulse durations apart. Arrangements must then be made for their separate signal processors' outputs to be brought into coincidence. A simple delay equal to the time separation of the two transmitted pulses must be introduced to the signals occurring earlier. Performance advantages of diversity operation are discussed in Chapter 7.

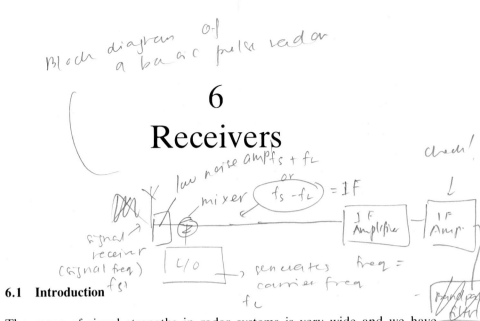

Block diagram of a basic pulse radar

6

Receivers

6.1 Introduction

The range of signal strengths in radar systems is very wide and we have already seen that the designer's aim is to reproduce signal inputs with as little added noise and distortion as possible. This is in the further cause of allowing extraction of all the *wanted information* contained in the signals. These aims give rise to numerous designs both of receivers and the modulation detectors which follow them.

Modulation can take many forms, among which the following are prominent in primary and secondary radar systems:

- Amplitude;
- Frequency;
- Phase;
- Time (pulse duration and spacing).

The history of receivers is worth a brief mention, for a significant change in receiver design technique occurred with the advent of transistor technology.

6.2 Tuned radio frequency (TRF) receivers and the 'super-het'

The earliest broadcast receivers consisted of a serial chain of amplifiers each having its own independently tunable passband. It was necessary, by careful art, dexterity and keen ear, to tune each to the frequency to be received. One can imagine it was sometimes more fun to 'tune in' a transmitting station than to listen to the broadcast material. The invention of the super-heterodyne ('super-het') technique took all this inconvenience away, leaving one tuning control only for the whole receiver.

In principle the super-het is delightfully elegant in its simplicity. The receiver incorporates an oscillator whose frequency is automatically arranged always to be a fixed value away from that of the carrier frequency of the signal to be received. The signal frequency and the oscillation frequency are put to a signal mixer. If the mixer has non-linear characteristics (i.e. its output amplitude is not a linear function of the input amplitude) then the mixer output will have signals whose frequencies are the sum and the difference between its two input frequencies. By design the difference between the oscillator and received frequency is always constant. This difference frequency (called the 'intermediate frequency' or i.f.) is amplified in a series of stages fixed-tuned to the i.f. and of passband wide enough to reproduce the modulations of the signal. The other component of the mixer output (carrier *plus* oscillator) is well outside the passband of the i.f. amplifier and is thus rejected.

This technique was almost always used when only vacuum tubes were available. Transistor amplifiers these days have such wide intrinsic bandwidth that interstage tuning of the i.f. amplifier is often dispensed with, the passband characteristic being formed by separate filters in series with the amplifier.

6.2.1 Dynamic characteristics

The quality and extent of signal processing which can be performed on the radar receiver's output is highly dependent upon the degree to which the receiver reproduces its inputs with fidelity. Receiver design is thus directed to this end. In pulse radar systems important information is contained in the signal's amplitude and duration. In pulse radars with moving target indication or detection, the phase of the signal must also be preserved. The first difficult design problem to solve is how to contain the large dynamic range of input signals.

It is not uncommon to find inputs ranging from noise level to 10^{10} times noise. When one realises that signals at the extremes of this range can be simultaneously in the system, the true magnitude of the design task is seen. Such a circumstance in a primary radar would be where a small aircraft was flying over a very large mountain at a range close to the radar: the radar must attempt to register the aircraft signal and reject that from the mountain. The term 'attempt' is used, for many systems would fail the test in the case cited above. To succeed, the radar needs to have the ability to preserve super clutter visibility (Super CV) or sub clutter visibility (SCV).

The first term, super CV, means that if a wanted signal is superimposed on an unwanted clutter signal, and the addition produces a greater value than the clutter signal alone, the system must preserve that difference. The degree to which it can is called Super CV. As an example, suppose a clutter signal S_c was of an amplitude 60 dB above noise, i.e. 1000 times greater (remember the consideration taken is amplitude not power), and that a target signal S_t of an amplitude 40 dB above noise is added to S_c. The target signal amplitude will be 1/10 that of the clutter ($S_c = S_t + 20$ dB). If the phases of the two were additive, the combined signal amplitude would grow to S_c + 10%, or 1100 times noise. Unless the receiver is capable of reproducing this difference at its output and subsequent processing makes it visible, the system would not have Super CV.

What prevents receivers providing Super CV? Unless the receiver has infinite dynamic range, there will always be a point beyond which any increase in input produces no change in output. The effect is known, not unnaturally, as 'limiting'. Indeed, sometimes the characteristic is such that it reduces output for increase in high level input ('fold-over'). In some systems which deliberately dispense with amplitude changes, limiting amplifiers are used to ensure no output change occurs for change of input above a certain level.

The second term sub clutter visibility (SCV) describes a property of the system which will show the presence of a signal which is *lower* in strength than competing clutter. Thus, if the system has an SCV of 30 dB it means that a wanted signal 1000 times less powerful than coincident unwanted signals will still be detected. Such properties are found in moving target indicator (mti) or moving target detector (mtd) systems.

Various ways are used to preserve dynamic range. A linear amplifier, whose characteristic is typified in fig. 6.1(a), is contrasted with a logarithmic amplifier, fig. 6.1(b). Although the latter makes better use of the available output range, the incremental gain is reduced and so small changes of input produce smaller changes of output than does a linear amplifier. Sometimes receivers combine the two to give the better incremental gain of the linear amplifer at low signal input levels and change to an approximate logarithmic characteristic at higher input levels to delay the onset of limiting.

6.2.2 *Swept gain*

A common method for making the best use of a receiver's dynamic range is to make its gain range dependent, on the premise that signal amplitudes in a radar system are themselves range dependent. For example, a mountain at short range will produce a signal greater than the same mountain at greater range.

The technique, illustrated in fig. 6.2 goes by various names such as gain time control (GTC), sensitivity time control (STC) (not to be confused with short time constant) and swept gain. It is common in modern radars to find the technique applied not as a variable gain in the receiver but as a variable attentuation of the receiver's input. The device most commonly used for this is the PIN diode, the 'P' and 'N' describe the normal properties of a semiconductor junction and the 'I' stands for an 'intrinsic' or pure layer of silicon separating the two. Its action is to provide attenuation at microwave frequencies proportional to an input control voltage. The attenuation range can be up to 50 dB.

An essential difference between the technique of swept gain and swept

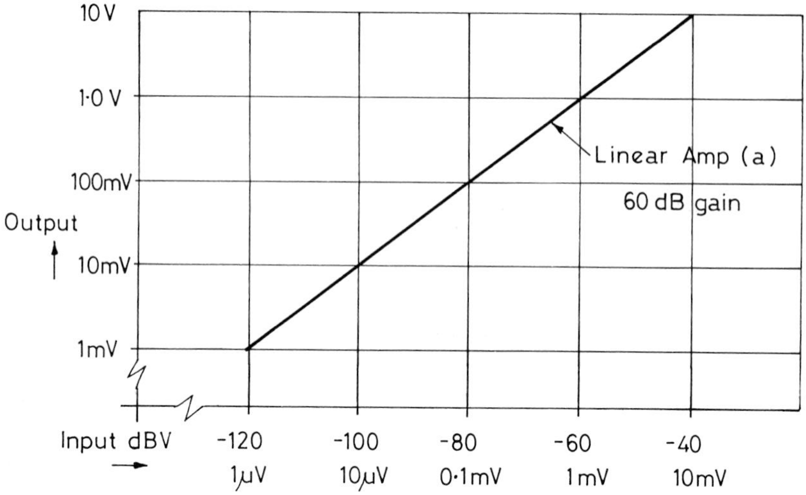

Fig. 6.1(a). Hypothetical input versus output curve for a linear amplifier.

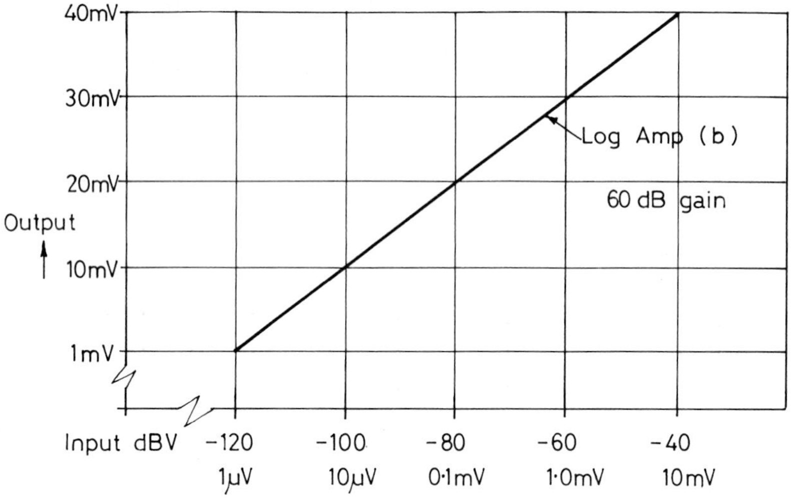

Fig. 6.1(b). Hypothetical input versus output curve for a logarithmic amplifier.

attenuation for the user is that swept gain will produce a variation in noise output from the receiver and hence produce less visible noise on the display – sometimes a 'black hole' in the ppi centre – whereas swept attenuation will not. Although either technique is valuable in containing a wide input range within the dynamic range of the receiver and in reducing small unwanted signals such as 'angels' (usually birds in small groups or large flocks), the rate of loss for wanted signals is the same as that for unwanted clutter.

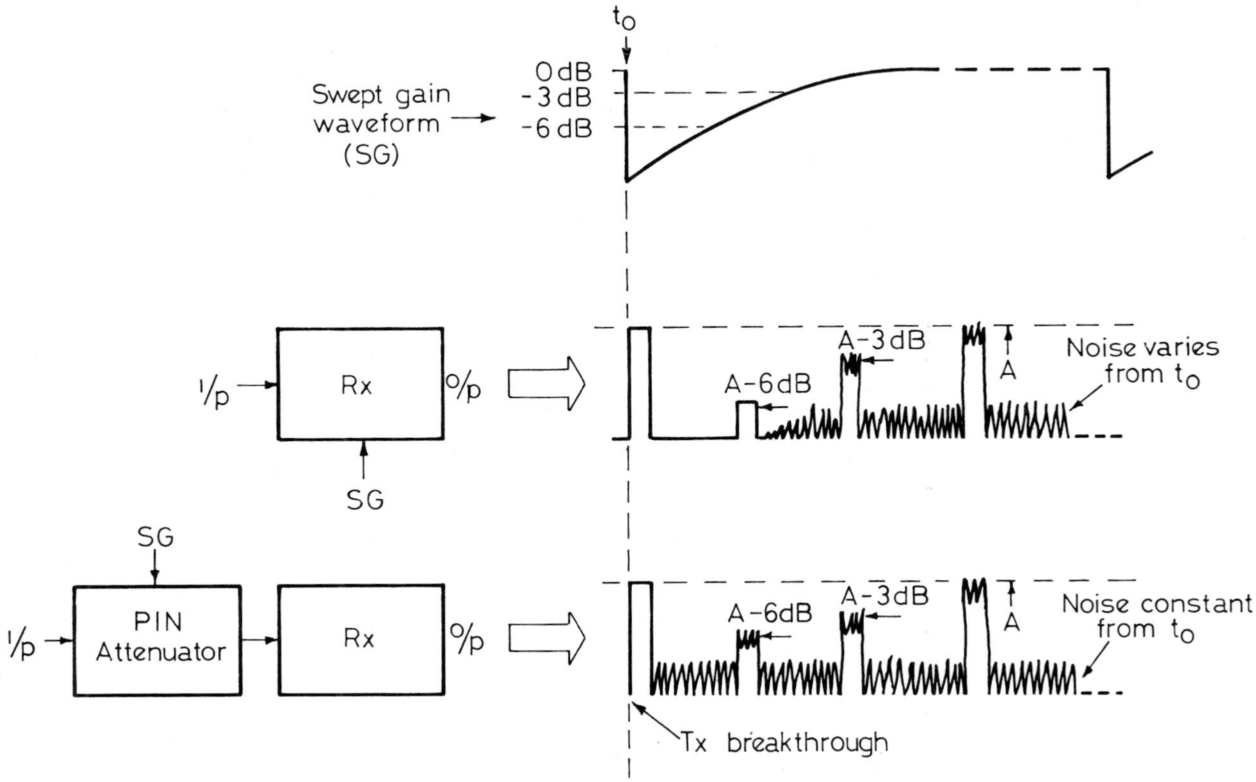

Fig. 6.2. Illustrating the action of swept gain and swept attenuation. The former acts upon receiver gain producing varying noise output. The latter acts upon signal input and noise is constant, being generated by the receiver input stage.

6.2.3 *Detectors*

In section 1.5 it was pointed out that the signal's modulation carries wanted information. This is extracted first by various appropriate forms of detector and later refined by processing. The modulations can be to all or any of the following system parameters:

- Amplitude;
- Frequency;
- Phase.

Amplitud modulatio

The first can be subdivided into a group of detectors which will give:

(a) The envelope of the signal's carrier.
(b) The envelope of modulation above a preset threshold.
(c) An output proportional to the signal duration.
(d) An output proportional to the signal's rate of rise or fall.

Those readers interested in the circuit techniques of such detectors will find useful data in reference 4.

For the purposes of this work the action of these detectors is illustrated in idealised functional form in fig. 6.3(a) – (f). The envelope detector in fig. 6.3(a) does just what its name suggests. It produces an output which is the envelope of the carrier frequency of the received signal. The threshold detector, fig. 6.3(b), is a simple variant of the envelope detector. It gives output for all values of input in excess of a set threshold. Sometimes the threshold is itself varied by other detection processes, e.g. by automatic sensing of mean noise level.

The time duration detector, fig. 6.3(c), gives an output proportional to the input duration. A variant, hardly a detector as such, gives an output up to input durations of a preset time and inhibits output for input durations in excess of the preset value. Such devices are used in pulse length discriminators and are useful in reducing clutter signals which are very often longer in duration than the transmitted pulse. The rate of rise/fall detector, fig. 6.3(d), is used in some SSR systems to recover the pulse duration and spacing information essential in SSR.

The frequency deviation detector, fig. 6.3(e), is commonly used for automatic frequency control functions. The value of f_0 is usually set by an oscillator whose frequency is accurately determined by a crystal source. The detector output is used in the correct sense to steer another frequency source to coincidence. Alternatively, in frequency modulated systems, the detector produces an output proportional to the frequency difference between the input and reference frequencies.

The phase sensitive detector (PSD), fig. 6.3(f), finds an essential place in radar systems with mti or mtd characteristics and provides the means of using the coherence of such systems. A reference frequency internal to the

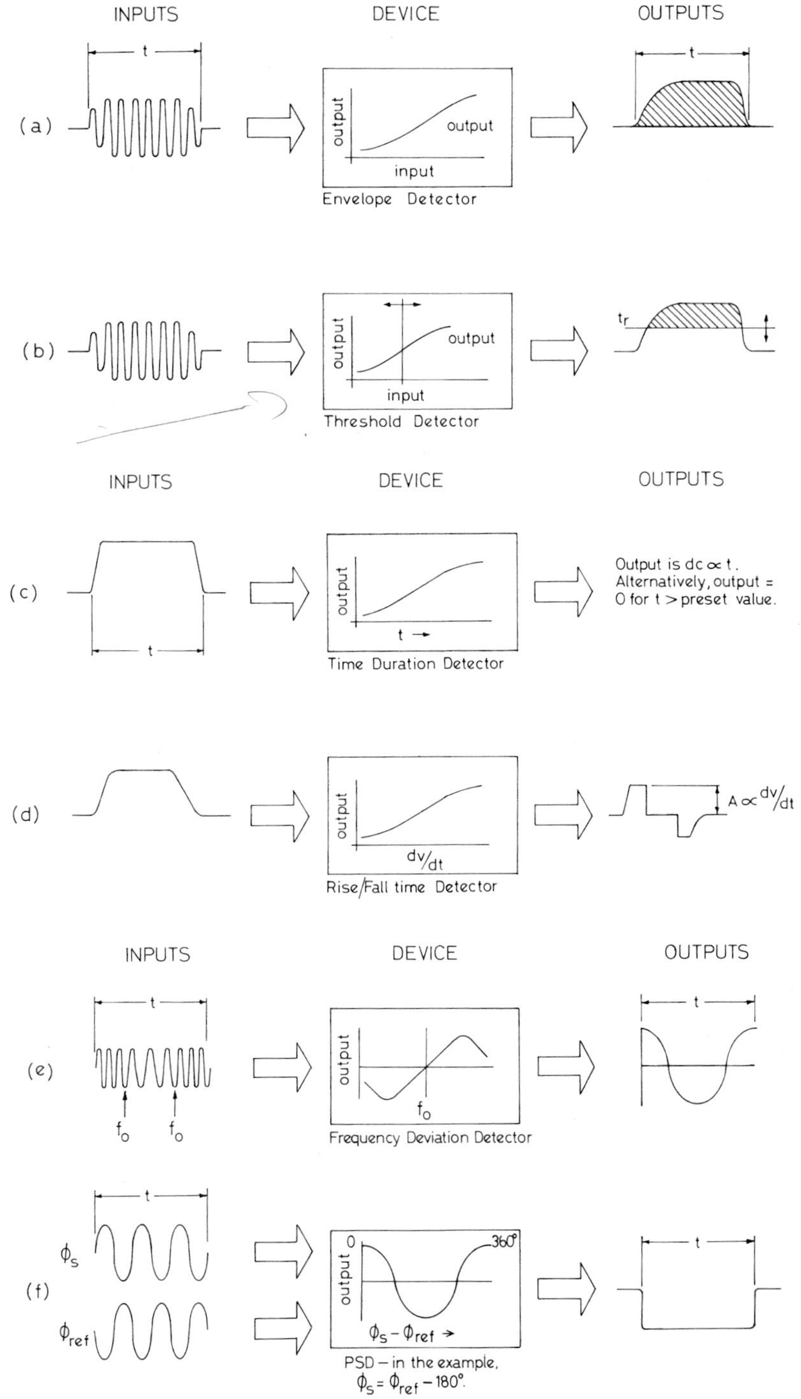

Fig. 6.3(a) to (f). Showing the action of various detection devices.

transmitter is used as one input to the PSD and the received signals form the other input. The output amplitude is proportional to the difference in phase between the two inputs. A point to note is the system's ambiguity to inputs greater than 360° difference and those equidisposed about the points of symmetry of the PSD characteristic. Solutions to problems created by this are described in part 2.

Part 2

Primary Radar

7
Theory and its practical impact

7.1 The elements of a primary radar system

The primary radar principle is a simple one: radiation from a known point is reflected by discontinuities in the atmosphere. Some of this energy is gathered at another known point, amplified, detected and displayed in a manner such that the observer can tell the location of the discontinuity. It is usual for the energy to be radiated and gathered by the same antenna, i.e. sent and received at the same point. This is the monostatic radar technique. It is possible, and in future may become common, to separate the radiator and receiver (or receivers), the system then becomes bistatic (or multistatic). The radar system can be represented as in fig. 7.1.

The modulator and power generator constitutes the radar transmitter. Modulation can take many forms, some of which will be described in later chapters. Most readers will be familiar with the most common type of modulation – that of amplitude, in the form of a pulse. Modulation can be looked upon as giving a 'signature' to the final transmission. In a pulse amplitude system, the pulse duration and repetition rate are part of the signature, both of which are used to good advantage in separating wanted from unwanted received signals.

The modulation is applied to a power generator. It would be more accurate, however, to call it a power translator, because in general its action is to take energy continuously from the electricity supply at a relatively low level and convert it to microwave energy which is released after modulation to the power radiator. The microwave energy and the modulation will be characterised by the spectrum of the output. An example has been given in chapter 1 of part 1.

The radiator will be recognised as the radar's antenna, designed to have special beam shapes in both azimuth and elevation planes to utilise the radiated energy to the best advantage of the user. We have seen that in the monostatic radar system, the same antenna is used for transmission and reception. Sometimes the beam shaping is made to be different for the two regimes.

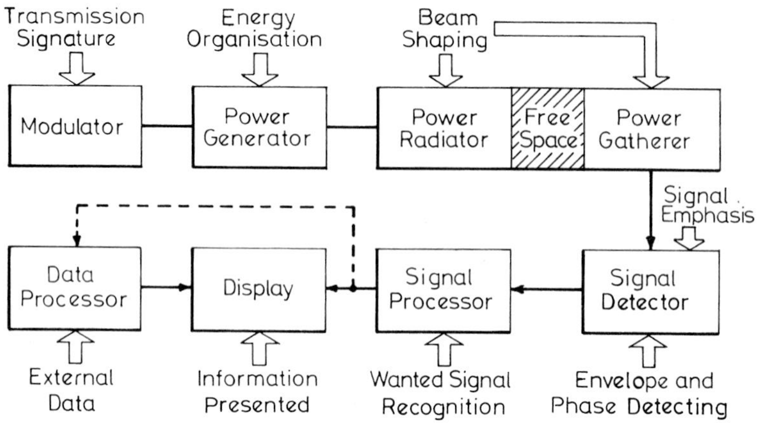

Fig. 7.1. The essential elements of a radar system and their functions.

Power gathered by the antenna is passed to a signal detector. But before detection takes place it is necessary to amplify the very weak signals received and give them emphasis with respect to system noise. This is where the 'signature' of the transmission is first used to advantage. Signals received by reflection of the radiated energy will have the same spectral structure as that of the power generator, but with slight changes imparted by the reflecting object. Incidentally, these slight changes are of great importance, as will be seen later. Knowing the spectral structure of the radiated power it is possible to amplify only those frequencies contained in the spectrum of the received signal. Frequency selective amplification of the pre-detection receiver thus confers a 'bandwidth' to the system.

The final detection process is usually performed after translating the signal's microwave frequencies to much lower values. Phase and amplitude data are extracted from the signal and passed to a signal processor. Within the signal processor, the detected data are used to recognise wanted and unwanted signals and to discriminate in favour of those wanted by the user. Finally, the wanted signals are displayed in a manner most convenient for the user, the most usual and familiar form being the plan position indicator (ppi).

It is becoming increasingly common to find the output of the signal processor being operated upon by a data processor whose purpose is to extract information from the signal processor output and convert it to a form more easily read from the display by the user.

Radar systems work in practice for three main reasons:

(a) The velocity of the radiation to and from the reflecting object is constant (the velocity of light).
(b) Sufficient effective radiant power can be generated to create measurable reflected energy.

(c) Receivers of adequate sensitivity are available to detect the resultant signals.

Because the velocity of propagation is constant, the distance between the radiator and reflecting object can be written as:

$$R = \frac{ct}{2} \tag{7.1}$$

where c = propagation velocity
t = journey time

The factor of two in the equation accounts for the energy's round trip from radiator to reflector and back again.

For practical purposes the velocity of propagation of radiation in free space is 3×10^5 km/s. Thus if the round trip time was 100 μs the range would be:

$$R = \frac{3 \times 10^5 \times 100 \times 10^{-6}}{2}$$
$$= 15 \text{ km}$$

Convenient approximate values to remember are that 1 km is equivalent to 6.6 μs round trip time and 1 nautical mile is equivalent to 12.4 μs round trip time.

7.2 The radar equation and its implications

This work is user-directed. Why, then, is it considered necessary to include such a section as this? It is for two reasons:

(a) Many users feel that their technical advisers are the only agents able to understand the complexities of radar performances, but this need not be forever true. What follows is an attempt at enabling users to understand the language of their technical advisers.
(b) Many users have the impression that radar performance can be calculated to fine limits with high accuracy. Although this is true, the accuracy has to be expressed in terms of *probability*.

By this presentation it is hoped to make a bridge between users and their technical advisers to their mutual advantage.

7.2.1 Two aspects of radar performance

It is usual to find radar performance described at a level of almost dismissive simplicity, being a matter of geometry and simple physics. A little more examination, usually avoided by users because of the mathematics involved, reveals deeper and more significant levels (in terms of understanding) rooted in statistical analysis.

The importance of this second level is such as leads me to tell of an encounter between designers, users, and their procurement executives. During this meeting, concerning radar performance, quite sophisticated 'user' spokesmen stated the operational requirement as '. . . if there is an aircraft in our operational area the radar must show us for sure it is there'. This was another way of saying that the radar should give 100% probability of detection. When it was pointed out that the range of any radar for 100% probability of detection was zero – not nearly zero but *precisely* zero – the meeting fell ominously quiet for a considerable time. And yet the statement was true. And until the laws of physics and the perception of mathematics change, it will always be true. There are many parameters in the radar equation which have to be expressed in statistical terms. One has been touched upon already: receiver noise. Others will come to light.

7.2.2 *The primary radar equation*

Using ordinary geometrical principles, it is possible to formulate the basis of a radar's performance in terms of range. At its root, the calculation to be made asks the question 'how far away can a target be before its detection ceases to be practicable'.

The first step is to describe how much power reaches the target in W/m^2. Imagine the target to be on the surface of a sphere of radius R and the power source to be at its centre radiating in all directions with equal strength (an omni-directional radiator). Call its intensity P_t. The power density at the target will be a proportion of the total which is spread over the surface area of the sphere. The total area of the sphere, from geometry is:

$$A = 4\pi R^2 \tag{7.2}$$

The power density will then be represented by:

$$P_{den} = \frac{P_t}{A} \text{ or } P_{den} = \frac{P_t}{4\pi R^2} \tag{7.3}$$

Note that doubling the radius (R) of the sphere increases the surface area four times and thus the power density in W/m^2 at the surface falls to a quarter.

However, the radiator is deliberately designed not to be omni-directional. Its directivity gives it gain emphasis in desired directions. If the target is illuminated by the antenna pattern's peak, the power density is magnified by the antenna's gain. Thus:

$$P_{den} = \frac{P_t G}{4\pi R^2} \text{ W/m}^2 \tag{7.4}$$

In all radar systems the power density at the reflecting object exists for a certain time. The product of power density and time is the 'energy on target' (watts × time/m^2). The radar performance is more conveniently described in terms of energy rather than power and this concept embraces all kinds of modulation, pulse shapes and the history of the pulse content.

For a simple pulse amplitude modulated radar, equation (7.4) can now be written in wider terms as energy on target $E_t = P_t\,T$ where P_t is the peak power in watts and T is the time during which P_t is present (the pulse duration) in seconds. The power incident upon the target will be reflected in many directions. Some of it will reach the receiving antenna. For most systems this will be the same antenna as used for transmission.

The reflected energy reaching the antenna ($E_r = P_r \times T$) will be in the form of a plane wave and the receiving antenna will intercept it and guide it into the radar receiver. The amount of received power can thus be seen to depend on:

(a) The *effective* as opposed to the geometric area of the reflecting target, call it σ square metres.
(b) The area of the receiving antenna, A.
(c) The range between target and antenna, R.

So we may write the received power, P_r, as

$$P_r = \frac{P_t \times G}{4\pi R^2} \times \frac{A \times \sigma}{4\pi R^2} \qquad (7.5)$$

The gain of the antenna is related to its area as follows:

$$G = \frac{A \times f \times 4\pi}{\lambda^2} \qquad (7.6)$$

where A = area in square metres
f = efficiency factor less than one
λ = wavelength in metres

G will be a magnification factor and is usually expressed in decibels (e.g. if $G = 30$ dB, i.e. $G = 1000$ times).

Substituting gain for area in equation (7.5) in terms of energy the following is obtained:

$$E_r = \frac{P_t T G}{4\pi R^2} \times \frac{G\lambda^2}{4\pi} \times \frac{\sigma}{4\pi R^2} \qquad (7.7)$$

so

$$E_r = \frac{P_t T G^2 \lambda^2 \sigma}{(4\pi)^3\, R^4} \qquad (7.8)$$

Since $E_t = P_t T$ and $E_r = P_r T$, the transposition of equation (7.8) to express range allows (mathematically) the deletion of the term T since it appears in both the denominator and numerator of the resultant equation which is the form usually seen as:

$$R^4 = \frac{P_t\, G^2\, \lambda^2\, \sigma}{(4\pi)^3\, P_r} \qquad (7.9)$$

It is dangerous, however, to take this expression as the true state of affairs. A number of other factors and circumstances have to be taken into account. They may be summarised as:

(a) The target effective area σ is not a fixed parameter. If the target is an aircraft, the effective area can fluctuate wildly and its value is dependent upon the aircraft aspect to the radar antenna.
(b) The antenna beam shape is curved and the maximum gain exists at its peak. As the beam sweeps across the target many samples are gathered with less than maximum gain.
(c) Propagation of the microwave energy is usually through the Earth's atmosphere and some of the energy is absorbed and some scattered by particles and gases in the atmosphere.
(d) The considerations so far have expressed the maximum range in free space, i.e. on the assumption that antenna beam shapes are maintained at their design levels in practice. Under these conditions energy reaches the target and is returned over only a single path, that is the line between the radar and the target. In many cases energy reaches the target not only by the direct path but by reflection. Particularly is this true of the elevation plane, because energy can also reach the target by reflection from the surface under the antenna (usually the ground or the sea). The distances involved usually allow direct and reflected energy to be considered as arriving simultaneously and so can interfere with each other. The effects were described in part 1, section 3.5.
(e) Integration effects: the radar system gathers a number of signals from the target as the antenna beam sweeps across it. Each signal has an organised history but is mixed with disorganised noise. In combining or adding these the net effect is to improve probability of detection over that for a given single signal to noise ratio.
(f) The ratio of the signal to noise power required to produce the practical level of detectability, usually denoted S_{min}.
(g) The degree to which the radar display and operator performance affect detectability.
(h) The degree to which the receiver's bandwidth affects the amount of signal and noise power passed.

It is not my intention to write a textbook on radar engineering – many already exist. What is intended is to make those works more readily understood. To this end the following expansion of the factors and circumstances listed above is included.

7.2.3 Target echoing area

The ideal radar target would be a perfect sphere, however most radar targets are of non-uniform shape. Take the case of an aircraft flying directly at the radar. Consider that the radar transmission lasts for 2 μs at a wavelength of 23 cm. Because the aircraft is many wavelengths long and wide this energy will be reflected from various parts of the airframe: nose, fuselage, wing leading edges, engine nacelles, tail fin, tail plane, flaps, etc. Each of these features is at a different distance (in numbers of wavelengths) from the nose. Each makes its own contribution of reflected energy along a line from the target to the radar antenna.

These contributions will at any instant add vectorially at the receiver input terminals. Because the aircraft is usually shorter in length than that occupied by the pulse of energy in space, some reflected energy from, say, the wing leading edges will be combined simultaneously with reflected energy from the nose. These contributions may be at various phases relative to each other. Taking extreme cases, if they are perfectly in phase, they will add; if in anti-phase they will subtract. It is thus easy to see that for the multiplicity of – call them scattering phase centres or 'scintilla' on the aircraft – the combined effect will fluctuate wildly as the aspect or disposition of the aircraft to the radar changes even by fractions of a wavelength. An example of this is seen in fig. 7.2, taken from the first classic book on radar systems engineering (reference 6).

In such circumstances the *effective* area of the target is the only tractable value to use in calculation and this itself varies as a function of the aircraft aspect to the radar. Typical values are given in table 7.1. The values given in

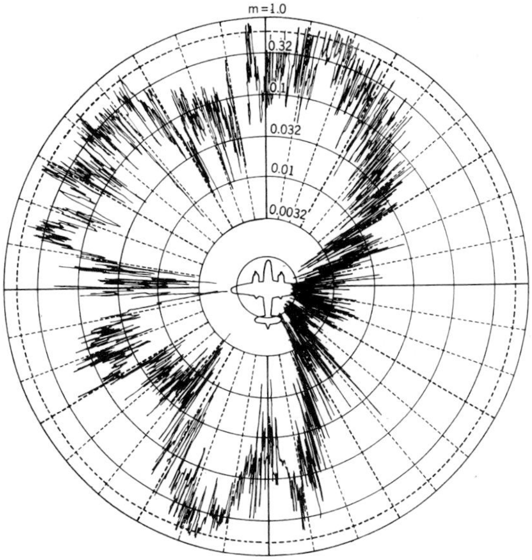

Fig. 7.2. Showing the massive variations in echoing area of an aircraft as aspect varies. General reduction on one side is caused by propellers being feathered. (*Reproduced by kind permission of McGraw Hill Book Co.*)

the table are averages of a set which have a statistical distribution above and below it. A commonly accepted form is the Rayleigh distribution. The signals returned from such targets are mixed at the receiver input with noise, which is itself a statistically varying quantity. Although the combined effect of the two leads to a distribution tending towards the Normal or Gaussian form, the target fluctuation re-imposes the Rayleigh distribution. Differences between the two distributions can be seen from fig. 2.6 in Part 1. This means that roughly 50% of possible values lie above and 50% below the average.

Table 7.1. Typical reflecting area in m^2

A/C	H/ON	SIDE ON	TAIL ON
Small jet fighter	2–3	10–20	2–3
2 engine jet	5–10	100–1000	5–10
4 engine jet	20–50	1000–2000	20–50

We usually wish to have higher probabilities than 50% for detection and so must make allowance for this. If the required P_d was 80% we would need to be assured that the echoing area produced the required signal to noise ratio (or greater) for more than 80% of trials. One way of translating this into the radar equation is to increase the value of the required received power P_r. More rigorously it would be true to say that the received signal to noise power

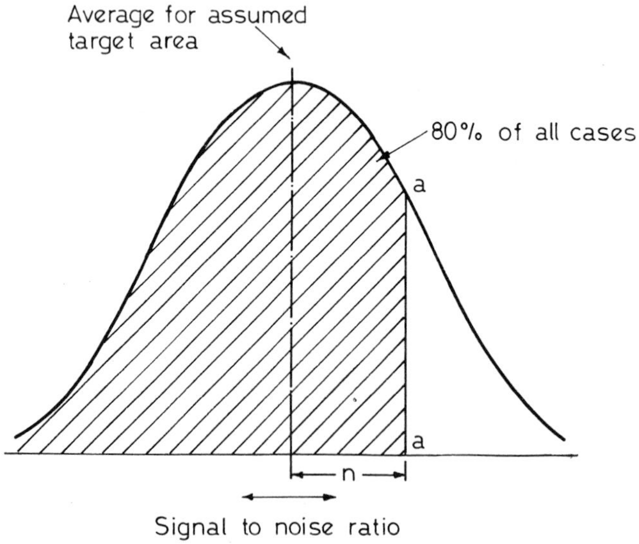

Fig. 7.3. For the combined statistical distributions of signal to noise ratio roughly half the population is above and half below the average. To include 80% of all cases requires an increase in signal to noise ratio.

ratio needs to be increased. However, it is often not possible to reduce noise or the system's bandwidth or the losses between antenna and receiver – these will have been engineered to be minimum in almost all cases.

An illustration of the need to account for the increased probability of detection by demanding a higher input signal to noise ratio is given in fig. 7.3, based upon the assumption above. The area under the curve represents the total population of possible values of signal to noise ratio obtainable from the compound of fluctuating target area and added noise. Taking the mean value achievable, for this model of distribution, is tantamount to specifying a 50% probability of detection, because half the signal to noise ratios would fall below the mean. We need to find that part of the curve which embraces the required probability of detection. Suppose this was 80%; the line a–a represents this. It can be seen that this corresponds to a requirement for an increased signal to noise ratio.

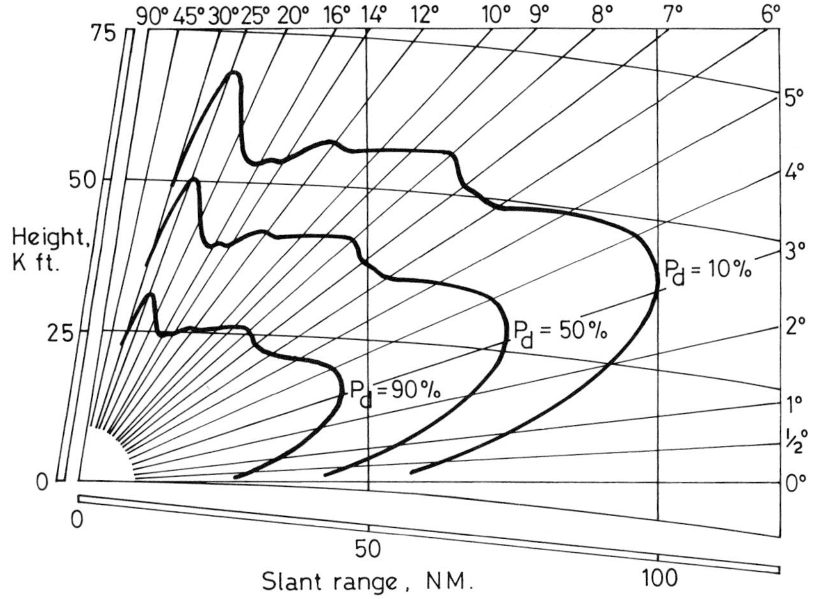

Fig. 7.4. Showing variations of range and vertical coverage with probability of detection (P_d) when all other parameters are unchanged.

It is appropriate here to stress another important corollary to the idea above. It concerns the way in which radar coverage diagrams should be read. They usually take the form of a range/height diagram, the shape being determined by the antenna's vertical polar diagram (VPD). The VPD does not show an area wherein targets are guaranteed to be seen. Equally the diagrams should not be interpreted to mean that targets outside the VPD area will not be seen. They are curves of *equi-probability* that a target of assumed echoing area will be seen to the stated range for a given set of radar parameters. A set of curves of a practical surveillance radar is reproduced in fig. 7.4 to show that the vertical coverage can vary as different probabilities of detection are taken.

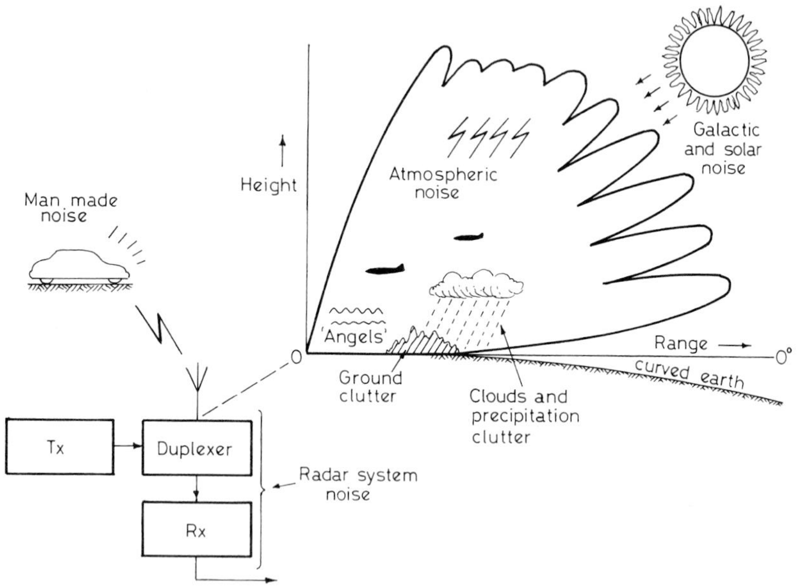

Fig. 7.5. The radar environment. Particular points are: Operation over a curved earth, shadowing by high ground, ground reflection effects cause vertical lobing. (*Reproduced by kind permission of Butterworths (ref. 4)*).

7.2.4 Antenna beam shape

The beam shape in azimuth is typically of the (sin x)/x form as shown in fig. 3.7. The radar design is usually made so that a number of radar returns are available from each target, giving a high probability of detection. That number is set typically at about ten. If the beamwidth of the antenna is 2° then there will be ten pulses per target, spread through this beamwidth. The gain of the antenna has a peak value used in calculating the peak range. Either side of this, the gain is less. As the beam is rotating, the value of gain for the antenna will be different for a given pulse at transmission than for the reflection of that pulse from the target. This is because the round trip time of the pulse is accompanied by antenna rotation. In the case taken the gain variation between transmission and reception is that equivalent to 0.2° of azimuth (1/10 of a beamwidth). It is usual to include a value of 1.6 dB loss for this effect.

7.2.5 *Atmospheric attenuation*

In a clear atmosphere with no precipitation, electromagnetic energy will in part be absorbed by oxygen and water vapour and will be dissipated as heat. Thus it suffers attenuation on its way from antenna to reflector and back again. This attenuation decreases as wavelengths increase. It is significant for radars whose range is required to be 100 nautical miles and above.

For instance, at zero elevation angle an S-band radar would suffer attenuation of about 3.5 dB round trip loss for a target range of 200 nautical miles. This is equivalent to −3.5/4 dB system loss (a range factor of 0.85). At higher elevations the loss is less because the radar rays are looking less obliquely through the gases. Tables for loss versus frequency and elevation are given in most textbooks. A comprehensive set appears in reference 6.

7.2.6 *Integration effects*

The considerations in radar range performance usually start with the 'single sample' case, i.e. the results of one transmitter pulse seen through the receiver bandwidth. In practice there will be a number of these samples received from an aircraft or isolated reflectors because transmissions are repeated and the scanning antenna has an effective beamwidth. Consider a radar having the following characteristics:

(a) Antenna rotation rate of 10 r.p.m.
(b) Effective azimuth beamwidth of 1.5°
(c) Pulse duration of 2 μs
(d) Sample rate (pulse repetition frequency) of 600 Hz.

An aircraft target travelling at 600 knots radially relative to the radar has displacement during the radar beam's passage across it. Calculation shows this to be the distance travelled in 25 ms (the beam's 'dwell time' on the target) and at 600 knots is equal to approximately 8 m. The pulse duration of 2 μs represents about 1/3 km in terms of radar distance. We can see therefore that the signals from the target within a single beamwidth's passage will overlap the same range increment almost completely (approximately 95% overlap). It would be possible, then to add such signals together (i.e. to integrate them) in an electronic system having a range-ordered store. In a perfect integrator the resultant signal to noise ratio would improve because the signal amplitudes would add directly and the noise contributions would add their average values. The nett result is an improvement $I_{av} = n$, where n is the number of pulses integrated. Equally the effect may be thought of as requiring a smaller input signal to noise ratio in the single sample case for a given overall probability of detection.

Obviously integration is never perfect and so-called 'integration loss' occurs. Note that this is not 'a loss because integration is performed' but *a loss of the full benefits* of integration. A good example of integration loss occurs in the 'real-time' ppi display. Most will be familiar enough with ppi construction to appreciate that integration takes place on the display screen itself by the storage of signal energy (given out as light) within the tube's phosphorescent layer viewed by the operator. If, in such a system, the antenna was stationary and an aircraft was to be illuminated by the radar, the returned signals would impinge upon the same small area of the display. As each new sample appeared, it would add to the previous samples. Gradually the phosphor would build up its brightness, resulting in greater target visibility. Thus we can call the ppi, and its observer, an integrating device. In practice such a combination is near-perfect. But the radar is not operated in this fashion; the antenna is rotating and so each successive sample is written onto a slightly different area of the ppi display.

The integration process is therefore imperfect and a good working expression for the loss incurred is at the rate of 1.0 dB per decade of samples, or pulses, integrated up to a number of thirty samples. Thus a system with an average of twenty pulses per beamwidth would not only gain from integrating these twenty pulses it would also incur an integration loss of 2 dB.

W.M. Hall, in his paper on prediction of radar performance (reference 7), also draws attention to another form of integration loss which is a function of the real-time ppi display and its adjustment. He points out that if the signal to be represented on the display has a range dimension small in relation to the display beam's spot size then extra noise can be introduced into the signal being displayed. The situation is described in detail in part 4, chapter 15.

7.2.7 Power losses

We have seen that the wanted signals have always to compete with the receiver's inherent noise. Before doing so the signal has to reach the receiver from the antenna. The journey from the antenna to receiver involves loss, as the signal has to traverse a number of elements en route such as waveguide, rotating joint, performance measuring components, diplexers, duplexers, connectors, r.f. switches, etc.

The radar designer's art has to be well exercised in order to keep such losses to manageable and minimum levels. Typically, losses of up to 5 or 6 dB are encountered in practice. This results in the signal at the antenna reducing to only about a quarter of its value at the receiver input terminals. Many of these losses are also encountered by the transmitted pulse which to a large degree shares the same route as the received signal on its way out into space. Such losses must also be accounted for in range calculation.

7.2.8 Display and operator effects

Many users will be familiar with the way a maladjusted traditional real-time ppi display can affect the radar's performance – too high a noise level reduces visible afterglow on wanted targets, too low a noise level reduces minimum detectable signal visibility.

Changing range scales can also affect detectability because the ppi spot size is fixed and changing the range scale changes the writing speed, giving differing values of screen excitation per target. Suppose we have a radar and its display set up to register a target as a rectangular arc of 1 mm in range by 4 mm in azimuth. If the range scale was halved, the target would now be registered as an arc of 2 mm in range by 8 mm in azimuth. It would be unusual for the tube's spot size to be changed and thus the target would be registered over four times the area of screen by the same sized spot. The integration effect of reinforcing target data by exciting overlapping portions of the screen would diminish and detectability would reduce.

There are, however, display techniques these days which overcome this by arranging the ppi writing speed to be held constant, irrespective of range scale. One such is described in chapter 14 of part 4.

7.2.9 *Transmitter power*

In the development of the radar equation (section 7.2.2) the sent and received power neglected the duration of the pulse. This was done for simplicity's sake since mathematically the pulse duration cancels out in the equation. Nevertheless, it is important to realise that what is being sent and received is energy and not just power.

It is almost a magical thing to me to realise that radar systems work with such little amounts of power. Consider a pulsed transmitter whose one megawatt output is on for one microsecond and off for one thousand microseconds. Each pulse releases only 1 joule of energy (a watt second, 1×10^6 watts $\times$ 1×10^{-6} seconds). Its mean power, averaged over a number of pulses, is one kilowatt. This small power enables one to detect aircraft out to ranges of greater than 100 km and up to 10 km height, in any direction. This represents a volume of greater than $100^2 \times \pi \times 10$ km (nearly one-third of a million cubic kilometres).

Imagine this in terms of heat. Suppose that instead of a microwave transmitter, a small electric fire has its energy directed into space by the antenna. The reflected heat is gathered, detected, analysed and displayed for you to see!

The term representing transmitter power (P_t) in the radar equation is the peak power. It is important to realise that the target detectability is affected by the *average* power delivered by the transmitter. Range performance is a function of the power on the target. Let us replace the peak power with the concept of average power (P_{av}).

$$P_{av} = P_t \tau f$$

where P_t = peak power

τ = duration of the peak power
f = repetition rate of the peak power (pulse repetition frequency)

We have already seen that the signal detectability is a function of the system bandwidth – the wider it is, the more noise it passes to obscure the signal. The designer aims at matching bandwidth (B) to pulse signal fidelity, producing a factor of $B\tau = 1$ (approximately). It is also (through integration effects) a function of f, the pulse repetition frequency. It is therefore easy to see that these factors can be, as it were, hidden in the usual form of the equation since the average power, $P_t\tau f$, in the numerator is counterbalanced by τ and f in the denominator.

7.2.10 The total effect

It might seem from the above that calculation of radar range performance is a fruitless task. Not so. The difficulties are many, but all may be overcome by calculation. The greatest difficulty, however, is in getting a group of people concerned for proper solution to the problem to agree upon the formulation. Experience has shown that those concerned with the long term day-to-day use of radar would prefer a system which gave 80% of its full performance for 100% of the required time, rather than 100% of full performance for 80% of required time and zero or unacceptable performance for the remaining 20%. Others, who have responsibility for ensuring specifications are met at system acceptance time, are more concerned that 100% of performance is achieved during acceptance trials.

One particularly contentious point is that associated with the target's echoing area and the manner in which it fluctuates. Earlier in this text the fluctuation was assumed to be of Rayleigh distribution. However, four types of fluctuations have been posited by Swerling (reference 8). All discount the fluctuation in signal strength imposed by the antenna beam shape. The four cases are really two, each with a slight reclassification forming the remaining pair.

The first case considers the echoing area to remain sensibly constant from pulse to pulse throughout the radar beam's passage across the target but that this value fluctuates from one antenna revolution to the next. The third case is similar but uses a different fluctuation distribution law. The second case considers the target echoing area to fluctuate from pulse to pulse during the target's dwell time within the beam. The fourth case is taken to be as for case two but uses the distribution law of case three.

The difference in practice between these cases is profound in its effect on calculated radar range, at probabilities of detection above 50%. Since most users are interested in probabilities of 80 to 90%, account must be taken of the effect. Suppose the requirement was to produce 90% probability of detection at 10^{-6} probability of false alarm in a system with fifteen pulses per target. Swerling's work leads to signal to noise requirements of:

- 12.5 dB for case one targets – typically aircraft
- 4.7 dB for case two targets
- 8.8 dB for case three targets – smooth shapes
- 4.7 dB for case four targets

The greatest difference is 7.8 dB, equivalent to a range factor of 0.64. That is, if the radar saw a case two target at 100 nautical miles it would only see a case one target out to 64 nautical miles for the same 90% probability of detection. A full exposition of the situation can be seen in section 2.8 of reference 1.

7.2.11 Performance advantage of diversity operation

In chapter 5 of part 1, the notion of diversity operation was introduced. Armed with the information presented so far, the effect of such operation can be understood. The fundamental advantage of this dual transmitter operation is that the effective mean power in the system is doubled (provided both transmitters have the same power and pulse characteristics). But unfortunately the looked-for increase in range and signal strength is not as large as one might at first expect. Doubling the mean power should (from the radar equation) increase the range by a factor of $\sqrt[4]{2}$ or 1.19, i.e. nearly 20%. But in integrating the signal outputs, a loss of 0.5 dB is incurred so the overall improvement in range falls to 15%.

It is not necessary to settle for this relatively small advantage, more is available by use of the 'diversity' mode of operation, of which there are three major types:

- Frequency
- Space
- Polarisation

By far the most common in radar systems is the first, and it is implemented in the following manner.

Frequency diversity

Each of the radar channels is operated not only at a small difference of time but at a slightly different frequency.

It is not usual to find the separation is greater than about 5%, since wider spacing could put the two channels outside one band of frequencies, losing the advantage of common equipment in both. Each transmitter is assigned its own frequency and has its own receiver and signal processor. As described above, the two channels operate in synchronism, a few pulse durations separating their transmission times. Signals output from the two processors are brought into time alignment by delaying the earlier of the two by the time separation of the transmissions.

How does this confer extra performance? We have now to invoke the principles of probability once again. Suppose a target was to be fluctuating and have a Rayleigh distribution about its mean value of 10 m^2. It has this probabilistic nature because of the reasons given in part 2, chapter 7: the vector addition of the contributions from the numerous 'scintilla' of the target give it a multi-spiked radiation pattern.

If the energy reaching the target was at a single frequency from both transmitters, the radiation pattern of the target (or its history of instantaneous echoing area) would be the same for both channels. The fact that the time of arrival of the two pulses was different would not affect matters, since the displacement of the target in a few microseconds would be insignificant. But since the frequencies are not the same, the target exhibits slightly different radiation patterns. The differences will be in the precise value of instantaneous echoing area seen at a given aspect angle. The radar channels will share a common line of sight to the target since they use the same antenna and hence share precisely the same target aspect angle. They will thus see, for any given transmission, independent instantaneous echoing areas. Positing a fluctuating target of 10 m^2 mean echoing area with Rayleigh distribution means that 60% of the time it will appear greater and 40% of the time less than the mean. Suppose the limit of acceptable performance of the radar allowed the target area to fall to 6 m^2. For the 10 m^2 target assumed, this would occur in 16% of cases (about 1 chance in 6). But there are two systems operating at frequencies different enough to create independent scattering diagrams. Thus when operated virtually simultaneously the chance of both failing to see a target of 6 m^2 or more is 1 in $6 \times 6 = 36$. It is this effect that creates a higher probability of detection in the frequency diversity mode.

Another way of looking at it is to take two polar diagrams, much as illustrated in fig. 7.2, and turn them into a gambler's spinning wheel. Put them both on the same spindle. Set them so that the aircraft axes are aligned and stick them together. Now spin the wheel. When it comes to rest at a nominated point, the probability that both patterns would have very low echoing areas is itself much lower than for each wheel independently.

The improvement expected of two different frequencies in a system aiming at 80% probability of detection is 4.8 dB or a range factor of $\text{antilog}_{10}(4.8/40) = 1.32$. Thus more than 30% extra range at 80% probability level is achieved. It is interesting to note that the improvement factor is greater if the single channel probability of detection increases. Equally, following the above arguments, using more than two frequencies confers yet more advantage in performance.

Space diversity

As its name implies, space diversity operation uses the separation of two (or more) sensors to gather data from a common target. Binocular vision is an example. It is not a technique commonly used in radar, most systems are 'one-eyed', having only one antenna! But it is often used in microwave communications links to counter the effects of variations in beam bending under anomalous propagation conditions. Coupled with frequency diversity, the so-called dual diversity mode (i.e. both frequency and space) confers much greater integrity upon the communication channel.

Polarisation diversity

It is possible to use this technique in radar systems; its principle is as follows. As described in chapter 1 of part 1, circular polarisation can be either 'right handed' or 'left handed' dependent upon which of the two orthogonal components of the electric vector is the earlier to be output. If the two radar channels are respectively left and right handed, the circular polarisation technique will still produce rejection of weather clutter in each channel. The same argument as advanced to explain the frequency diversity improvement factor is offered to account for the measured improvement resulting from the use of polarisation diversity: that is, differences in instantaneous echoing area resulting from differences in vector addition of all the small contributors to the reflected energy from the target.

Experimental measurements at L-band show that the effect can be usefully present but at a lower level than that achieved by frequency diversity. In radar systems designed to give data on weather clutter as well as other targets and which therefore have to include circular polarising elements, the polarisation diversity effect is almost a free by-product.

7.3 The radar environment

The radar environment is illustrated in fig. 7.5. It shows a cross-section of a surveillance volume plotted as a vertical polar diagram in a height/range form. Three features are immediately obvious:

- The radar operates over a curved earth.
- It can't see through mountains.
- The vertical pattern is modulated by ground reflected energy into a series of lobes.

Various sources of signals are shown. A number of these can be at various times both 'wanted' and 'unwanted'. Sometimes they are simultaneously present in the radar system, which should be designed to give users the ability to select these at will, as their operational tasks demand.

The antenna has sidelobes in both azimuth and elevation planes. These gather unwanted signals (even if they are from wanted targets). They are 'unwanted' because the concurrent azimuth data define only the main beam bearing, which is different from that of sidelobes. Antenna design attempts to

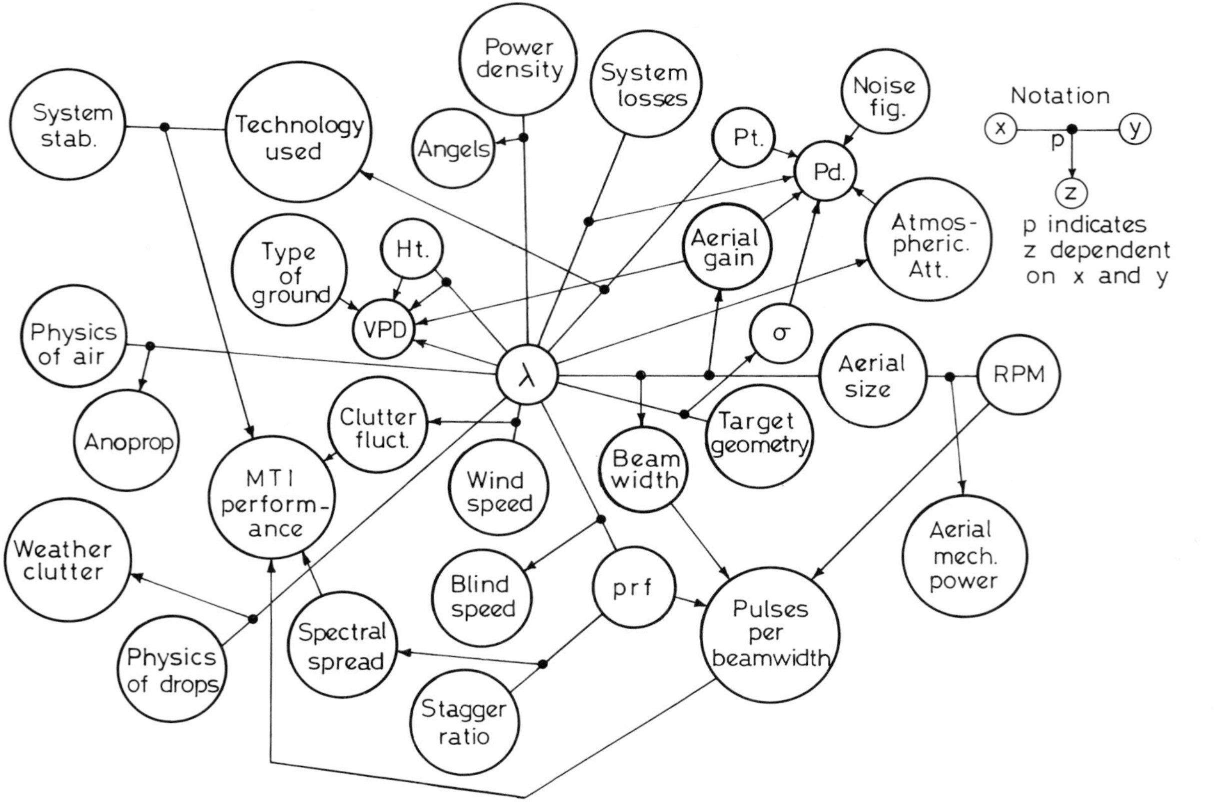

Fig. 7.6. Map of dependencies in radar. Legends: P_t = transmitted power, P_d = probability of detection, H_t = height of antenna phase centre above ground, VPD = vertical polar diagram shape, σ (sigma) = target echoing area, RPM = antenna rotation speed, p.r.f. = pulse repetition frequency, λ (lambda) = wavelength.

discriminate in favour of wanted targets but since these can coexist with the unwanted (e.g. aircraft and weather clutter signals), signal processing has to bear the burden of discrimination. The radar design itself requires the choice of wavelength.

The longer it is for given shapes, the less will be the clutter gathered. Clutter echoing areas tend in general to be much greater than aircraft or ships. However long wavelengths demand bigger antennae for a given beamwidth and also lead to bigger electronic equipment. This choice of wavelength is central and repercusses into practically every aspect of design. The following section attempts to convey this idea.

7.3.1 The importance of wavelength

A hint of the issues at stake was given above – the relationships between beamwidth and wavelength. There are numerous others which will be appreciated from a study of the remainder of the text in part 2. The situation is represented in fig. 7.6 which is drawn as a 'map of dependencies'. It is by no means complete, but is sufficient to show the central importance of the choice of wavelength. The notation indicates 'dependency'. What those dependencies are can be found in radar theory and physics.

7.4 Check list

Gathering all the factors and effects together into a radar equation is now possible. It reads as follows:

$$R^4 = \frac{P_{av}\, G\, A\, E_a\, \sigma\, n\, I_n\, F_p}{(4\pi)^2\, K\, T\, F_n\, (B\tau)\, f_p\, (S/N)_1\, L_{sys}\, L_a}$$

where P_{av} = average pulse power of transmitter
G = antenna gain
A = antenna area (m^2)
E_a = antenna efficiency (a factor less than 1)
σ = target echoing area (effective), m^2
n = number of signals integrated
I_n = Integration efficiency (factor less than unity)
F_p = antenna propagation factor (lobing effects)
K = Boltzman's constant (1.38×10^{-23} J/degree K)
T = ambient temperature in degrees K
B = receiver bandwidth (Hz)
τ = pulse duration (seconds)
f_p = pulse repetition rate (Hz)

$(S/N)_1$ = signal to noise ratio required at receiver output (single sample case for given false alarm and detection probabilities)
L_{sys} = system losses incurred by transmitted and received power en route from and to equipment and antenna
L_a = atmospheric attenuation (2-way)
F_n = receiver noise figure

This form of the equation, albeit unusual, allows discussion on all aspects governing the range performance. It is hoped that it gives a useful check list for users in discussing performance with engineers.

8
The wide variety of primary radars

8.1 Narrowing the field

Primary radar technique is employed to serve a large number of users across a wide range of operational requirements. Table 8.1 is reproduced from the *Electronics Engineer's Reference Book* (reference 5) to show the breadth of scope covered. The list is not exhaustive but sufficient to indicate how the range of operational requirements leads to a wide variety of equipments and techniques. Although all of the foregoing applies to all types of primary radars, the following text will concentrate upon operational requirements fulfilled by ground based systems, itself a very large group. As suggested earlier, most primary radars are typified by those found in the world of air traffic control.

From the table, the following types of radar are abstracted:

(a) En-route radars (E.R.R.).
(b) Terminal manoeuvring area radar (TMA).
(c) Approach control radars (ACR).
(d) Airfield surface movement indicators (ASMI) (sometimes called airfield surface detection equipments).

These four groups exemplify the very different characteristics required of most ground based radar equipments and allow concentration upon how their differing requirements highlight differing electronic designs. A small number of characteristics implied by the operational requirements mark differences in the technology required to meet them. They are:

- Data renewal rate (e.g. antenna rotation rate).
- Resolution capability.
- Mean power requirements.
- Range performance.
- Complexity of processing.

The operational requirements of the four radar groups above infer the properties

Table 8.1

Operational role	Typical wavelength	Typical peak power and pulse length	Deployment	Characteristics
Infantry manpack	8 mm	F.M.–C.W.	Used for detecting vehicles, walking men	Has to combat ground clutter and wind-blown vegetation etc. Hostile environment. Battlefield conditions
Mobile detection and surveillance	3 cm	20 kW at 0.1 μs to 0.5 μs	Used in security vehicles moving in fog, also military tactical purposes	Needs high resolving power for target discrimination. Hostile environment. Mobile over rough ground
Airfield surface movement indicator	8 mm	12 kW at 20–50 ns	Used on airfields to detect and guide moving vehicles in fog and at night	Resolving power has to produce almost photographic picture. Can detect walking men
Marine radar (civil and military)	3 cm	50 kW at 0.2 μs	Used for navigation and detection of hazards. Longer wavelength for aircraft detection, i.e. 'floating surveillance' radars	3 cm needs good resolution and very short minimum range performance. All have to contend with sea clutter and ships' movement
	10 cm and 23 cm	(as TMA)		
Precision approach radar (PAR)	3 cm	20 kW at 0.2 μs	Used to guide aircraft down glidepath and runway centreline to touchdown	Very high positional accuracy required together with very short minimum range capability and high reliability
Airfield control radar (ACR)	3 cm	75 kW at ½ μs	Usually 15 r.p.m. surveillance of airfield area. Guides aircraft into PAR cover or on to instrument landing system	Needs anti-clutter capability. 10 cm has MTI. Good range accuracy. High reliability required.
	10 cm	0.4 MW at 1 μs		

Airborne radar	3 cm	40 kW at 1 μs	Used for weather detection and storm avoidance	Forward-looking sector scanning. Storm intensity measurement system usually incorporated
Met. radar	3 cm 6 cm 10 cm	75 kW to ½ MW ½–1 μs	Surveillance and height scanning gives range/bearing/height data on weather. Also balloon-following function	Good discrimination and accuracy. Rain intensity measuring capability
Air traffic control–terminal area surveillance (TMA)	10 cm 23 cm 50 cm	½ MW to 1 MW 1 μs to 3 μs	Surveillance of control terminal areas. Detection and guidance of aircraft to runways and navigational aids. 60n miles range	Needs good discrimination and accuracy MTI system necessary. Display system incorporates electronic map as reference
Air traffic control–long range	10 cm 23 cm 50 cm	½ MW to 2 MW 2 μs to 4 μs	Surveillance of air routes to 200n miles range. Monitoring traffic in relation to flight plans	Has to combat all forms of clutter. MTI essential. Circular polarisation necessary except at 50 cm
Defence–tactical	10 cm 23 cm	½ MW to 1 MW 1 μs to 3 μs	Mobile or transportable. Used for air support and recovery to base	As for TMA radar plus ability to combat jamming signals
Defence–search	10 cm 23 cm	2 MW to 10 MW 2 μs to 10 μs	Used for detection of attacking aircraft monitoring of defending craft including direction for interception. Usually a height finder operated together with search	As for ATC long range plus ability to combat all forms of jamming. Forward stations report data to defence centre

(Reproduced by kind permission of Butterworths (Ref. 4).)

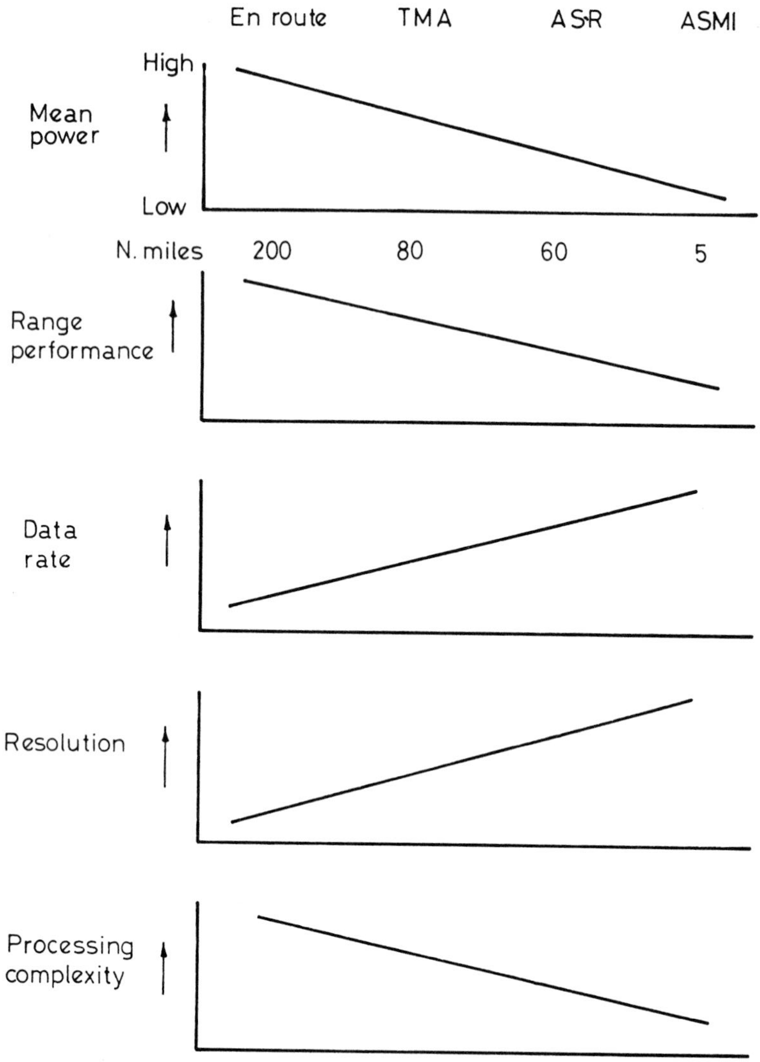

Fig. 8.1. General requirements in various radars, En route, TMA (terminal manoeuvring area), ASR (airfield surveillance radar) and ASMI (airfield surface movement indicator).

of the radar system to meet it as shown in fig. 8.1. 'High' and 'low' are used to express trends rather than absolute values and the scales are non-linear. From this picture it is possible to see clear distinctions. There is a need to produce high mean powers for E.R.R. and TMA radars at a relatively low data rate and resolution. Range performance has equally to be high but is in accord with high mean radiated power. At the other end of the scale, those systems concerned with targets close to the end of their missions need to give the user a high data rate and resolution but because target range required is small, make small demand on mean power, thanks to the laws of physics, sensitive receivers and the constant physical size of the target.

An important distinction needs to be made among the whole group concerning clutter rejection. En-route radars need clutter rejection within about 50 nautical miles of their range and protection against clutter coming in as a result of anomalous propagation. The group of radars from ASR and below – precision approach radars, coast watching, marine radars, etc. – need to know where ground clutter is and so do not generally need ground clutter rejection systems.

Although fig. 7.6 shows the relationships intrinsic to the radar art to be complex, we can say that data rate is predominantly dependent only upon how fast the antenna rotates. The range performance can be taken as 'that which is required' from the combination of all other parameters. This leaves only three topics requiring attention for the whole group:

- How to get enough mean power?
- How to get enough resolution?
- How to get enough clutter rejection?

As another instance of the perfidy of the radar business it is better to start with the second question, regarding resolution because it affects the others.

8.2 Resolution

The resolution capability of a radar governs the limits within which two wanted targets can be distinguished, one from the other. The most usual type of surveillance radar has two degrees of freedom to determine this:

(a) Pulse duration.
(b) Azimuth beamwidth.

A three-dimensional radar has an extra degree of freedom in the elevation domain, governed by its elevation beam structure. Sticking to the usual two-dimensional surveillance radar, it can be seen from fig. 8.2 that the smaller the resolution cell, the better is the ability to discriminate between two individual targets. But because the antenna pattern is necessarily three-dimensional, it gathers unwanted clutter signals in a volume much greater than that occupied by the wanted target. The only way to improve matters is to reduce the size of the resolution cell. In a two-dimensional surveillance system required to examine a vast volume of space including altitudes of up to 100 000 ft with a data rate of about seven reports per minute, the recourse is to use as short a pulse duration and as narrow an azimuth beamwidth as possible. The other method of breaking the elevation domain into segments (i.e. by multiple elevation beams – the three-dimensional technique) is very expensive albeit very rewarding in radar performance.

It will be realised that the smaller the resolution cell becomes, the better is the wanted signal to unwanted clutter ratio down to the limit where the target is physically larger than the cell.

An example of extremely high resolution in an ASMI (airfield surface movement indicator) or ASDE (airfield surface detection equipment in US parlance) is illustrated in fig. 8.3. Permanent features can be filtered out by processing so that moving objects are clearly shown against the background of a synthetically generated runway, taxi-way and parking area network.

The two-dimensional surveillance radar has then a resolution cell having a cross-sectional area of two dimensions to consider – those of distances equivalent to azimuth beamwidth and pulse duration. It also has a third dimension – the vertical extension of its cross-sectional area through the vertical polar diagram.

Filling the resolution cell with the requisite power and sorting out the contents of each cell, to retain the wanted signal and reject the rubbish, are the next topics.

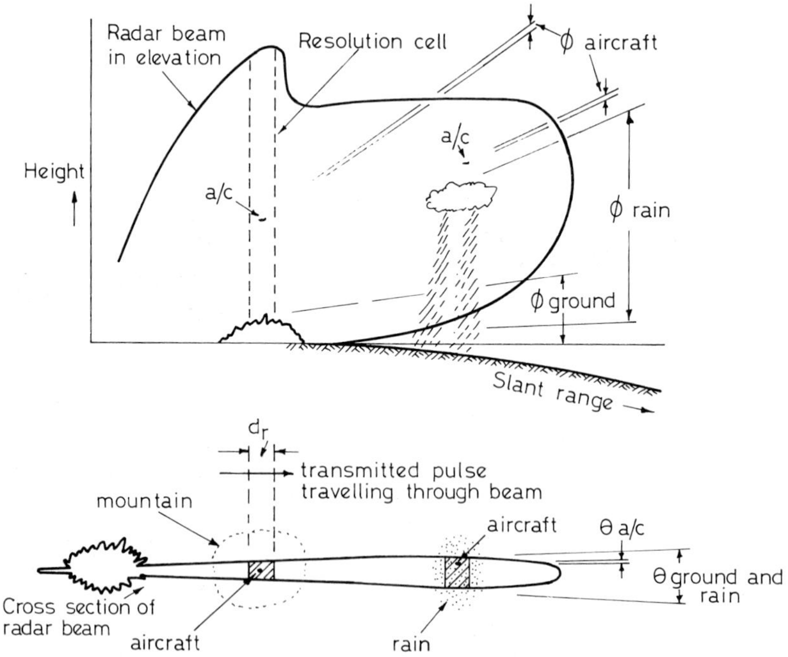

Fig. 8.2. Illustrating the 3-dimensional nature of the resolution cell. Volume is pulse duration (d_r) × linear equivalent of azimuth beamwidth (θ) × linear equivalent of vertical beamwidth (∅). Wanted targets are generally much smaller than the resolution cell.

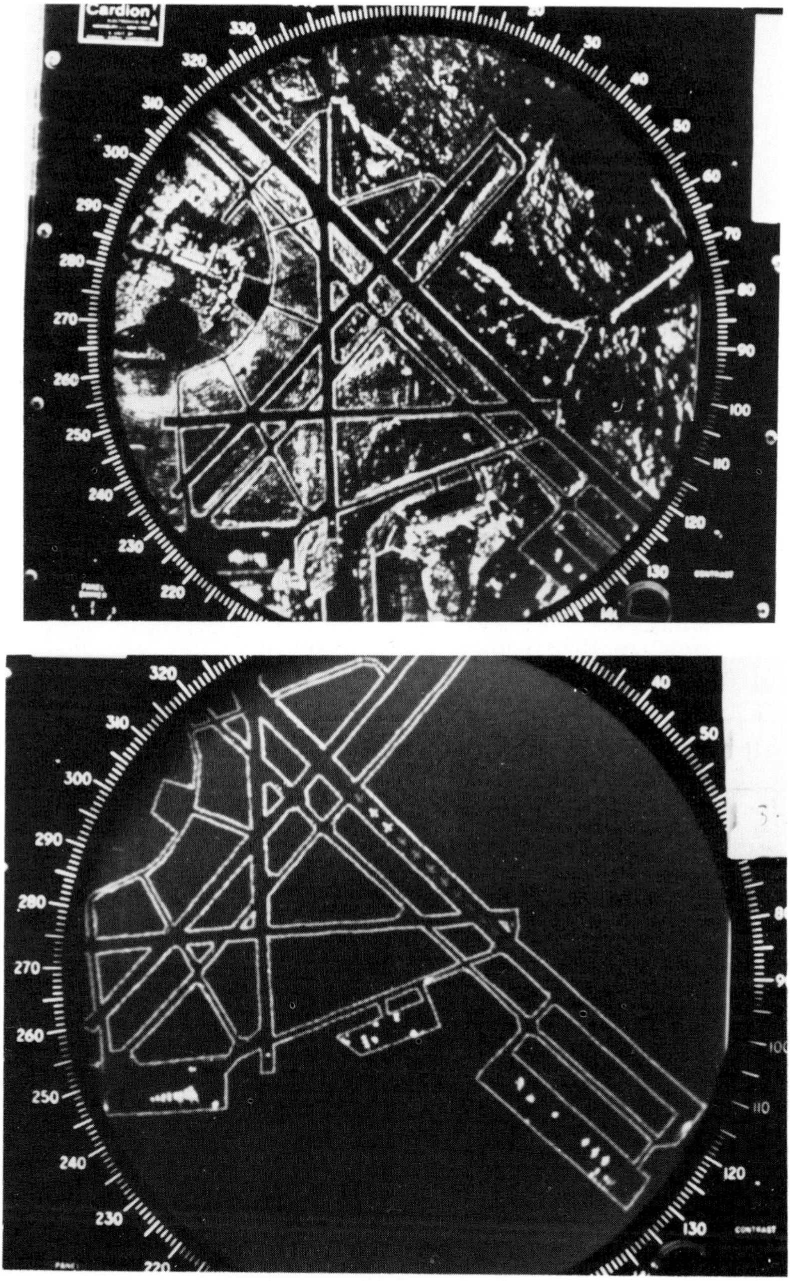

Fig. 8.3. High resolution of ASDE equipment achieved by very short transmitted pulses and very narrow azimuth beamwidth. Processing can remove selected fixed features leaving targets on runways, taxiways, etc., clearly visible. The aircraft taking off is a DC9. (*Photos courtesy of Cardion.*)

8.3 Filling the cell – transmission techniques

In part 1 the subject of output devices available was briefly touched upon. It is considered helpful to expand upon it because the vices and virtues of each govern their suitability for various roles in the group of radars selected. The term 'suitability' is emphasised, for no output device is intrinsically 'better' than any other.

8.3.1 The magnetron

Modern magnetrons still operate upon the original principle developed in the U.K. by Randall and Boot in 1940. Continual development has vastly improved their life in service, stability and output purity.

The device is illustrated in fig. 8.4(a) and (b). A cylinder is mounted between the poles of a powerful magnetic field, the lines of force being parallel to the cylinder's longitudinal axis. The cylinder, as shown in cross-section in fig. 8.4(b), contains a cathode which, when heated, releases electrons towards the anode when sufficient voltage differences exists between the anode and cathode. The anode is made in the form of cavities of equal size and shape. Their proportions at microwavelengths govern their resonant frequency, that is, they behave as tuned circuits. The interaction between the electron stream and the strong magnetic field gives the electrons a curved path as they try to reach the anode. Its a bit like playing table tennis in a very strong cross-wind, the ball being an electron and the magnetic field the wind.

This curved path takes electrons past the mouth of more than one cavity. The effect of this is to excite the cavities to resonance as a result of the electro-magnetic field generated at the cavities' mouths. Since the cathode is itself cylindrical and the cavities evenly disposed around it, the whole group of cavities begins to oscillate. They share common coupling by the electron stream and hence power generated by the group can be extracted from any one of the cavities. This is done by inserting a coupling loop into one of the cavities and transferring its gathered power into a waveguide by a properly matched probe. The output power is extracted from the electrons in the stream which finally comes to rest at the anode, the electrons having given up their kinetic energy in their travel.

The magnetron has a wide range of output power and frequency of operation from millimetres to 23 cm wavelength, and with peak output powers up to 5 MW at the longer wavelengths. The significant limit of their capability is that of mean power output. This is set by the physical size of the cathode and its ability to withstand heat created by some of the electrons which, having taken up energy from the field in the space between cathode and anode, return to the cathode instead of escaping to the anode. Incidentally, this heat is usually sufficient to maintain the cathode at its

Fig. 8.4(a). A high power L-band magnetron. The complete assembly is shown on the right. The magnet is on its left. Precision engineering of piece parts is self-evident. (*Photo courtesy of English Electric.*)

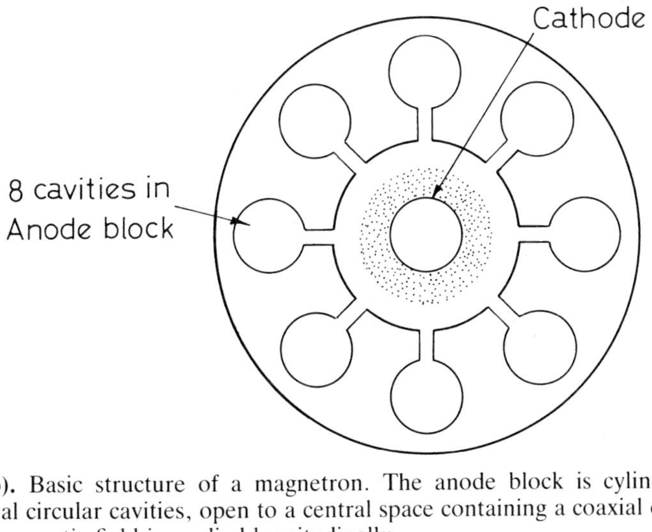

Fig. 8.4(b). Basic structure of a magnetron. The anode block is cylindrical with longitudinal circular cavities, open to a central space containing a coaxial cathode. A powerful magnetic field is applied longitudinally.

proper working temperature and the cathode's heater current is only required until this 'back bombardment' has established itself. Mean powers available from commercially available types range from about 10 kW at 23 cm falling linearly to about 50 W at 8 mm.

8.3.2 The klystron

Although klystrons are used at low power both as oscillators and amplifiers (see reference 4, chapter 34) we are concerned with those used as high power microwave output devices. In these, four or five resonant cavities spaced along the long tube's axis are used to govern the speed of travel of electrons down the tube. The arrangement is illustrated in fig. 8.5 and a photograph of one such as used in Marconi Radar's 600 MHz en-route radar is presented in fig. 8.6.

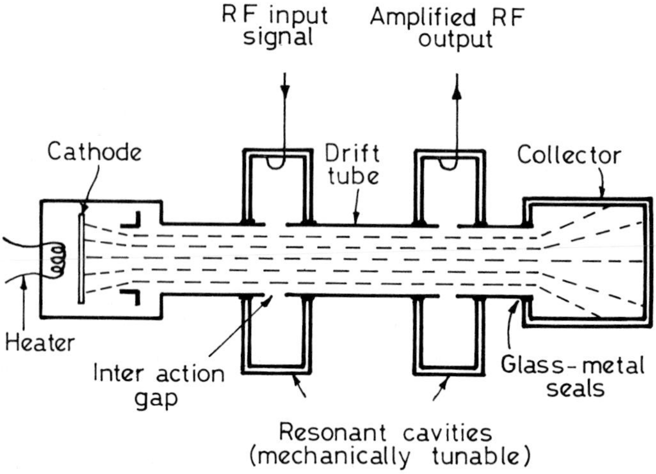

Fig. 8.5. Simplified diagram of a high power klystron. Fields across the interaction gaps modulate velocity of electrons causing 'bunching'. The effect magnifies as electrons progress to the collector.

The klystron's action is as follows. A high voltage is applied across the cathode and collector. Electrons released from the cathode are drawn through the long tube. At the tube's mouth and throughout the tube's length the electron beam is focussed by an external magnetic field provided by sets of flat coils concentric with the tube. The beam's electrons are accelerated by the potential applied between the cathode and an anode at the start of their journey down the tube. The sealed glass tube through which the beam passes has a series of hollow resonant cavities mounted concentrically around it.

Each can be tuned by symmetrically altering its physical size by mechanical means. The first cavity is excited by an externally generated low level r.f. signal introduced by a loop coupler.

The field created across the beam by the cavity's excitation causes the electrons to be either accelerated or slowed down, dependent upon their position in the cavity's length. The beam energy is therefore organised or modulated into bunches in sympathy with the oscillations within the first cavity. This very high frequency bunching is accentuated by the subsequent cavities which are tuned to the input frequency. They are caused to resonate by the bunched field passing through their middle or interaction space. As the beam progresses through the so-called 'drift' tube past these successive interaction spaces, the change of amplitude in the beam caused by bunching becomes very large, far greater in magnitude than the original modulating input. At the final cavity, whose exitation is very great, output power at r.f. is extracted, again by loop-coupler, and fed to either coaxial feeder or waveguide, dependent upon the frequency and power output levels.

Power gains of up to 40 dB (10 000 times) are possible at normal radar

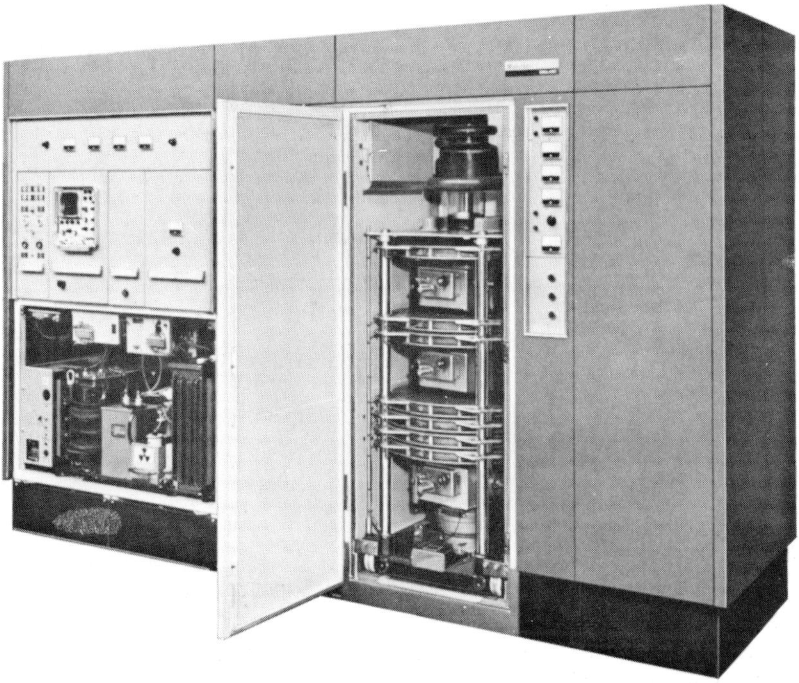

Fig. 8.6. A 600 MHz high power klystron, used in Marconi radar types S264A, S650 and S670. The klystron assembly fills the open-doored cabinet. (*Photo courtesy of Marconi Radar.*)

frequencies. But the klystron's chief value as a radar transmitter is its ability to sustain high mean power levels in its amplifier role. When pulsed (i.e. their mean power is spread into 'on' and 'off' periods) it is possible to obtain peak powers of between 0.5 and 20 MW. Klystrons are commonly used at ultra high frequency in CW mode for television transmission at power outputs of up to 60 kW. Continuous-wave use implies that the mean and peak outputs are nearly the same.

8.3.3 The travelling wave tube (TWT), sometimes pronounced 'twit'

The TWT, like the klystron is a linear beam tube. Its construction is shown in schematic form in fig. 8.7(a). As in the klystron, a cathode releases electrons which form a beam under the influence of a focussing axial magnetic field and the attraction of an anode. Within the glass envelope of the tube is a 'slow wave' structure. In early designs it took the form of a helix – a drawn-out coil – supported throughout the tube's length so that it was coaxial with the tube. In modern tubes this helical form is replaced by slow wave structures more easily manufactured under control.

At the cathode end of the tube the slow wave structure is energised by an r.f. input signal. As the electrons pass through it, the interaction of the beam and field created by the input signal causes the electrons to change speed, as in the klystron, causing 'bunching'. Because the beam and bunched wave are synchronous at the input frequency, the travelling wave created induces higher and higher voltages in the slow wave structure as it progresses down the tube. The amplified input signal at its end is cavity-coupled to the outside world, energy being extracted by a coupling loop into a coaxial feeder or waveguide. As in the klystron, the electrons (now with practically all r.f. energy expended) impinge upon a collector. The residual energy is dissipated as heat.

A great variety of TWTs are available and, like the klystron, are able to be used as pure CW amplifiers or to be pulsed – the normal mode of operation in radar. Up to 1 MW of output can be obtained in pulsed TWT amplifiers. A typical low power assembly is illustrated in fig. 8.7(b).

8.3.4 Unambiguous range

Transmitter design includes making a choice for these various types of output device. The first step in this choice is usually made by deciding to adopt the simplest solution to a fundamental problem: how to get range information.

Radars almost invariably need to show target range (there are exceptions such as police radar speed traps which only need one's identity and speed!). This has to be unambiguous, a single and reliable value. After each pulse has been transmitted the system must go into receive mode long enough for

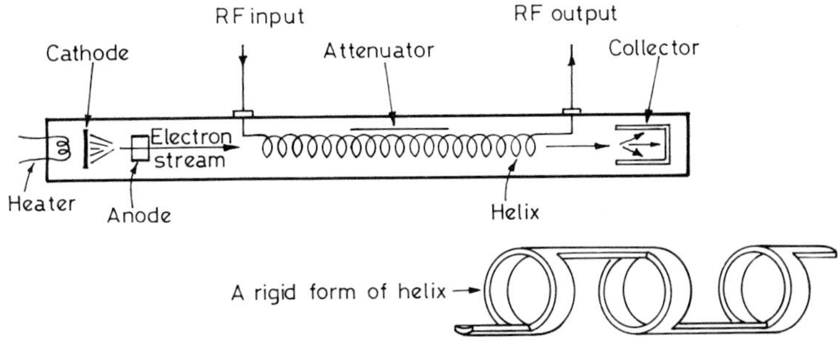

Fig. 8.7(a). Simplified diagram of a travelling wave tube (TWT). Interaction between the electron stream and field excited by the helix modulates electrons to increasing extent as they progress towards the collector.

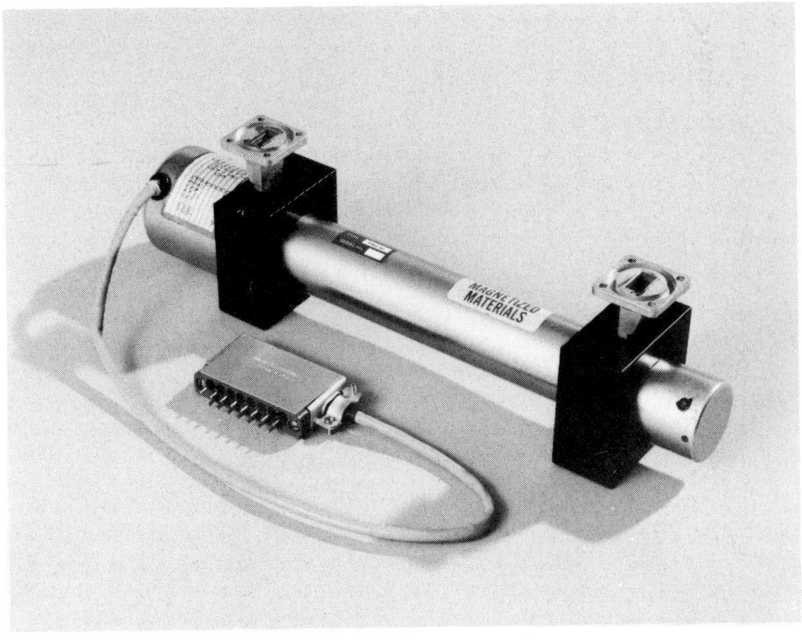

Fig. 8.7(b). A modern version of a 3 cm TWT. (*Photocourtesy of English Electric.*)

echoes to return from the longest displayed range. If this were not done, the target range would be ambiguous, as in the following example.

Suppose a radar was designed to have 100 nautical miles range. The round trip time for this (neglecting display and processing 'overheads') would be 100 $\times$ 12.36 = 1236 μs. The necessary interpulse time would therefore have to be at least this value and the p.r.f. would be its reciprocal, i.e. 809 Hz. Suppose

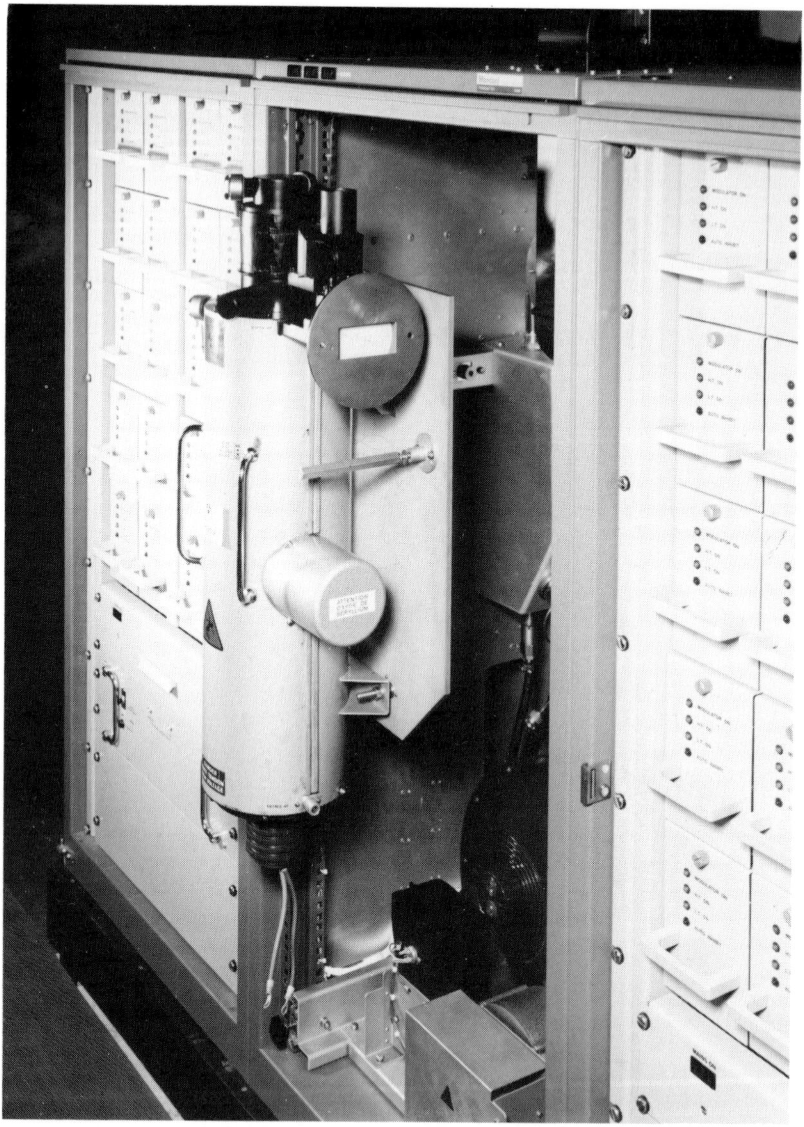

Fig. 8.8. High power TWT shown in its servicing position in Marconi radar type S512. It operates in S-band giving 2.5 kW mean power. (*Photo courtesy of Marconi Radar.*)

also that a target was at 60 nautical miles range. Its echo would arrive 60 × 12.36 μs after transmission, i.e. 743 μs later. If the system was made to operate at twice the p.r.f., i.e. 1619 Hz, and the display triggered at each transmission, the target would appear to be only 10 nautical miles distant because the display would be indicating only half the required 100 nautical miles, i.e. 50.

If the display was still triggered at the required rate (809 Hz) and the transmitter fired at twice this value the ppi would show a bright ring at 50 nautical miles (half the full tube radius) representing transmitter breakthrough and any targets within 50 nautical miles would be indicated twice with a separation of 50 nautical miles. Techniques do exist which permit very high p.r.f.s and long range performance, but they are far from simple.

The need to produce unambiguous range information in the simplest fashion dictates the interpulse period. The intrinsic radar range requirement calls for a transmitter mean power to match it. The transmitter device will have two parameters which determine how it will best be operated. They are:

- Peak power capability.
- Mean power capability.

Suppose the requirement was to produce 2 kW mean power from a device with a peak power capability of 1 MW with interpulse period of not less than 2 ms. The pulse duration (t) is easily calculated from the relationship:

$$P_{mean} = P_{pk} \times \frac{t}{T} \qquad (8.1)$$

where t = pulse duration in seconds
T = interpulse period in seconds

From (8.1) above we have

$$t = \frac{P_{mean} \times T}{P_{pk}} \qquad (8.2)$$

For the example taken,

$$t = \frac{2 \times 10^3 \times 2 \times 10^{-3}}{1 \times 10^6}$$

$$t = 4 \times 10^{-6} \text{ s or } 4\ \mu\text{s}.$$

But suppose the system was also to have a range resolution of 1/10 nautical miles. This demands pulse duration of not more than 1.2 μs. In the example above, reducing the pulse width would result in less mean power, for the peak power cannot be increased in compensation. If the transmitter device was a magnetron, then it would have to be of a type that had at least 2 kW mean power capacity *and* a peak power capability of 4/1.2 MW or 3.3 MW. Such considerations must be taken by the designer.

Another set of choices is available to the designer. They are to use either a klystron or TWT, instead of a magnetron, coupled with the technique of pulse compression. This is a means of providing the necessary mean power by using

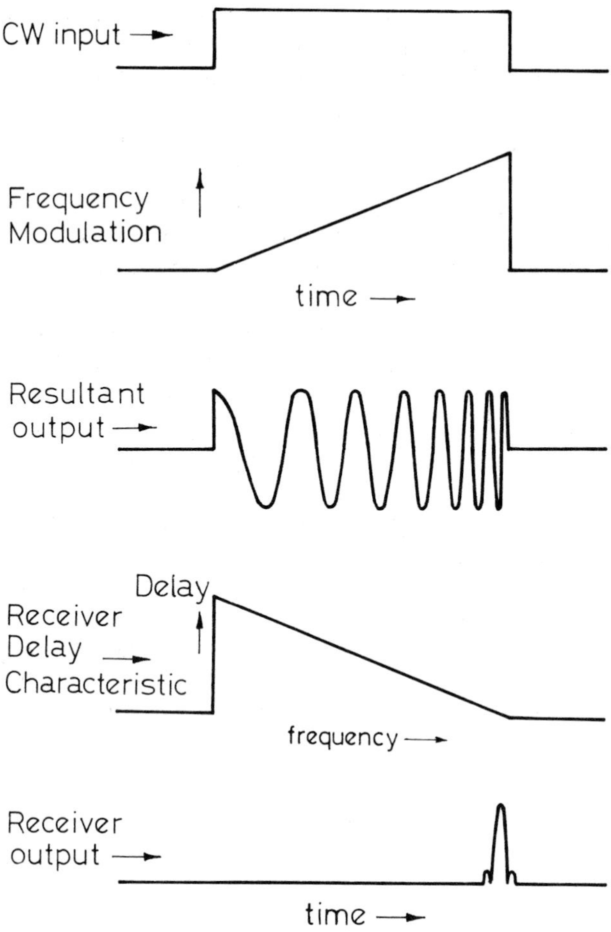

Fig. 8.9. Illustrating the technique of pulse compression. Long transmitted pulse is frequency modulated. Received signals undergo frequency-dependent delay compressing the output into very short pulses.

a long pulse in transmission and compressing the received long pulses into a duration suitable for the required range resolution. It operates as follows. We have seen that an essential difference between the self-oscillator and power amplifier systems is that the latter accept a 'gated' CW input and the amplifier magnifies this pulse of CW to produce a higher powered output pulse. In a pulse compressed system the input pulse has its frequency modulated (usually at a linear rate) throughout its duration. The waveform of the output is as shown in fig. 8.9.

Upon reception all pulses will have this history but before they are used in signal processing they pass through a special delay device. The delay is made to be frequency dependent in such a way that those frequencies in the long pulse arriving first are delayed most, those arriving last, least. The frequency ν delay characteristic is the exact inverse of the modulation history. This has the effect of compressing the received pulses into a much shorter time than their input duration. It can be considered as a means of increasing the effective peak transmitted power in the ratio of the input to output pulse duration ratio. For example, a 20 μs transmitted pulse which was compressed by 1 μs would effectively increase the peak output power by a factor of twenty.

Some unfortunate by-products come with the technique. The radar cannot receive the reflected pulses until transmission is finished, otherwise the sensitive receiver would be blown up by its own transmitter's power. In any case the signal components cannot emerge from the delay device in their compressed form until they have effectively suffered the delay time of the device. Thus the system's minimum range is limited to the equivalent of the transmitter pulse length. In the example above, a 20 μs pulse duration would limit the minimum range to 20/12.36 nautical miles = 1.6 nautical miles. In systems requiring 0.5 nautical mile minimum range, a short pulse without frequency modulation must be transmitted at a different frequency from any in the long pulse and at the lower peak power of the output device. Signals at this single frequency are treated in the normal fashion. The extra complexity of such a system is obvious and is illustrated in fig. 8.10.

8.4 Choice of transmitter techniques

In recent years heated arguments have been put for and against the choice of self-oscillators and power amplifiers in primary radar transmitters. It is hoped that what follows will generate a little light to go with the heat! I have already published articles directed to this end in the context of air traffic control radars. These have a special set of operational requirements which make choice relatively easy (see reference 9). Table 8.2 shows the output devices available for use.

If one considered the whole gamut of radars exampled in table 8.1, the choice of transmitter device is governed by a small number of parameters:

(a) The wavelength permitted.
(b) The required resolution.
(c) The mean power required.
(d) Whether the system is to be coherent or not.
(e) Whether frequency agility is needed or not.

We can dismiss two of the devices shown in table 8.2 as being, for general use, non-runners. The amplitron on grounds of its low gain (typically 10 dB)

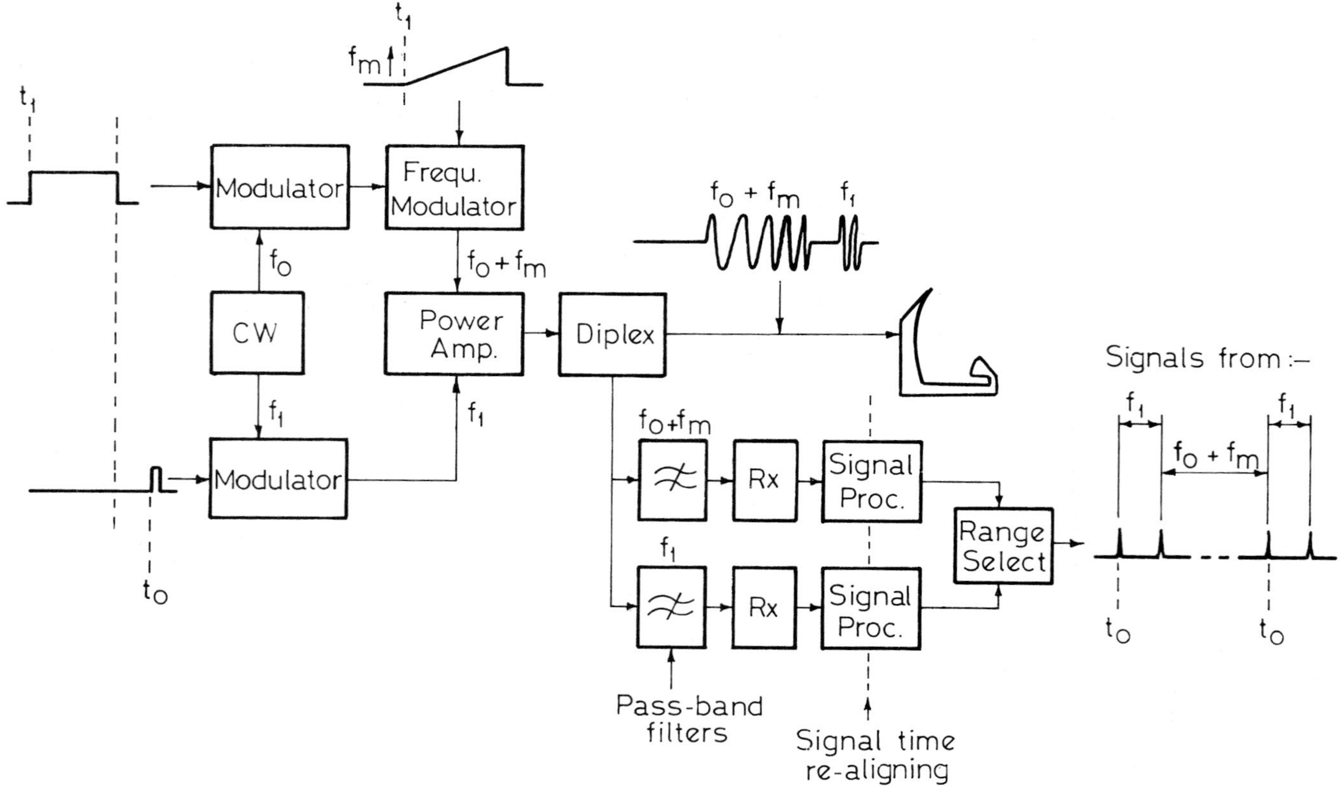

Fig. 8.10. Functional schematic diagram of a practical pulse compression system. A short pulse at f_1 has to be introduced to cover the range equivalent to the long pulse duration. Note the need for complex processing.

Table 8.2

Cross field devices		Linear beam devices	Solid state devices
Magnetron (SO)	Resonant cavity (SO)	Klystron (Dr)	Multiple transistor modules (MTM) (Dr)
	Coaxial (SO)	Travelling ware tube (TWT) (Dr)	
Amplitron	(Dr)	Hybrid TWT/klystron – twystron (Dr)	

(SO) = Self-oscillator
(Dr) = Driven power amplifier

and the twystron on the grounds that its mean power is usually far in excess of normal requirements. The coaxial magnetron is a recent development of magnetron technology using the same principle of operation as its predecessors but having much longer lifetime – of the same order as its competitors, the linear beam tubes. Figure 7.6 gave an indication of the degrees of freedom and constraints the designer has. The importance of satisfying the requirements of resolution has also been discussed. All the devices remaining from table 8.2 are capable of producing coherence, an essential property of moving target indicators and detectors. A singular parameter is left for discussion: that of frequency agility.

8.4.1 Frequency agility

This term is used to describe the ability of the transmitter (indeed the whole radar system) to operate successfully using one of a set of different frequencies during a small number of repetitions. The main reason for this requirement is to protect the system from the effects of 'jamming'. If a military radar used only one frequency, an enemy would be able to measure it, together with its transmitter signature (pulse duration, frequency modulation, etc.). The enemy would then be able to radiate signals on this measured frequency which swamp the radar's receiver or totally confuse the radar's signal processing, thus reducing its operational effectiveness, sometimes to the point of its being totally useless.

There are a number of ways of managing the frequency agility. Obviously a change of frequency at each successive transmission is the most effective, particularly if the choice is randomly selected from a large group of values within the radar's operating band. As will be seen later, some signal processing systems will fail under these conditions and a small number (typically four to eight) of transmissions are needed at one frequency before changing to another. These two modes of operation are known respectively as 'pulse-to-pulse' and 'burst-to-burst' agility.

Although a version of the magnetron, called the 'multipactor', has a degree of frequency agility its variation is limited to about 2% of its operating frequency.

The linear beam tubes and multiple transistor modules can have a very great range of frequencies, easy to control since the devices are amplifiers and the frequency of their input signals at low power level can be accurately governed at great speed. Thus unless frequency agility of a high order is a requirement the choice of transmitter devices is between all those listed in table 8.2.

8.4.2 Cost

Linear beam tubes (klystrons and TWTs) are very vulnerable to mechanical damage. Alignment of their internal structures has to be maintained to close tolerances. If mechanical shock takes alignment out of tolerance the tubes are liable to total destruction when operated. They are very expensive; typical costs (1985) are in the £50 000 bracket. They have an 'in-service' life of around 30 000 hours.

Magnetrons are comparatively rugged. Modern versions of traditional designs cost around £5000 and have a service life of some 5000 hours. Coaxial magnetrons cost in the region of £15 000 but have a much longer life of around 20 000 hours. The inference from this is:

(a) If frequency agility is required, then a linear beam tube or MTM amplifier is necessary. If not, a magnetron would produce much lower capital and life cycle costs.
(b) If the pulse compression technique is used, again costs are much higher. The increased complexity is also a disadvantage.

Table 8.3 is presented to show some of the characteristics which need consideration in the design of a transmitter and the consequences of choosing a particular technique. The 0 to 3 weighting given is of the author's own devising and the judgements are directed at a medium performance air traffic control radar. Although others may choose different values for the weighting of individual characteristics, the table as given illustrates two important points. First that MTM transmitters are fast becoming contenders in the race and secondly that discounting frequency agility, the coaxial magnetron is still the strong contender particularly for those users who put store by technical simplicity.

Table 8.3

Characteristics	CFO	TWT	Klystron	MTM
Lifetime	3	3	3	3
Cost of tube or output devices	2	1	1	1
Efficiency	2	1	1	2
Size of Tx equipment	3	2	1	2
Cost of Tx equipment	3	1	1	1
Cooling complexity	2	1	1	2
Output device protection	3	1	1	3
Mechanical handling	3	1	0	3
Shelf storage problems	3	1	1	3
Coherence	2	3	3	3
Complexity of transmitter	3	1	2	2
Device failure mode	0	0	0	2
Signal processing complexity	3	1	3	1
Electric stress	2	1	1	3
Field experience	3	2	3	1
Frequency agility	0	3	3	3
	37	23	25	35

As has already been pointed out in reference 9, the proviso is always that the design of the modulator driving the output devices be optimised.

9
Signal processing

9.1 Introduction

Attempting a definition of signal processing gives a hint of the way misunderstanding between user, procurement authorities and designers comes about. It would be easy to say that signal processing is the means of enhancing wanted target data and rejecting the unwanted. But the system can receive signals that are sometimes wanted and at other times not. As an example I cite a case, many years ago, wherein an atc approach radar was designed with mti – it worked very well for its day. As required, it rejected signals from fixed targets and accepted those from moving targets. Some of the latter were generated by rain and heavy clouds. The user said ‘please can you invent a means of seeing aircraft signals even if they are obscured by signals from rain coincident with them’. After some ingenious work on the circular polarisation technique, the user specification was met and he was happy – if it rained the new design cleared the radar screen of signals received from rain and with very little loss of coincident wanted aircraft signals.

Not many months went by before the user said ‘please, we think we’ve asked for the wrong thing; your new design has resulted in the Airport Operations Manager being inundated by complaints from pilots that the air traffic controllers keep sending them into rain storms’. What the user should have asked for was the means of preserving both the wanted aircraft target data and the weather target data and allowing their separate and controllable display. This example is given to show that it is not only the engineering branch of procurement authorities that are at fault if the user is dissatisfied with equipment he gets. Quite often the user is at fault in being insufficiently precise in specifying his needs. Let us content ourselves with a definition of the aim of signal processing as ‘To extract the maximum data from input signals so that they may be categorised and selected for display by the user’.

The following text will assume that one or a number of detection processes described in part 1, chapter 6, have been performed.

9.2 First steps in removing clutter

The simplest situation requiring resolution is that where the wanted signal is 'in the clear', i.e. the only competition it has for detection is noise. In chapter 2 of part 1 it was pointed out that target detection is achieved by checking that a series of signal detections has occurred at the same range over a sufficient number of repetitions. In a radar system with a 'real-time' plan position indicator (ppi) the viewer makes his own decision as to whether this criterion has been met. By 'real-time' is meant that for every new input, an output is produced simultaneously (or sensibly so). If so, the target is seen as an arc of brightness whose range dimension is equal to the number of resolution cells occupied by the target. The azimuth dimension is set by the antenna's horizontal beamwidth and how much of it is filled by the target. If the target is a point source it will result in a range dimension governed by the transmitted pulse duration, how much of its energy has been received and the receiver bandwidth. In section 3.4 it was seen that this depends upon a number of other parameters such as antenna beam shape in azimuth, target size, range and receiver bandwidth matching factor.

Because the process of detecting targets 'in the clear' is one of range correlation it can be used to reject the clutter classed as 'asynchronous interference'. This type of interference can be caused by receiving energy reflected from objects illuminated by a transmitter other than one's own; but predominantly they result from one's own or 'home' station receiving other stations' transmissions directly as either station's antenna looks at the other.

The amount of interference received will be a function of the path attenuation and the product of the two antennae's effective gain along the line of sight. Obviously the interference will have a maximum value when the two antennae's main beams are looking at each other: since it would be highly unlikely that each had exactly the same rotation rate, this case is a very rare one. The most common circumstance is where the interference has two peaks of strength – one for each time either beam sweeps past the other station's position. At other times the effective mutual antenna gain is via the two antennae's sidelobes. If the two stations have different pulse repetition frequencies and hence different interpulse periods, consecutive interference pulses received will be at a different apparent range. Thus, over a beamwidth's history such signals will not correlate in range and thus will not form a 'target-like' response on the ppi. Nevertheless if they are not inhibited from the display they can be a great distraction to the operator. (See fig. 2.5(b) in part 1.) A simple system often used to effect interference suppression is illustrated in fig. 9.1.

Two versions of the detected signals are input to a coincidence gate. One is delayed precisely by the 'home' station's interpulse period (T) and the other input is direct. The logic of the coincidence gate is such as to allow b to emerge only if accompanied in time by an a input. Thus only those inputs

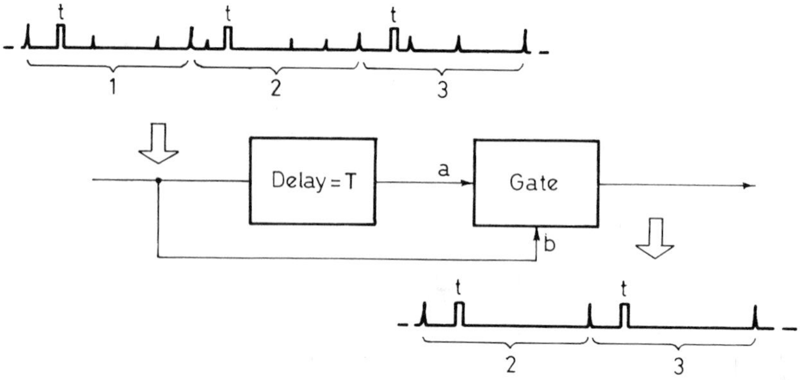

Fig. 9.1. Simple interference suppressor. Three successive interpulse periods are shown as input containing wanted signals (t) among others at irregular intervals. By introducing a delay (T) equal to the radar interpulse period, only signals with this regular interval are allowed to output.

separated by the 'home' station's interpulse period can emerge. There is of course a probability that the interference will accidentally coincide but this is calculably rare. Another less welcome effect in many equipments is that of a group of signals within a beamwidth from a wanted target the first will be lost, as will the last, because they have no concurrent partners to form coincidence.

In some systems the delay mechanism is an analogue device which uses ultrasonic delay lines (a column of water, mercury or a block of quartz).

The detected envelope of the received pulse signal is used as modulation to a carrier frequency of between 10 and 30 MHz, dependent upon design criteria, and launched into the delay medium as a pressure wave of ultrasound. This will emerge at the delay column's end at a fixed time later, dependent on the propagation velocity of sound in the delay medium.

For instance a column of water (whose sonic propagation velocity is about 1400 m/s), would have to be about 1.4 m long to produce a delay of 1 ms (the equivalent of some 80 nautical miles range). The pressure wave is then converted, by a crystal transducer, back into an electrical signal for amplification and detection into its original form. This chain of events involves great loss. Much careful design goes into ensuring the user doesn't suffer from it. Currently it is much more convenient and economical to use digital techniques rather than analogue. These commonly employ shift register circuits to effect delay. They operate as follows. The radar range to be processed is divided into a large number of equal and small increments. Typically a ten-bit binary structure is used, yielding $2^{10} = 1024$ increments. Suppose the range to be represented was 100 nautical miles, each increment would then be equivalent to 100/1024 or about 1/10 nautical miles.

Each increment is allotted its own digital storage circuit. The whole battery of 1024 is, through the marvel of micro-miniaturisation, able to be mounted in spaces the size of postage stamps. It can be operated as a range-ordered store in the following manner. The device is meant to store data which record 'presence' or 'absence' of a signal at particular ranges. It does this by using a system of 'clocking'. Each stage of the register accepts short duration clock pulses generated externally whose interpulse periods are regular and generally arranged at a rate equivalent to radar transit times of 12.36 μs/nautical miles or multiples of this.

In the example taken, a range of 100 nautical miles served by a ten-bit register would lead to clock intervals of $(12.36 \times 100)/1024$ μs = 1.207 μs. The clock oscillator would thus be 1/1.207 MHz or 828.5 kHz. The store would begin its life completely empty of data. It would be activated at a time equivalent to zero radar range and receive, as input, any detected signals and the first of the train of clock pulses. If there was a signal present during the time represented by the first range increment, the first clock pulse would cause the first storage element to be set in its 'signal present' state. At the arrival of the next clock pulse, this stored data ('signal present') would be automatically transferred to the second storage stage. The following clock pulse would pass the data from the second to the third stage and so on to the last storage element, at which time the 'clock' is stopped.

The original signal is established in store at the register's end and any other signals which were present in the radar period from zero to the equivalent of the store length would be stored in the appropriate storage cell of the 1024. For instance, if there were signals present at 10, 26, 50 and 80 nautical miles in a given radar period then the store would have its 102nd, 267th, 517th and 823rd range cells indicating 'presence'. This follows from:

$$n = \frac{R}{R_{max}} \times N$$

where R = actual range in nautical miles
R_{max} = maximum range represented
n = number of the occupied cell
N = number of range increments

If two such registers were to be used in parallel, it is easy to visualise how their contents could be compared cell by cell after successive radar periods, and output only given when the nth cell in each register had 'signal present' states. By arranging the clock pulse to be generated at true radar rate, real-time sequences can be reproduced. However, we shall see that such digital techniques have great flexibility in not being tied to real-time as are analogue delay devices. Useful information on the engineering form and working of such devices may be found in reference 4.

The simple case taken above applies to digital processes where signal amplitude data are not required. However, there are many instances where amplitude data are vital to the signal processing. Nevertheless, the same technique can still be used but with a stack of range-ordered storage units, one layer of the stack for each of a number of binary digits into which the sampled signal amplitude is coded.

No, it is not complicated and it works like this. An analogue to digital (A to D) converter is used. Suppose the dynamic range of signal amplitudes to be handled was designed to be 0 to +4 V (including noise). By expressing this range of values as a twelve-bit binary word, there will be 4096−1 (i.e. $2^{12} - 1$) increments of amplitude, each of 4/4096 volts or approximately 1 mV. At any instant the detected signal output to be digitised can have a value between 0 and +4 V. The A to D converter will express the value as a digital word as shown in table 9.1.

Table 9.1

Input value	Binary expression								
	1	2	3	4	5	6	---	12	Binary bit
	(1)	(2)	(4)	(8)	(16)	(32)	--	(2048)	Decimal value if bit present
7 mV	1	1	1	0	0	0	---	0	
	(1)+(2)+(4)=7								
19 mV	1	1	0	0	1	0	---	0	
41 mV	1	0	0	1	0	1	---	0	
4095	1	1	1	1	1	1	---	1	

Note: The 4096 values include zero

The A to D converter is organised to sample the input signals by using micro-chip technology in much the same way as the shift register. By associating a separate shift register store to each bit of the 'amplitude coding' A to D converter, it is possible to store the complete time and amplitude history of detected signals. An example of such a system in operation is shown in fig. 9.2; it is typical of values expressing only radar receiver noise. Strange to think that most of the time, this is all that is present in the system.

9.3 Thresholding techniques

There is a large class of radars whose prime role is to detect moving targets and for whom signals returned from fixed ground echoes represent a hindrance. There are others in which signals from weather or sea clutter need to be discriminated against. A very effective way of reducing such clutter signals down to noise level is to apply adaptive thresholding techniques – adaptive in the sense that the thresholds automatically adjust themselves to

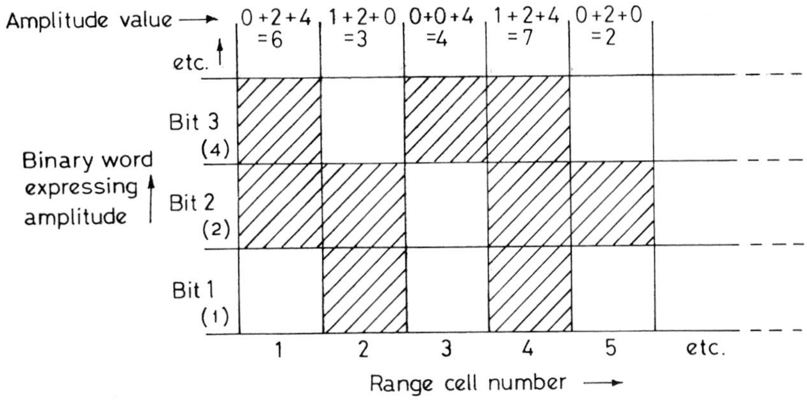

Fig. 9.2. Analogue to digital (A to D) conversion. In the example shown a noise signal in the time equivalent to range cells 1 to 5 is 6, 3, 4, 7, 2 units decimal. Binary bits expressing these are shown shaded. Thus for cell 1, decimal 6 ≡ binary 011; for cell 2, decimal 3 ≡ binary 110, etc.

the clutter environment as it changes in both short and long time periods. The thresholding techniques to be described confer super clutter visibility but not sub clutter visibility.

9.3.1 Temporal threshold

In this process, the whole surveillance area is divided into range/azimuth cells of equal dimension. Suppose, the range domain is broken into 1024 equal increments and the 360° of azimuth into 64 equal angles. The resultant matrix has 65 536 cells. As the antenna rotates, the receiver output (signals and noise) is envelope detected range increment by range increment and the peak value for each measured and stored as a digital word, in range and then azimuth order. Thus after one antenna rotation the store contains the peak signal amplitudes found in each of the cells. The peak amplitudes are then used to set a threshold for each cell, against which the signals in each on the next antenna revolution, are compared. The threshold usually has a small extra margin added to account for the almost inevitable fluctuation in clutter amplitude found from revolution to revolution of the antenna. Any signal exceeding the threshold is output to the display or plot extractor.

In order to account for clutter signals which gradually fade or grow over periods longer than about ten revolutions of the antenna, a system of integration is used whereby the margin is increased or decreased to follow such fluctuations. This integration over long periods permits transitory signals such as aircraft to be shown if they exceed the clutter amplitude plus its margin in a given cell. The aircraft signal's dwell time in one cell is insufficient for it to affect the threshold setting to any significant degree. However slow moving rain or weather clutter will produce effects which will cause the threshold to be raised or lowered, depending on its tendency. Thus all fixed and slow moving clutter will be removed automatically from the system's output, as will noise, for the margin set will be higher than noise level.

One such system in operation is illustrated in fig. 9.3(a) and (b). It will be realised that the matrix of cells containing clutter can themselves be used as indicators of clutter areas. The technique is one of many sometimes known by the name which sounds like 'seefar' (CFAR) which stands for constant false alarm rate.

The term can be understood from the following example. If the radar receiver output contained weather clutter signals at, shall we say, twice that of the receiver noise level, their character would be somewhat like that of noise itself. If displayed without any processing it is easy to imagine that there would be an increased number of clutter plus noise 'events' which exceeded

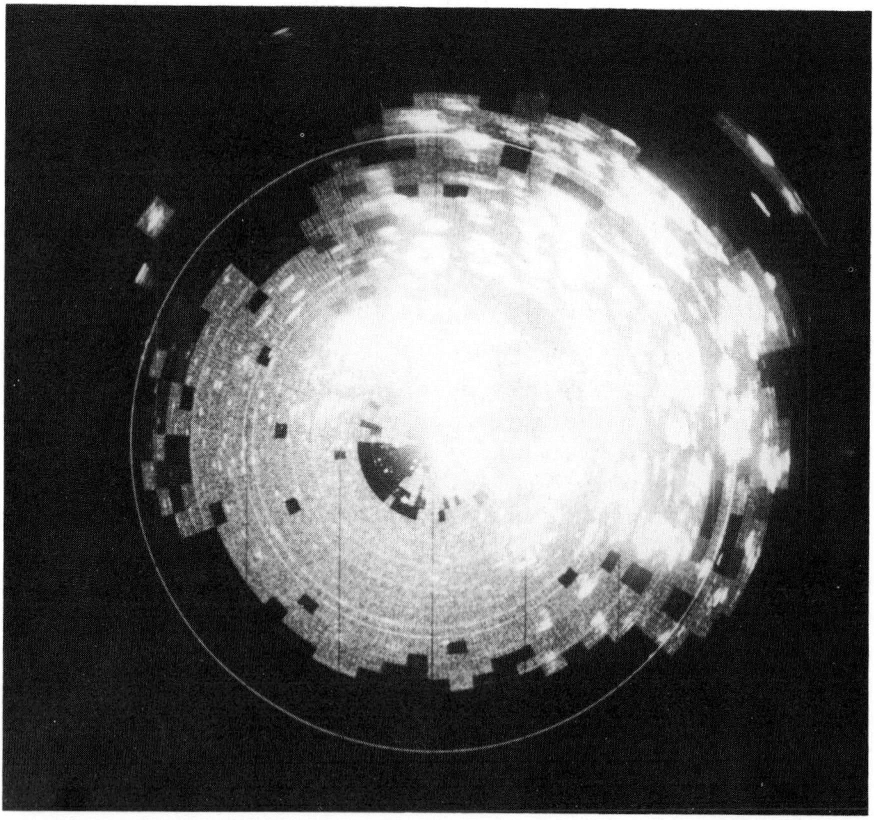

Fig. 9.3(a). Threshold set to low level, thus noise events result in 'cell setting'. Isolated groups of cells 'un-set' are indicated in the noise background. Individual aircraft signals are those with no 'cell' background. (*Photo courtesy of Marconi Radar.*)

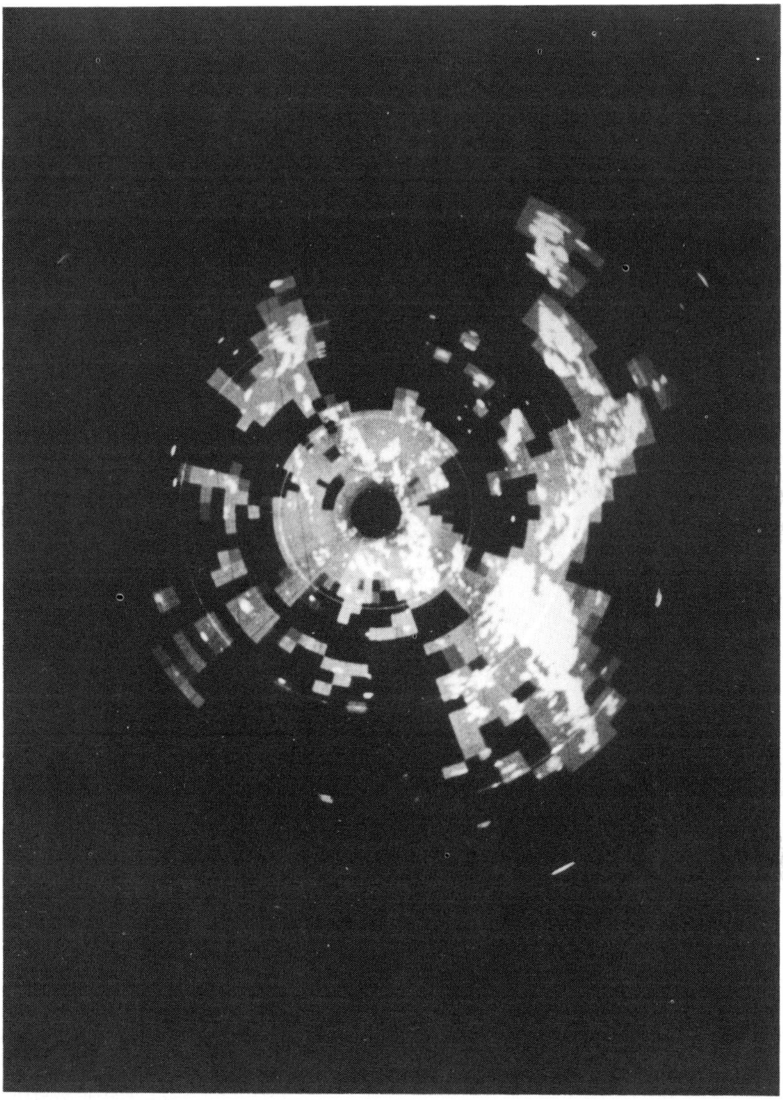

Fig. 9.3(b). Threshold level and margins set at higher level than in Fig. 9.3(a). Clutter areas are clearly indicated by the cell structure. (*Photo courtesy of Marconi Radar.*)

the radar's (or the viewer's) detection criterion. Thus in this cluttered area the false alarm rate would increase above the average. The thresholding technique described above reduces the noise-like fluctuations to a common and invisible level set by the receiver noise and thus restores the false alarm rate to a constant value without impairing super clutter visibility.

9.4 Logarithmic amplification and STC

A common technique for reducing weather and sea clutter is to use a logarithmic amplifier in the receiver chain and to pass the detected video signals through a short time constant (S.T.C.) circuit consisting basically of a simple capacitor and resistor arrangement. The action of the S.T.C. circuit performs the function of mathematical differentiation and its output is thus proportional to the rate of change of its input. The system is illustrated in fig. 9.4 together with waveforms showing the effect. It is seen that by careful choice of the time constant of the S.T.C. circuit components, the effect on short duration pulses such as from point targets is relatively small, whereas its effect on extended clutter such as from mountains, high ground and weather is profound.

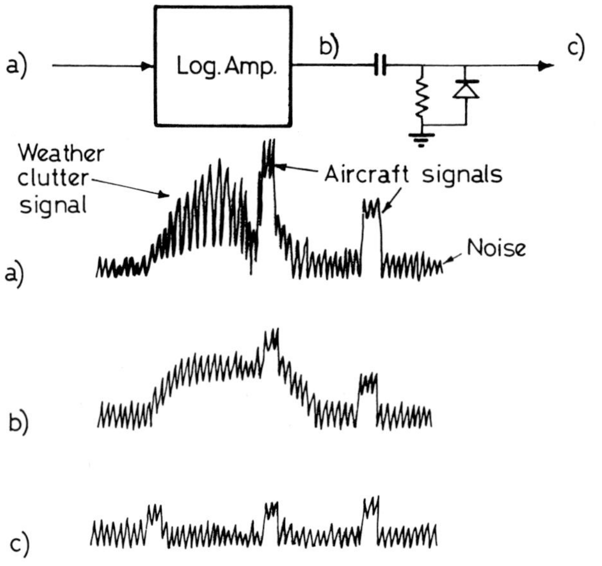

Fig. 9.4. A simple 'Log – S.T.C.' method of reducing clutter. It exhibits super, but not sub-clutter visibility.

9.5 Moving target indicators (mti and mtd)

A.T.C. radars, designed to detect aircraft, have to incorporate processing which separates signals received from moving objects from those generated by fixed objects. This is necessary because significant areas of cover will contain signals from both at the same instant. What characterises the two types? The first obvious answer is displacement with time. How is this simple difference between fixed ground clutter and moving targets to be sought? Let us first consider using the technique of cell storage described in section 9.2. We have seen it possible to break the surveillance area into very small cells and examine the contents of each at successive antenna revolutions. Thus we should be able to detect aircraft as they go from cell to cell. Yes, it is possible to have what one might call 'area mti' using this principle and indeed it is done. But the system has no sub clutter visibility (SCV). To obtain SCV it is necessary to use more than just target amplitude data. The phase information contained in the signals must be used as well. Consider some practical values. Take a typical atc radar having a pulse repetition frequency of 600 Hz and a pulse duration of 2 μs. The range dimension of the radar's resolution cell will be 2/12.36 nautical miles or about 1000 ft (300 m). An aircraft moving at 200 knots radially to the radar would have displacement, during a radar sampling period of only 15 cm or so. Even if the radar's range incrementation and accuracy were to be as small as 1/64 nautical miles (30 m approximately), it can be seen that measuring the range change between samples is far too insensitive a technique for use as a moving target indicator.

We begin to see now the need for coherent radar systems, for with their ability to maintain phase reference from one sample to the next, i.e. from one transmission to the next, it is possible to use the wavelength of transmission as a 'vernier' of range.

Consider the two targets shown in fig. 9.5(a). One is a fixed feature, the other a moving aircraft. The range to either can be expressed as

$$R = \frac{n\lambda}{2} + \frac{d\lambda}{2} \tag{9.1}$$

That is, a fixed integral number of half wavelengths plus a fraction of a half wavelength. Imagine that the antenna is stationary; the signals returned from the fixed target for each transmission will always be the same value in these terms, i.e. $(n_1\lambda/2) + (d_1\lambda/2)$ will be constant for each sample. For the aircraft

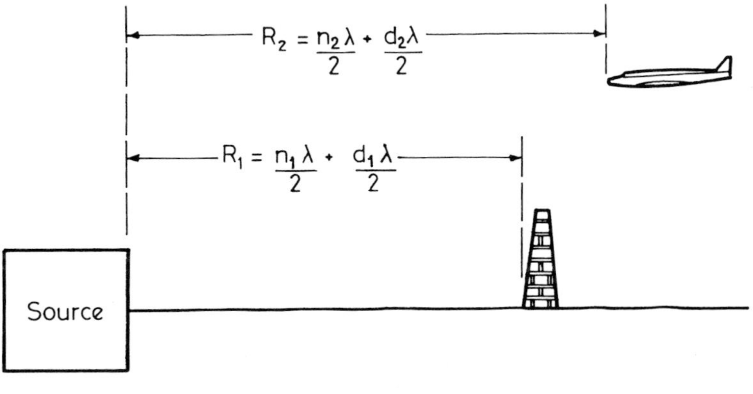

Fig. 9.5(a). Range expressed in terms of half-wavelengths for fixed and moving targets.

however, the value of $n_2\lambda/2$ or $d_2\lambda/2$ will vary from sample to sample because of the aircraft's movement.

A coherent radar system is designed to have an extremely high order of stability; enough to enable these small differences of distance between samples to be detected by use of the system's internal phase reference. A phase sensitive detector such as illustrated in chapter 6, part 1, is used. One arm of the detector accepts the reference oscillation as input, the other accepts the received signal. Phase comparison is made usually at the intermediate frequency of the radar system. A typical phase detector characteristic is shown in fig. 9.5(b). It shows how its output amplitude and polarity change as a function of phase *difference* between the two inputs. Because of the stability of the coherent system, the range continuum can be regarded as made up of a joined series of these phase detector characteristics in space with no overlap or gaps between them. For those who wonder why the phase relationships established at the microwave frequency are held when translated into an intermediate frequency, Appendix 1 contains the mathematical proof.

The example given in fig. 9.5(a) shows a fixed feature of clutter (perhaps a

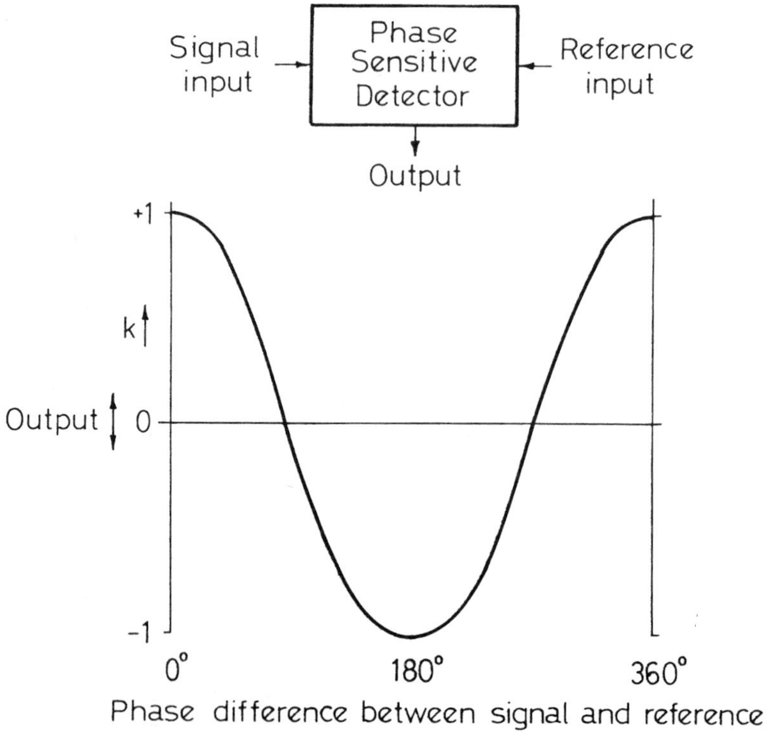

Fig. 9.5(b). A phase sensitive detector (PSD) characteristic. (Derivation of its co-sinusoidal shape is found in Appendix 1.)

chimney stack) and a moving aircraft. The action of the mti detector will be as follows. As the antenna beam illuminates the chimney the first pulse will be received after its double journey. It will have an amplitude dependent upon the parameters of the radar equation (power, gain, target size, etc.) and its rf content will have a phase relative to the radar's coherent reference oscillation.

The intermediate frequency version of the pulse is passed to the phase detector carrying both the amplitude and phase of the original signal. It emerges with its envelope modified in amplitude depending on the phase difference between the pulse and reference inputs. If we regard the phase detector curve as giving a modulus (k) to any input and that k can have maximum and minimum values of $+1$ and -1, the input pulse of amplitude A will emerge as $A \times k$. So a pulse whose phase relative to the reference is zero will emerge with amplitude $A \times +1 = A$. If the pulse had 180° phase relative to the reference it would emerge with amplitude $A \times -1 = -A$. For 90° phase difference it would become zero, for $A \times 0 = 0$. It is seen that the phase detector translates the input signal into a form where its amplitude and polarity are phase dependent.

The second pulse to enter the system will have exactly the same range (i.e. neither $n_1\lambda/2$ or $d_1\lambda/2$ will have changed). Thus the phase difference between the signal and the coherent reference will be the same. The only difference between the first and second pulse outputs from the phase detector will be those originally present at the phase detector's input caused by the difference in antenna gain at the instant of target illumination. In mti systems this is designed to be as small as possible, sometimes by preceding the phase detector by a limiting amplifier to prevent the antenna's horizontal polar diagram from modulating the amplitude of such signals as the beam sweeps across the reflecting object.

As for the signals from the aircraft, at each successive sample (i.e. each transmission and resultant reception) the phase between the received signal and the reference will change from sample to sample. If it was travelling radially at 200 knots and the sample rate was 600 per second (i.e. p.r.f. = 600 Hz) then the displacement between samples for a wavelength of 23 cm would be $(200 \times 6080)/(60 \times 60 \times 600)$ ft/sample = 6.74 inches or 17 cm, which is 17/11.5 half wavelengths, i.e. 268° phase difference between each sample, since half a wavelength represents 180°.

The net effect for these two cases is shown in fig. 9.6(a) and (b). The fixed target produces the same *phase difference* between reference and input signals and hence the same value of modulus for each sample. The moving target produces a *non-zero difference of phase* between its input signal and the reference. Also if its velocity (speed and direction) is constant throughout the antenna beam's illumination time, the difference of phase between samples will be constant, depending on the sample period, the wavelength and the target velocity.

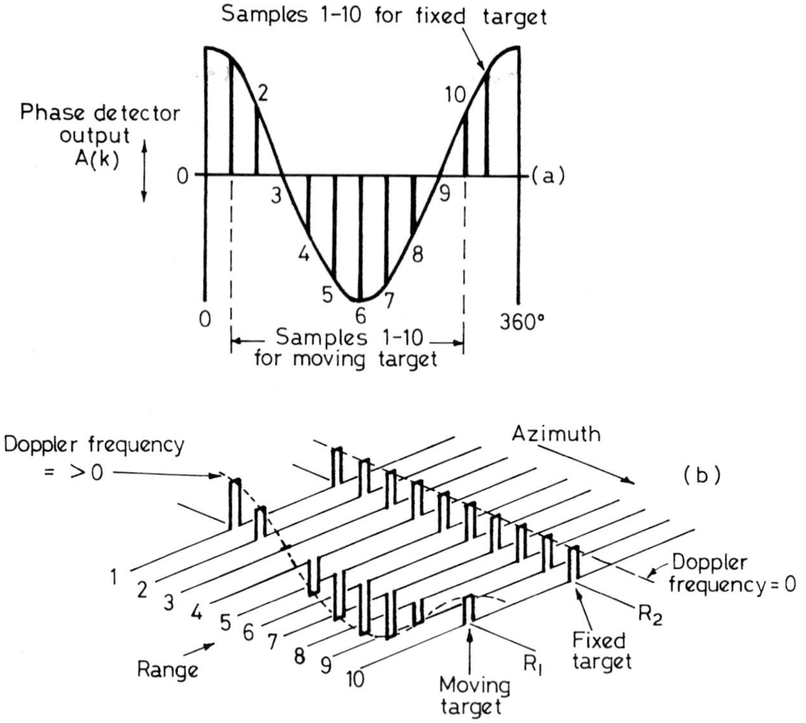

Fig. 9.6. Phase detected outputs for 10 samples in a beamwidth of moving and fixed targets at R_1 and R_2 respectively. Phase for moving target varies across samples 1 to 10 evincing a doppler frequency of > 0. Constant phase of samples for fixed targets give doppler frequency of zero.

From this we see a direct analogy to the classic doppler effect and the aircraft signal's doppler frequency can be calculated along the following lines. In part 1 'frequency' was defined as the rate at which a cycle of events takes place, a 'cycle' in our case being 0° to 360°. So a rate of change of phase of 360° per second is equal to 1 cycle per second or 1 Hz. By the same token a series of events at the rate of 3600 degrees/second has a frequency of 3600/360 cycles per second or 10 Hz. In the example above where the target travels radially at 200 knots, the sample time being 1/600 seconds and the wavelength 23 cm, we have seen that the phase change between samples is $1.4\lambda/2$ or 268°. This represents 268 × 600 degrees/second. Thus the equivalent frequency is (268 × 600)/360 Hz = 447 Hz.

The doppler frequency in mti is apparent to the user as the rate at which the signal amplitude from the phase detector changes. From fig. 9.6 (a) and (b) we see that a fixed target has zero doppler frequency and moving targets have non-zero doppler frequency values.

9.5.1 *Cancellation techniques*

Having sensed the difference between fixed and moving targets it is necessary to make it manifest or apparent to the user. In principle it is easy to do. In a simple 'two-pulse' cancellation system, the output from the phase detector is put into two signal paths simultaneously. In one path the signals undergo a delay exactly equal to the sampling period (i.e. one interpulse period, the time between transmissions). In the other path no delay is introduced. The delayed signals and undelayed signals go into a 'difference-taking' circuit. If the times of arrival of the two are the same and their amplitudes are equal their difference will be zero and no output results. If their time of arrival is the same and their amplitudes differ, the difference will be non-zero and it gives this non-zero signal as output. The system is illustrated in fig. 9.7.

The events n, $n+1$, $n+2$ ... $n + m$ are each separated by the interpulse period T, since they are the result of successive transmissions and all timing is relative to transmission time, t_0. The resultant, shown as a normal oscillogram

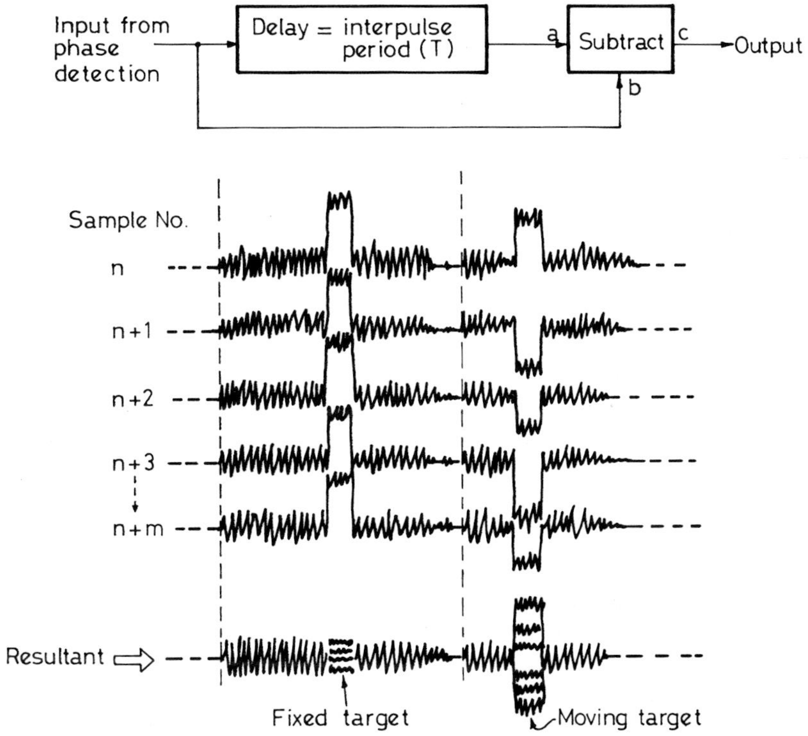

Fig. 9.7. Simple illustration of the mti principle. Successive subtractions of phase detected outputs at sample periods of T reduce fixed target signals to noise level, but preserve moving target signals above noise.

of actual signals, illustrates how the fixed target is reduced to noise level and the moving target fluctuates as successive samples yield different amplitudes caused by the target's minute movements between samples.

An interesting sidelight upon this is to consider the degrees of stability required in mti systems. Values for the reference oscillator of about one part in 10^9 are not uncommon, with time coincidences or delay stabilities in terms of nanoseconds. The phase will be modulated if the path length from receiver to target and back varies; that is the very principle upon which mti works. This means that the antenna assembly has to be extremely rigid since any vibration of the antenna (particularly the 'feed' portion which is usually supported away from the centre of rotation) would change the path length and spoil the hard won coherence of the system.

The simple two-pulse canceller shown in fig. 9.7 can be extended to produce more effective clutter rejection. Two versions are illustrated in fig. 9.8(a) and (b). Figure 9.8(a) represents a four-pulse canceller using the subtraction or difference taking method. The second is a variant wherein a 'summation' method is used. The weighting given to each of the summing arms is set by coefficients of amplification.

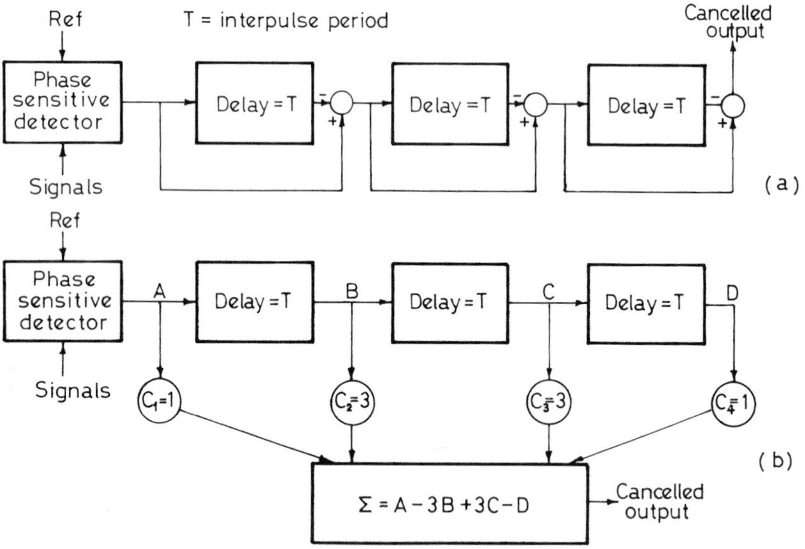

Fig. 9.8(a) and (b). Two forms of 4-pulse canceller; (b) is the more modern.

I cannot forbear drawing attention to the audacity, the sheer optimism of design engineers in their thinking it actually possible to make such accurate and finely grained measurements of distance by the means described above. It may have escaped the reader's attention that what a pulse doppler mti does is to measure target or clutter distance to fractions of a centimetre and at distances of hundreds of kilometres repeatedly and accurately. One instinctively dismisses the possibility, put in these terms, and yet it is done.

A simple mti system has a significant drawback. It suffers from three sorts of 'blindness':

- Blind speeds.
- Blind phases.
- Tangential blindness.

9.5.2 *Blind speeds*

Regular blind speeds occur in a system with a single regular sampling rate (i.e. a fixed p.r.f.). If the moving target travels at a speed which gives it a change of range between samples which is an exact multiple of a half wavelength then successive samples would fall at exactly the same point on the phase detector curve and each output from the phase detector would have the same polarity and amplitude value. That is, it would appear not to be moving. Also its doppler frequency would exactly equal the sample rate. Another way of looking at it follows from the relation above where the range can be written as

$$R = \frac{n\lambda}{2} + \frac{d\lambda}{2}$$

A target moving at the rate of an integral number of halfwavelengths per interpulse period produces no change in the term $d\lambda/2$. It is this term which the phase measuring 'vernier' of the coherent mti system uses to tell fixed from moving objects. This vernier is ambiguous in range because it will not detect changes in the $n\lambda/2$ term.

There is a simple analogy which readers might find helpful. Imagine that a soldier is marching towards you in the far distance – far enough away against an even background for you not to be able to see his change of range, only his 'left–right, left–right' movements. If you were to look at him as does a radar, by blinking open your eyes for a brief moment, say for half a second, and to close them for a much longer time, say five seconds, you would see a series of snapshots of the soldier. At each sighting you would see his arms and legs at differing angles. But if your blink rate coincided with his 'left–right, left–right' marching rate, he would always appear to be in the same posture at each sighting and you would be led to believe him to be stationary. So it is with a pulse doppler mti. If the sample rate equals the target's doppler frequency (or a multiple) the target appears stationary.

9.5.3 *Blind phases*

There is another effect in mti systems which can cause loss of signal and although not as severe as the blind speed effect, it is bad enough to receive special treatment. The effect is brought about by the phase detector characteristic's symmetry. We have seen that moving targets produce different amplitude outputs from the phase detector at each sample and it is this difference which makes them visible to the user.

It is entirely possible that two successive samples in the phase domain lie equally above and below the point of symmetry of the phase detector curve. In such a case, even though the target is moving, the phase detector output for these two samples would be the same amplitude and polarity and would produce zero output. The circumstance is exampled in fig. 9.9.

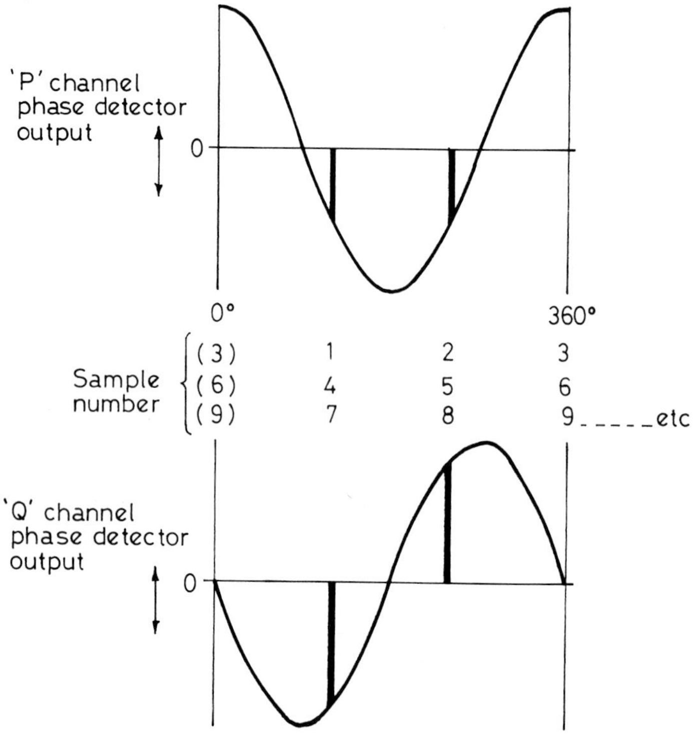

Fig. 9.9. Two phase detector characteristics with one reference input displaced by 90° relative to the other. A series of nine samples from a moving target are indicated. Their disposition produces 'blind phases' in the 'P' channel but not in the 'Q' channel.

To overcome the effect, a pair of phase detectors is used. In one phase detector the reference oscillation is used directly and in the other a phase delay of 90° is introduced. The input signal is applied to each simultaneously and thus the two phase detector characteristics have the same shape but effectively they become 90° out of phase. It is thus impossible for two successive samples to be symmetrically disposed in the two curves at any one time.

This technique is known as 'phase and quadrature' mti (quadrature because of the 90° phase shift) and the two channels are referred to as the 'p' and 'q' channel respectively. Each channel feeds its own cancellation system so that fixed clutter continues to be cancelled in each. A moving target, even if at a blind phase of one channel, will still produce output from its partner.

9.5.4 *Tangential fades*

One of the biggest drawbacks to any mti system is, so to speak, a self-inflicted wound. Because the system cancels signals from fixed clutter then, if the radial displacement of a moving target is zero (or as in the case of blind speed apparently so) the signals from it will also be cancelled. In chapter 7 the different types of target fluctuation were described. The most common are those whose effective echoing area does not change significantly from pulse to pulse (Swerling, cases one and three). The notion of the vector summing of energy reflected from various parts of the target was also introduced, the radar operating upon the resultant amplitude and phase of the numerous 'scintilla' of the target.

Another way of saying that targets meet case one or three criteria is that the summation of the reflected energy produces little change in amplitude or phase from sample to sample. Thus such targets, if they move at right-angles to the radar's scanning beam (i.e. they move at a tangent to the beam), will produce the same value of $(n\lambda/2) + (d\lambda/2)$ for each sample, so they appear as though they were not moving and thus will be treated as a fixed piece of clutter. In theory then, if such a target was to fly a precise orbit around the mti radar, it would be invisible. Operators used to mti systems will know that it is almost impossible to direct a craft to do this to demonstrate a tangential fade. It only happens when you don't want it to!

All doppler mti systems suffer from tangential blindness. To what degree will be seen from the following text.

9.5.5 *Overcoming mti weaknesses*

The blind speeds effect explained above is seen to be caused by the system's use of a fixed sample rate, i.e. a fixed pulse repetition frequency. If we look at the overall mti response for the simple two-pulse canceller system shown in fig. 9.10 it takes the form, familiar to many, of a series of half sine waves. The nulls are equispaced and represent the doppler frequency as multiples of the sampling rate or alternatively target radial velocity equal to a half wavelength (or exact multiples) movement between samples.

We see also that output is at a maximum for a velocity equal to $\lambda/4$ between

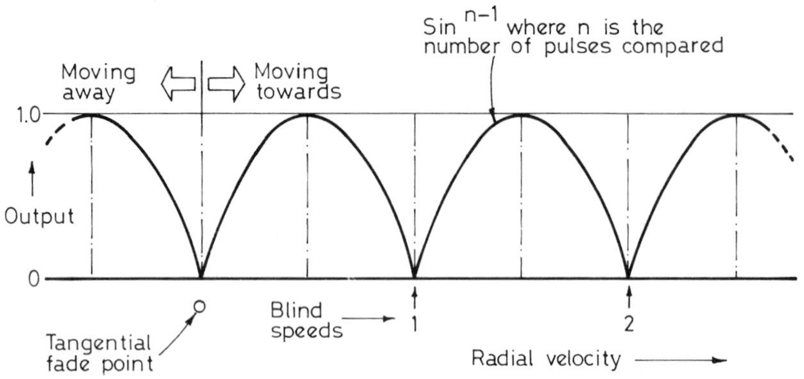

Fig. 9.10. Overall mti output characteristic of a 2-pulse cancellation system with regular sample rate (fixed p.r.f.). Output falls to zero when target doppler frequency is a multiple of the sample rate.

samples which gives rise to a doppler frequency of half the sample rate (p.r.f./2). The blind speeds occur according to the relationship

$$V_b\,(n) = 0.0097 n\lambda f \qquad (9.2)$$

where V_b = blind speed in knots
n = any integer
λ = wavelength in cm
f = p.r.f. in Hz

To the user, the nulls represent failure of the system to provide him with available data.

9.5.6 'Staggered' pulse repetition frequency

Suppose we were to make the sample rate alternate between two different fixed values. Each interpulse period would have its own blind speed and they would first coincide at their lowest common multiple. Thus, since the target can have only one velocity at one time, if it was 'blind' for one of the two p.r.f.s, it would be visible at the other. This would be true up to the velocity where both blind speeds coincide. The average performance of their combination would be a compound of the two separate mti characteristics. An example of a pair of p.r.fs in this so-called 'staggered' p.r.f. system is shown in fig. 9.11. The two interpulse periods assumed are 2000 μs (p.r.f. = 500 Hz) and 2666 μs (p.r.f. = 375 Hz). This is equivalent to a 'stagger ratio' of 375:500 or 3:4.

It is clear that the two independent characteristics, each having its own series of half sine waves, do not coincide at zero until three times the first blind speed of the higher p.r.f. or four times the blind speed of the lowest p.r.f. A number of points are significant to the user:

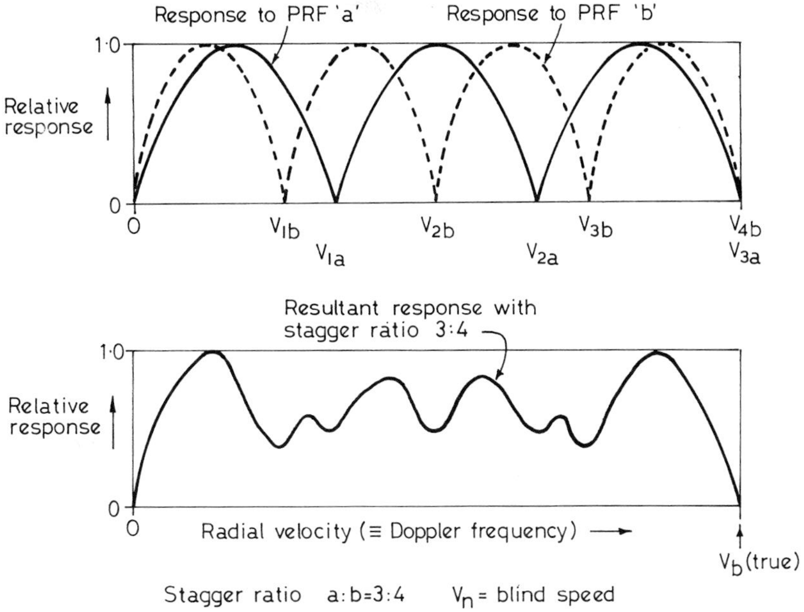

Fig. 9.11. By alternating between two sample rates the overall characteristic has blind speed (V_b (true)) at much higher doppler frequency (and hence radial speed) than either sample rate. Intermediate losses are incurred.

(a) Most systems operate in real-time and produce one output pulse for each new input pulse. Thus in a simple 2-pulse canceller only one of the two interpulse periods is in play at one transmission. As a consequence, over a beamwidth's history, a target flying at the blind speed of one of the p.r.f.s would produce no mti output every time that particular p.r.f.'s sample rate was used. So if there were potentially twenty pulses per beamwidth, only ten would produce an output for a dual p.r.f. stagger. This is illustrated in fig. 9.12.

The effect can be understood as follows. Suppose the first pulse into the system was to be at point one in the phase detector curve. Suppose now that the target travelled at the blind speed of the p.r.f. whose sample period was that between the first and second pulse. The second pulse would also produce the same polarity and magnitude of signal as the first. They would have no difference and hence would be cancelled. The third pulse would lie on a different point of the curve because the sample time would have now changed. As a consequence the target would have moved by some other value than half a wavelength in this other sample time. Remember that the wavelength is presumed constant. The cancellation circuit will now be comparing amplitude and polarity of pulse number 2 with pulse number 3. This produces an output because the amplitudes/

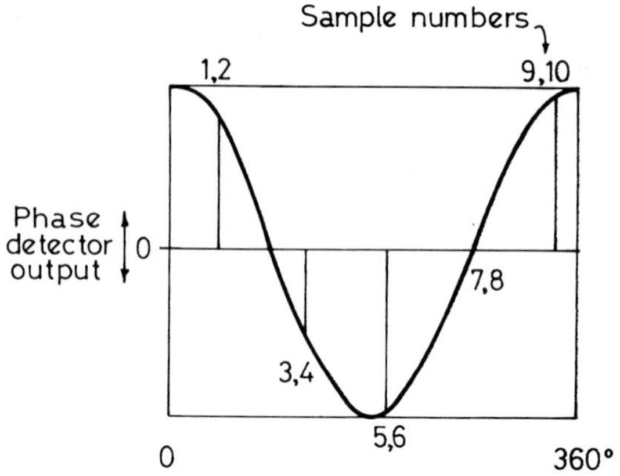

Fig. 9.12. Phase detected outputs in a staggered p.r.f. system with two sample rates where the target moves at the blind speed of one and alternate pairs of samples produce no change in PSD output.

polarities are different. But now the sample period reverts to the original value and if the target continues at the original speed, the fourth pulse will have the same amplitude and polarity as the third and so on.

(b) Fixed clutter echoes will still be cancelled because no matter how short or long the sample period is, the same phase difference between reference and received signals will persist because the wavelength is constant and the range $(n\lambda/2) + (d\lambda/2)$ will be constant.

(c) It follows from (a) above that the 'average' curve which is the arithmetic sum is only real in the sense that it represents what happens after integration over a number of pulses. The integration is usually performed by the display phosphor (successive pulses overlapping on the screen) or in the operator's eye and mind.

(d) The symmetrical dips between zero and the first 'true' blind speed become deeper as the stagger ratio becomes smaller. That is, if the stagger ratio was 10:11 then the vestigial blind speed nulls represent a greater loss. The smaller the stagger ratio the higher is the first 'true' blind speed. These two points are illustrated in fig. 9.13. They are calculated for an L-band radar operated at 600 Hz.

9.5.7 Technical implications

In order to effect cancellation in staggered p.r.f. systems it is necessary to switch extra delay into the canceller 'delayed' channel every time signals from the shorter interpulse period are to be dealt with. If this was not done then the arrival of signals at the 'difference taker' circuit would not be simultaneous. In analogue cancellation systems, where the delay medium is an ultrasonic

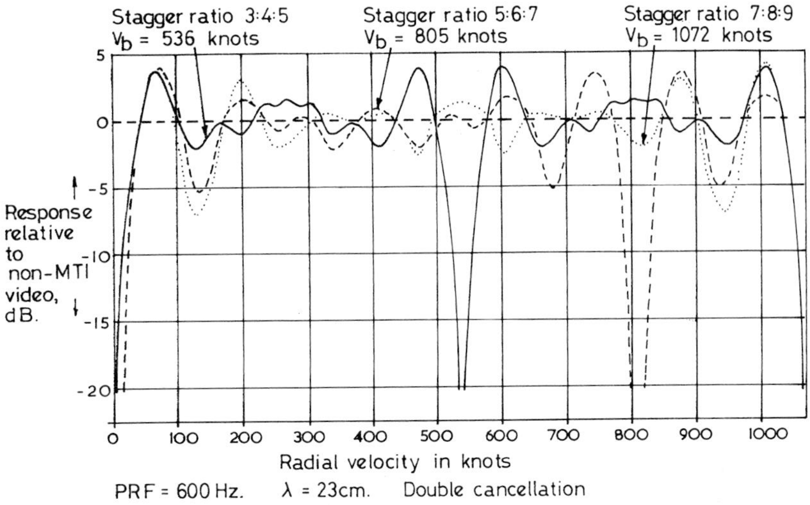

Fig. 9.13. Showing the effect of various p.r.f. stagger ratios for a fixed wavelength and mean p.r.f.

delay line, the electronic switching at alternate periods makes for complexity in equipment: duplicate sets of analogue delay circuits have to be provided, an expensive move which greatly increases maintenance time and reduces reliability.

Fortunately digital techniques have made possible much greater flexibility in staggered p.r.f. facilities together with greatly enhanced mti performance. The major advantage conferred by digital technique is the ability to use the 'shift register' or modern equivalent types of circuit as a delay medium. They operate in the same way as described in section 9.2. The delay and data access times are now no longer constrained to 'real-time', as is the case with analogue delay devices. Delay time can be as long as one wants to wait; a set of data can be input and stored until next week if desired. Naturally the data input is in the real-time domain – unalterably, since returned signals travel at the speed of light and are stimulated after one's own transmission. Having been stored in a digital delay medium, the data can be played out at a rate which is equal to, faster or slower than the input rate and the play-out start time is also controllable. Thus, in a staggered p.r.f. system the extra delay required can now be provided by delaying only the time at which the data store is addressed. So instead of a complex chain of ultrasonic delay columns, their transducer drivers, detectors and gain compensating amplifiers, the designer needs only to use a simple digital circuit which inhibits the clock pulses addressing the store for the required delay time. This simplicity allows very complex stagger ratios to be mixed either as a known sequence of up to six (or more) interpulse periods or as a completely random set. The latter is usually reserved for systems liable to be jammed.

9.5.8 *Digital mti*

From the above it might appear that digital mti systems are significantly different from their predecessors, but the differences are by no means fundamental. They give better radar performance, greater stability, higher reliability and wider flexibility. Stability, because the need for many continuously variable controls ia avoided, i.e. 'knob twiddling' can be done away with. Reliability, because most devices operate at low electric stress and so breakdown rates are extremely low; individual 'modules' of the kind used in digital mti have actual and predicted failure rates of many tens of thousands of hours. Flexibility, because digital techniques allow blocks of data (stored as '1's and '0's) to be moved around either as serial or parallel streams with great ease.

Furthermore, because the circuits are all solid-state and computer-like devices, the designer's problems are greatly eased. Most importantly for the user, reliability and stability are enhanced to a high degree. The basic principles of operation remain unaltered. A simple block diagram of a digital mti is illustrated in fig. 9.14.

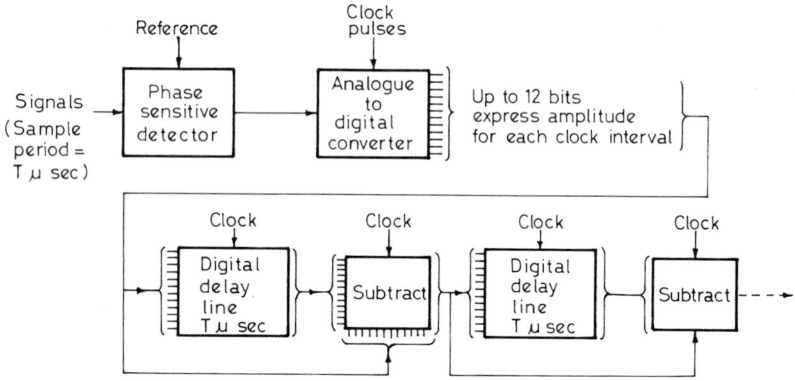

Fig. 9.14. A digital form of mti equivalent to that in fig. 9.8(a).

The system operates in the same manner as illustrated in fig. 9.7; delayed and undelayed signals are compared. The technique takes the bipolar analogue signals from the phase detector and translates them in an A to D converter into digital words expressing their amplitude through the range continuum at each instant the regular clock pulses demand. The amplitude domain is divided into as many parts the designer chooses; eight, ten or twelve bits, dependent upon the degree of cancellation required. A ten-bit word will express the signal amplitude as 1024 equal small increments. If the peak signal from the phase detector was 1 V, then each increment would represent about 1 mV. The clock pulse interval would be designed to match the minimum received pulse duration expected. If the pulse duration was 1 μs, it would be common to use a clock interval of about one quarter of this (to give four range samples per pulse, i.e. the clock oscillator source would have a frequency of 4 MHz).

The digital delay line (or lines) would then have storage cells in the range domain for each of the ten bits' amplitude structure and each of these 'columns' of ten would be arranged as a row of range cells, one row for each clock or range interval up to the range to which the mti is required. It would be common to find the range domain expressed as a twelve-bit word giving 4096 equal clock intervals. If, as exampled above, the clock interval was $\frac{1}{4}$ μs, the range time would then be (4096/4) = 1024 μs, equivalent to (1024/12.36) nautical miles = 83 nautical miles. The design flexibility is immediately obvious: range and amplitude quantities can be subdivided easily at the designer's dictation. The necessary devices for A to D conversions, storage, clock generation, digital amplitude comparison, etc., are all readily available from microcircuit suppliers (provided they haven't been swallowed by video games manufacturers!). The engineering problems are thus mainly those of packaging, optimum modularity and maintainability.

Figure 9.14 indicates that the delayed and undelayed comparison can be continued ad infinitum. In practice it is rare to find more than three stages in cascade. The repeated cancellations are done in order to reduce clutter to lesser and lesser levels; a single cancellation leaves residue caused by antenna scanning and clutter fluctuation (see fig. 9.15(a)). The improvement in cancellation performance is not however free from unwanted side effects, which will emerge later.

Since the digital technique allows, to a degree, escape from real-time, it is possible to design variants upon the cancellation technique illustrated. The most common called the 'integrate and dump' technique follows from the arrangement shown in fig. 9.8(b). As illustrated, the canceller will give an output for each new input made. For fixed clutter, ignoring any amplitude changes brought about by the variation of antenna gain as the beam sweeps past, A = B = C = D. Under these conditions the output of the summing network will be zero, but only if all components are present. As the antenna beam illuminates new fixed clutter features, their cancellation will be incomplete until four successive samples of the same piece of clutter have been taken, i.e. after four successive transmissions. If the cancelled output was displayed direct, with no subsequent processing, one would see remnants of clutter caused by this incompleteness. However, since these remnants or clutter residues are always in the same place, they would continually integrate in individual cells of a following temporal threshold system and thus become invisible.

The 'integrate and dump' method essentially adds together the results of cancellation over a number of repetitions, typically six or eight. Range-ordered stores (commonly the identical type of circuit used to form the delay sections) are used for this in the following manner. Integration over 8 repetitions is assumed. The summing network's output is directed, range increment by range increment into these stores successively until all eight have been addressed. When the last store has been served, the contents of each store's corresponding range cells are added together. If they exceed a given threshold an output pulse is generated. Thus if the radar system beamwidth was such as gave sixteen pulses per target and, in a given case, all generated a usable signal, the integrate and dump output would consist of two pulses – one because of the successful integration of the first eight and the other for the second eight. Obviously in the case of cancelled fixed clutter, the threshold of detection in the integrate and dump system is set so that no output is generated when clutter residue (caused by the incomplete cancellation mechanism described above) is present.

The method is essentially one directed to improved target detection but not to its display. Two effects unfortunate for the user occur. The first is that targets are displaced in azimuth by half a beamwidth (or the equivalent of eight transmissions), however this can be corrected by offsetting the display and antenna north references. The second is more important. A real-time display shows targets by the excitation of a small area of the display tube's phosphor by the closely spaced pulses received repetitively at one range and in one beamwidth. Thus the small arc of sixteen repeated pulses at one range allows the target to be clearly seen. If these sixteen were reduced to only two as in the integrate and dump system and these widely spaced apart then visibility would be severely reduced. It is thus needful to take the process further and to use each of the integrate and dump output pulses to generate, artificially, the missing pulses. The user now has to suffer very obtrusive false alarms, for if the integrate and dump output is present by a false alarm, it will generate seven contiguous pulses for display which will look very like a real target response.

MTI improvement factor (I)

Although still the subject of much philosophical debate among designers the most satisfactory method so far devised of enabling engineers to characterise the performance of an mti system is to use the notion of improvement factor – 'improvement' being that of the visibility of moving targets in simultaneous clutter. It is usually expressed as a decibel quantity, e.g. an I factor of 20 dB means a hundred-fold improvement in moving target visibility over competing clutter. It is another way of describing Sub-clutter Visibility (scv). How is it calculated?

Take as a model your desire to be protected from rain. Suppose the available technology allows you to make an umbrella, but the fabric used is not completely water-proof – it has holes in it and lets 1% of the rain through. As far as keeping you dry in rain, the umbrella has an effectiveness of 100 to 1: its improvement factor is thus 20 dB (assuming 'wetness' is directly proportional to number of raindrops!). If you found a new technology which gave fewer holes in the fabric (say four times less) then the umbrella's improvement factor would go up by four times to 400 to 1 or 26 dB. The mti improvement factor works in the same way, except that it comprises a number of sub-factors, not just one as in the model above. The important sub-factors are described in reference 6, chapter 17.

Moving target visibility in clutter is largely limited by incomplete cancellation of the clutter itself. The sub-factors expressing this have two prominent members:

(a) I_{scan}: The fixed clutter samples should all be the same for complete cancellation. They vary from sample to sample because the scanning beam's shape imparts differences to each.
(b) I_{stab}: The 'system stability' is imperfect, containing small but significant errors in coherence, transmitted pulse to pulse amplitude variation, frequency modulation within the transmitted pulse, timing errors (jitter), instability of reference oscillators, etc.

I_{scan}. The effect is illustrated in fig. 9.15(a) which shows a beamwidth's history of signals from a tall mast and exhibits the amplitude changes caused by the antenna beam shape. For complete cancellation, each sample should have the same amplitude. Attempts are made to effect this by passing the signals either through a limiting amplifier or through one having a characteristic which reduces output variation (a 'compression' amplifier). It can be seen that in either case, there will be contiguous samples at the beam's edges whose difference is non-zero. This difference represents 'clutter residue'. It can be reduced by repeated cancellations but only at the expense of widening the zero doppler pass-band and thus reducing visibility of targets with low radial velocities. The shape of the overall pass band is $\sin^{n-1} \theta/2$ where n is the number of pulses compared in the canceller and θ is the phase difference per sample between signal and reference.

I_{stab}. Modern technology has allowed surveillance radars to be designed with system stabilities of about 50 to 60 dB. Discussion on this is given in reference 9, concerning differences between types of transmitter.

Overall effect

The assumption is made that all the sub-factors in the overall I factor (I_{total}) are independent and the value calculable from:

$$\frac{1}{I_{total}} = \frac{1}{I_1} + \frac{1}{I_2} \dots + \frac{1}{In}$$

It can be seen then that system design is aimed at reducing all sub-factors to the same low level to prevent one from dominating the rest. For instance, if the value of I_{scan} was set to 40 dB it would seem a great waste to serve the system with an I_{stab} value of 60 dB. The difference implies that the transmitter/receiver is (60–40) = 20 dB = 100 times better than required! The battle for higher values of I_{total} continues.

Time varying weights (tvw)

Among the sub-factors reducing the total improvement factor are two which are directly related to the effects of staggered p.r.f. They are:

$$I_{stagger(scan)} \text{ and } I_{stagger(clutter)}.$$

By use of the technique of tvw both can be virtually eliminated. Complete explanation of the working of tvw requires complex mathematical analysis of the canceller's response to the input signal's spectrum, which becomes disturbed by staggering p.r.f. However the action of tvw can be readily understood from the following. Consider the sub-factor $I_{stagger(scan)}$.

Suppose the mti was a non-staggered p.r.f. four pulse canceller of the form shown in fig. 9.8(b). The co-efficients (C_1 to C_4) within the canceller would be as shown, resulting in an output of value:

$$A - 3B + 3C - D$$

If all samples were of equal amplitude, the resultant would be zero. Even though a fixed target generated equal outputs from the phase detector, the scanning antenna imparts amplitude changes because of its non-linear beam shape as shown in fig. 9.15(a). If it were linear and the samples were regular in time, all would be well, as can be seen from the example in fig. 9.15(b).

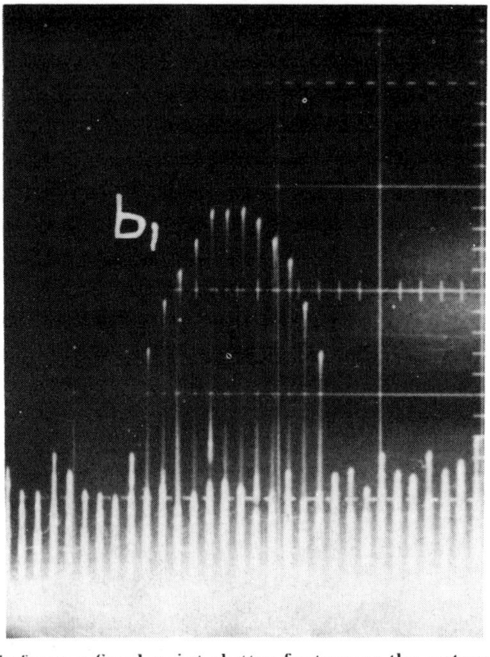

Fig. 9.15(a). Signals from a fixed point clutter feature as the antenna sweeps across it from left to right. Amplitude modulation is caused by antenna azimuth beam shape. (Low amplitude pulses are caused by transmitter breakthrough.) (*Photo courtesy of Marconi Radar.*)

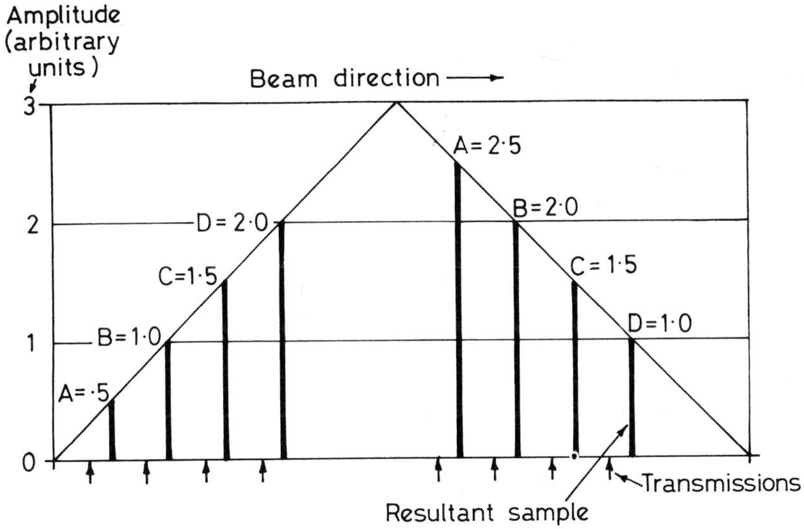

Fig. 9.15(b). A hypothetical radar beam shape modulating the amplitude of four successive samples from a point target in a 4-pulse canceller. The sequence A to D corresponds to those in fig. 9.8(b) (i.e. D was the earliest in time).

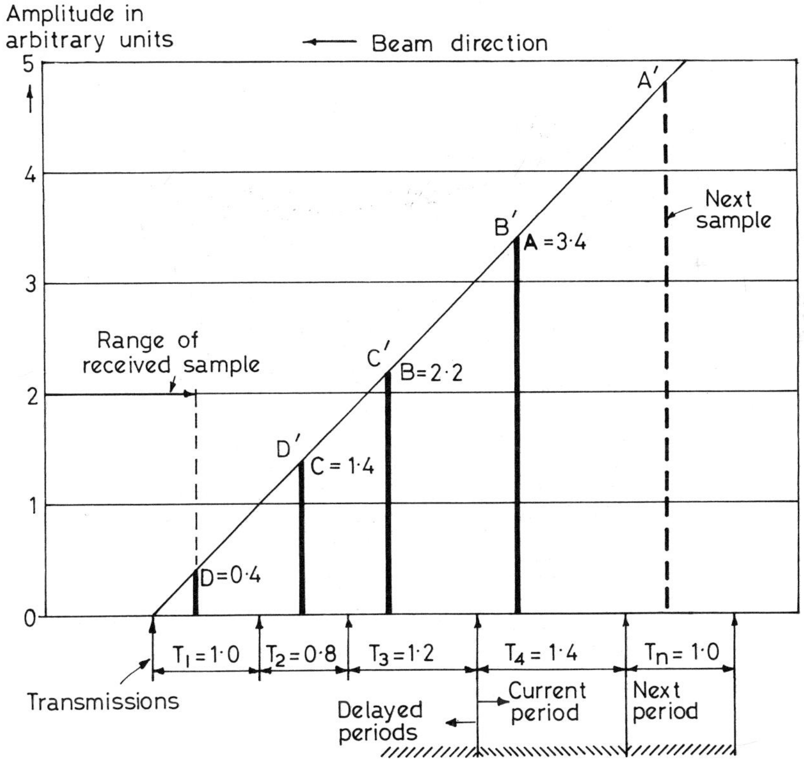

Fig. 9.15(c). Showing how amplitude differences between four successive samples in a 4-pulse canceller are disturbed by a 4 period p.r.f. stagger system. Time varying weights (C_n = K in fig. 9.8(b)) must be applied to correct for this.

Here, the beam shape is represented as symmetrically triangular with four successive samples upon which the canceller's summing network operates. For the values shown, the samples entering the beam (those on the right) will produce the result:

$$\begin{matrix} 2.5 & - & 6 & + & 4.5 & - & 1 & = 0 \\ \text{(A)} & & \text{(3B)} & & \text{(3C)} & & \text{(D)} & \end{matrix}$$

Those leaving the beam (on the left) will produce the result:

$$0.5 - 3 + 4.5 - 2.0 = 0$$

Imperfections result when the beam shape is non-linear for obvious reasons, but the beam shape is a close approximation to linear over the azimuth represented by four successive samples.

Suppose now that the system was to have a four period staggered p.r.f. with ratios 8:10:12:14. The samples are now disturbed in time as shown in fig. 9.15(c). If the co-efficients of the summing input network were kept at A=1, B=3, C=3 and D=1, the resultant output of A−3B+3C−D would be:

$$3.4 - (3 \times 2.2) + (3 \times 1.4) - 0.4 = 0.6$$

The desired result is zero so, patently, the stagger system has spoiled cancellation. The way to redress the balance is to make the co-efficients vary as the time periods between samples change.

Another way of looking at it is to restore the 'centre of gravity' of the four samples to the same point as in a non-staggered system. This can be done by operating upon the co-efficients of samples B and C alone. Following the ideas put forward in reference 6, the co-efficient to be used to cancel the four samples of which A is gathered in the current period T_4, can be calculated from:

$$K = 1 + \frac{(T_1 + T_3)}{T_2} \tag{9.3}$$

In the example taken

$$K = 1 + \frac{(1.0 + 1.2)}{0.8} = 3.75$$

Substituting this value into the cancellation equation gives

$$\text{Output} = 3.4 - (3.75 \times 2.2) + (3.75 \times 1.4) - 0.4 = 0$$

For the next transmission (T_n), whose interpulse period will now be $T_4 = 1.0$, a new set of four samples labelled A′ B′ C′ D′ in fig. 9.15(c) will be gathered, three of which remain the same as before (D in fig. 9.15(c) will drop out and be replaced by C). A new co-efficient will now be required of value

$$K' = 1 + \frac{(0.8 + 1.4)}{1.2} = 2.833$$

So as each of the four interpulse periods becomes T_4, a separate value of K is required. These are calculated by the designer and programmed into the summing network's input stages for automatic operation in synchronisation with the stagger programme itself.

9.5.9 Doppler filtering and Moving Target Detection (MTD)

Most of the foregoing descriptions of mti have been in terms of what happens when we take samples of the signals from illuminated targets and process them in time sequence. Time is the only thing the designer gets for nothing; except time! I hope that the reader has now enough information to realise that in mti systems the overall response characteristics have three domains. They are:

(a) Range (time to target and back).
(b) Amplitude (radar sensor and target parameters).
(c) Phase (the last remnants of the range expressed as fractions of a wavelength).

We are now concerned with the last domain, values for the first two being given from zero range to the calculable maximum range. We have seen that for fixed wavelength and sample rate (p.r.f.), the target radial velocity has a direct equivalent doppler frequency (see equation (9.2)). Fixed targets give rise to signals having zero doppler frequency and moving targets to those having non-zero doppler frequency. It would seem sensible then to say that if we had a filter designed to distinguish between these two the game is over and won, but the situation is not as clear-cut as that. The overall mti system response of a modern six-pulse stagger system is shown in fig. 9.16. It can be properly regarded as a pass-band filter response, showing how the output amplitude varies with the doppler frequency of the input signal. Looked at from the user's viewpoint it says:

'All targets (provided they return enough energy to be detected) above a radial velocity of about 5 knots will be of sufficient strength for me to see them'.

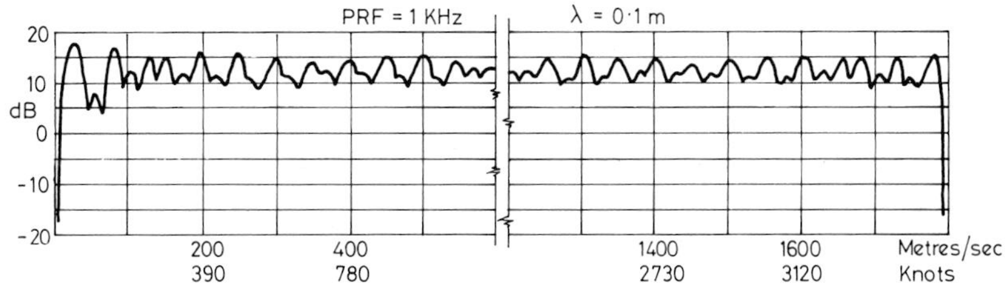

Fig. 9.16. A practical mti characteristic of a 6-period staggered p.r.f. system with mean p.r.f. of 1 KHz and wavelength of 10 cm.

Notice that the doppler filter has a bandwidth. That is, a certain band of doppler frequencies will be more or less rejected; the closer they are to zero, the greater the rejection.

So far we have considered 'fixed' and 'moving' targets. Mountains and hills are pretty fixed. Very often they have trees upon them – they are pretty fixed too. But the trees have branches and twigs and leaves. When the wind blows, it makes them move. Because mti systems are sensitive to very small movements of reflecting surfaces (fractions of centimetres between samples) the movement of wind-blown vegetation often spoils cancellation of erstwhile fixed clutter. This circumstance leads to a helpful distinction between types of target. There are those which have movement *and* displacement, e.g. aircraft, vessels, vehicles which follow tracks. There are those which have movement *but no* displacement, e.g. hovering helicopters, moored vessels in a heaving sea, wind-blown vegetation. In a sense, sea clutter and rain over large areas come into this category.

All of these circumstances can be described in terms of the doppler frequency components of their signals. For instance, a wind-blown wooded hill will give rise to a signal which has amplitude set by the normal radar parameters (power incident upon the target effective echoing area, etc.) and with a spectrum of doppler frequencies centred upon zero and extending either side. This extension is caused by the movement of vegetation as the wind blows the tree's branches and makes the leaves flutter in random directions. In the case of signals from rain, hail or snow they will have a doppler spectrum whose spread of frequencies is wider and centred on a value equivalent to its radial velocity, in turn governed by the wind velocity. The spectral spread will be dependent upon how much internal turbulence is present in the clouds from which the rain comes.

A hovering helicopter will give rise to a special spectrum. Because its main body is stationary the main doppler component will be at zero but the rotor blades will create spectral lines away from zero at values dependent upon the blade speed.

Examples of the doppler content of some of the above are shown in fig. 9.17. Remember that the diagram shows amplitude as a function of doppler frequency. All could appear simultaneously in a given radar beamwidth and if at the same range, will thus occupy the same radar resolution cell. Under these conditions it is easy to appreciate that the doppler components of the different types of signal may be all that remain to distinguish one from another. Thus it appears that if the signal processor was designed as a set of doppler filters side by side this distinction would be possible. MTD is based on this.

An idealised set of such filters is shown in fig. 9.18(a) and their manner of use and value to the user is illustrated in fig. 9.18(b). They can be looked upon as a means of detecting targets in a dimension other than range and amplitude. Since each filter output is separate, and all are simultaneously

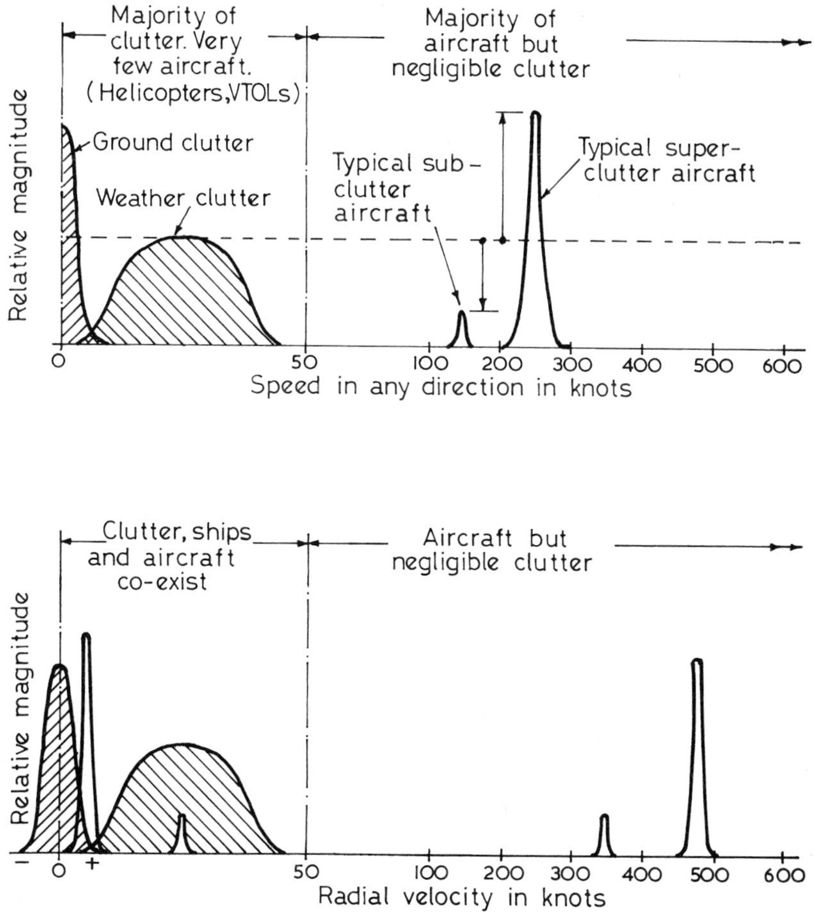

Fig. 9.17. Typical clutter and target signal scenarios. Note the difference of description of the horizontal axes; a target with a speed of 250 knots can have radial velocities from 250 knots down to zero.

available, then even if signals from clutter and aircraft were at the same range and hence in the same resolution cells, the doppler component of the aircraft would appear at one filter output and be distinguished from the clutter which appears at the output of different filters. Obviously the wanted target could still have radial velocity which was low enough to create doppler frequencies equal to clutter but most of the time they are much higher. Weather clutter, since it is wind-blown, will have doppler frequencies contained at the equivalent of wind velocities and it is possible to posite filter characteristics which match all the required conditions in an ideal fashion. Such a set is shown in fig. 9.19.

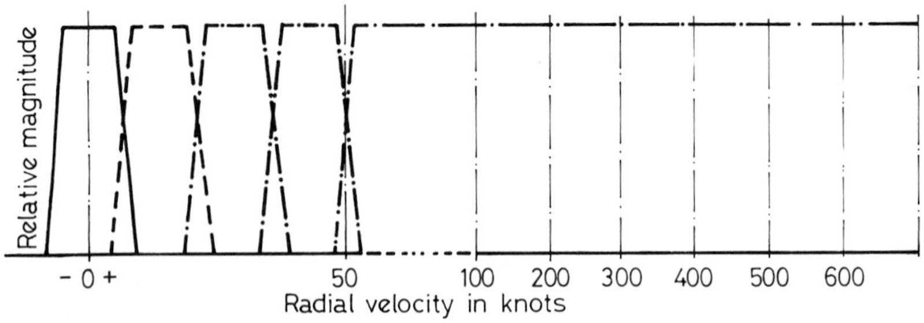

Fig. 9.18(a). An idealised set of doppler filters for an atc radar with reference to the scenario of fig. 9.17.

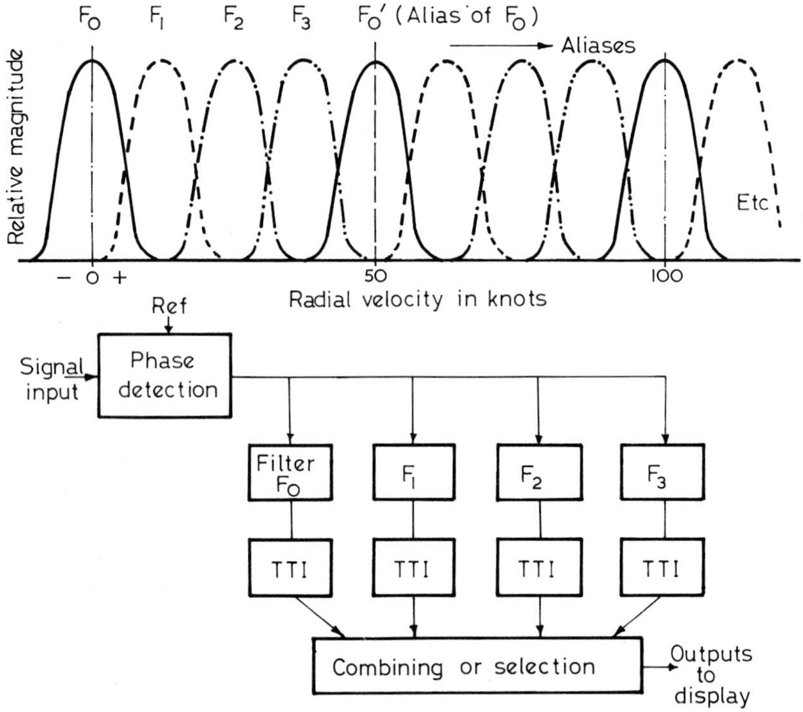

Fig. 9.18(b). A practical realisation of the ideal. The characteristics of filters F_0, F_1, F_2 and F_3 repeat themselves in the doppler frequency (radial velocity) domain. Each filter is followed by a Temporal Threshold Integrator.

Constructing doppler filters

In a coherent mti system, targets moving at a constant radial velocity generate signals having a constant phase shift from sample to sample if the sample rate is fixed. A useful extension of the ideas given at the beginning of this chapter

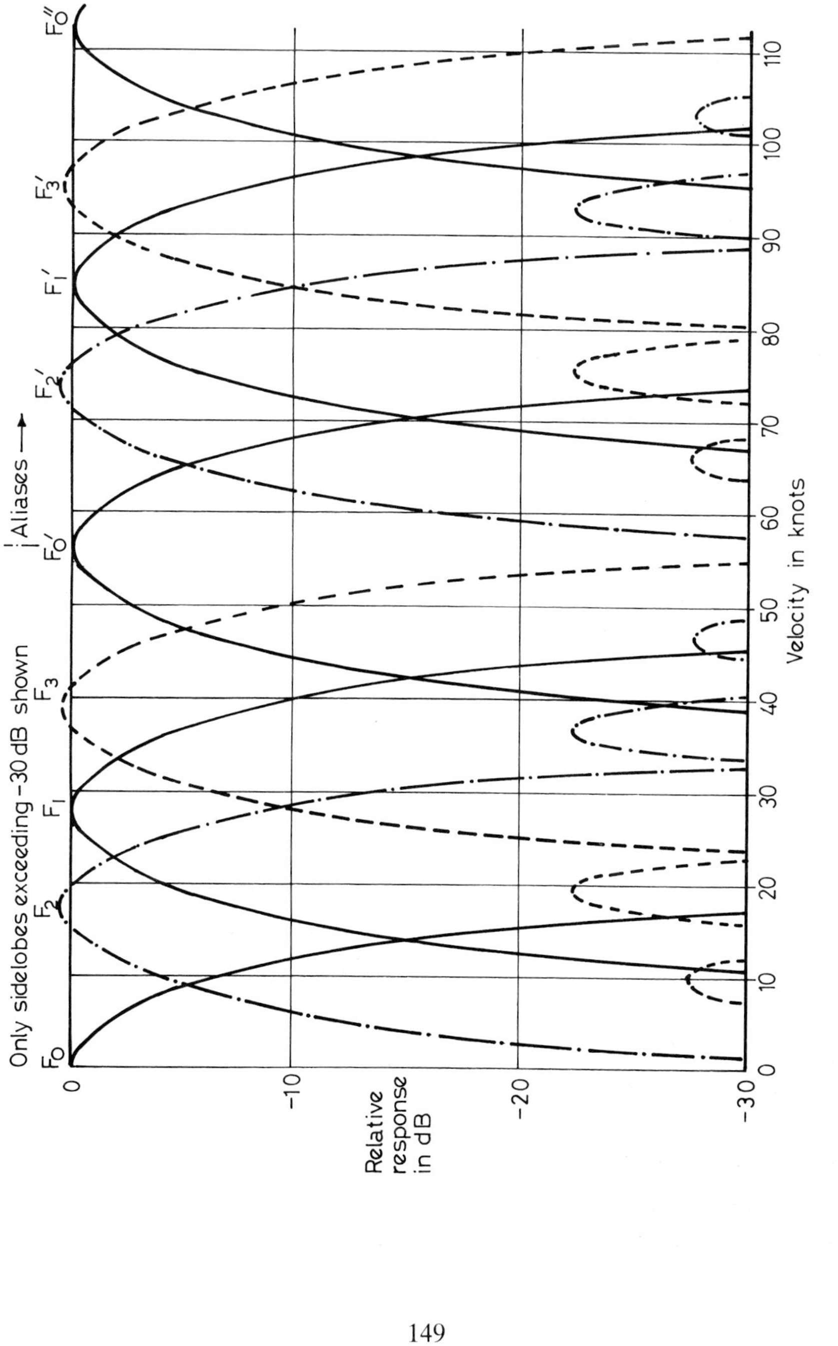

Fig. 9.19. A practical set of four filters showing pass-band overlap and sidelobes.

is to consider the signal relationships to the coherent reference as purely vector quantities.

In the case of the fixed target (i.e. clutter) the 'constant phase shift between regular samples' is zero, giving a vector ϕ_c relative to reference. For the moving target each sample is ϕ_t different from its predecessor, relative to the coherent reference. This is illustrated in fig. 9.20. The individual samples are shown as having the same amplitude for clarity's sake, and thus antenna beam shape modulation has been neglected.

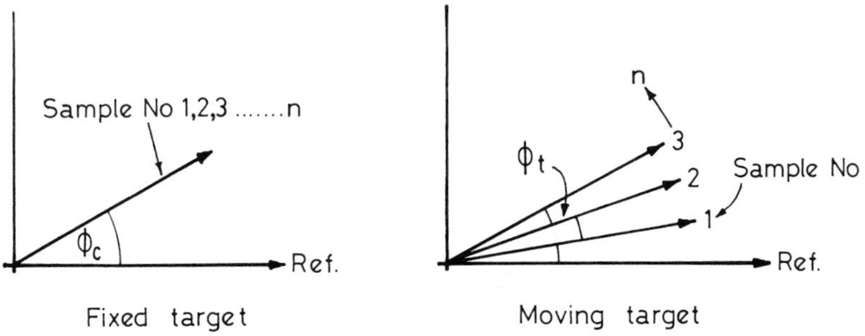

Fig. 9.20. Vector representation of successive samples of fixed and moving targets in a coherent system.

Suppose we had a set of circuits which would measure and store four successive vectors and add them together. Take as example a moving target which generates 30° phase shift per sample. Their summation would be as shown in fig. 9.21. If each vector was of unity amplitude the resultant (E) would be 3.34. Suppose further that as each successive sample was received it was made to rotate by 30° back into the same angle as the reference. Their summation would now be the maximum possible, i.e. 1+1+1+1=4. If this magical rotation was done it would be the equivalent to maximising detection of targets having a doppler frequency equal to 30° phase shift between samples. How to do this vector rotation?

The designer invokes the principle of vector multiplication. By this device, two vectors $r_1\theta_1$ and $r_2\theta_2$ give a product of:

$$(r_1\theta_1)\ (r_2\theta_2) = r_1r_2\ (\theta_1 + \theta_2) \qquad (9.4)$$

That is, their resultant modulus is the product of the two separate moduli at an angle which is the sum of the two individual vector angles.

Using a more complex set of digital mti circuits of the kind described in section 9.5.8 of this chapter, such an operation is performed. The designer can choose what rotation angle he desires and can thus construct a battery of circuits which maximise doppler frequency detection across the whole of the expected range from 0° to 360° in whatever increments desired.

Consider the example cited above where a rotation of 30° was given in one

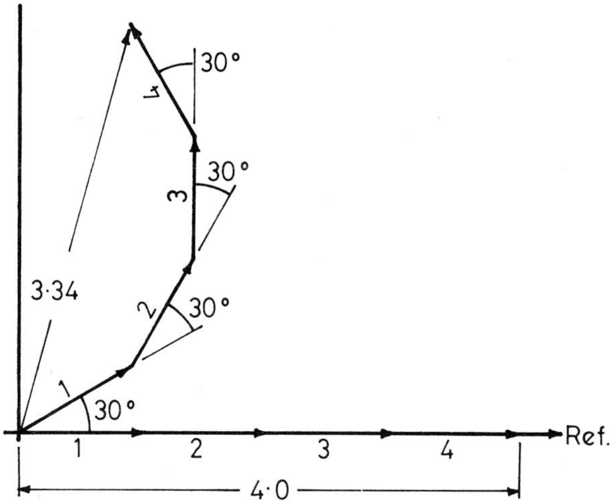

Fig. 9.21. Four successive samples in a coherent system. They show 30° phase shift between samples. If each was rotated back by this amount upon reception their sum would be maximised.

such set of circuits. What is the effect on signals exhibiting 45° phase shift per sample? Each successive sample would be rotated by 30° and so the net result would be as shown in fig. 9.22. Instead of producing an output (E) of 1+1+1+1=4 it becomes

$$E = \sqrt{(x^2 + y^2)} \tag{9.5}$$

where $x = (1 \times \cos 15°) + (1 \times \cos 30°) + (1 \times \cos 45°) + (1 \times \cos 60°)$
$= 3.04$, and
$y = (1 \times \sin 15°) + (1 \times \sin 30°) + (1 \times \sin 45°) + (1 \times \sin 60°)$
$= 2.33$

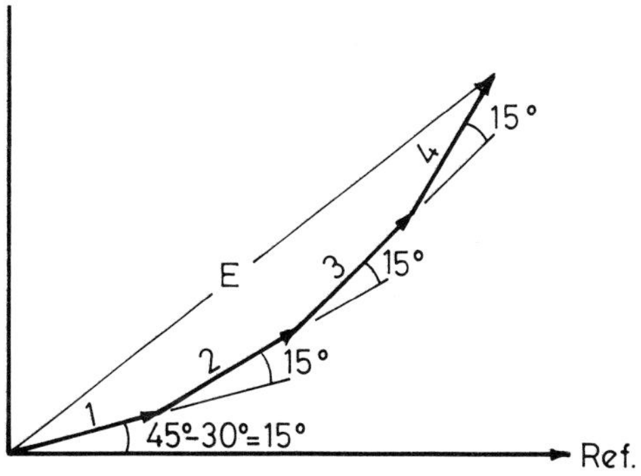

Fig. 9.22. Showing the resultant of four successive samples in a coherent system, each rotated back by 30° when their phase shift caused by target movement between samples is actually 45°.

Therefore from equation (9.5), $E = 3.82$.

These calculations are performed by the 'set of circuits' described above. Their individual actions are complicated but need not concern us beyond understanding the basic principle involved.

As the input sample phase shifts diverge from the 'vector rotation value' it is seen that the resultant output amplitude falls away from maximum. The set of circuits thus behaves like a pass-band filter having a maximum response at the doppler frequency equivalent to the phase rotation value chosen. What is the passband characteristic?

The reader may have noted a marked similarity between the principle described above and that in part 1 used to explain the process of antenna beam formation. The two cases are almost identical and given an infinite number of samples lead to a common amplitude characteristic having the form $y = (\sin x)/x$. This is exactly the shape of an antenna beam formed from a linear array with an infinite number of elements and, as in the case of the antenna, the pass-band or doppler filter characteristic has 'sidelobes', but now in the frequency domain.

In passing it is interesting to observe that since both the antenna beam and doppler filter are developed by the same principle, the antenna could be considered as a 'space filter'. Its individual elements (spaced at equal fractions of a wavelength apart) are the counterpart of the individual samples (four in the model above) taken of the target's doppler history. In both cases the contributions from individual elements in the system are processed to form a final output. In the antenna and the doppler filter it is the result of vector addition.

In antenna beam shape design it is possible to reduce sidelobe amplitudes by giving unequal weights to the contributions of the individual elements in the array. So it is with doppler filters: by giving unequal (but carefully designed) weights to the contribution of each of the four samples, sidelobe levels can be reduced in exchange for main beam shape.

9.5.11 A bank of doppler filters

Using the principle above, it is possible to design a set of filters side by side in the doppler frequency range to be covered. A number of interesting effects, important to the user, can be seen. The most significant is the effect called 'aliasing'. Each of the four filter passbands (F_0, F_1, F_2, F_3) are duplicated in the doppler frequency domain. This duplication (indeed continual replication) comes about because the phase detector characteristic by which the signal's phase is sensed cannot extend its measurement beyond 360° of phase, it is ambiguous outside this range. In other words a phase shift of 15° between samples cannot be distinguished from a phase shift of 15° + 360° – there is no 'rev counter' in the vector summation circuits!

The effect is directly analogous to that of mti 'blind speeds'. Thus, each filter has a set of 'aliases' spaced at intervals of the system's blind speed and which are a direct result of the need to preserve a constant sample rate (i.e. a fixed repetition period). The effect is not without its attractions, though, because it means the whole of the required radial speed range can be covered by a small number of filters, each having its aliases stretching into infinity. Although the set of filters is shown operating with positive radial velocities there is a sense in which the filters are capable of accepting 'negative' doppler frequencies. This notion is easy to understand if one considers targets moving towards the radar as generating a positive doppler frequency and those moving away, negative values. Just think of the audible doppler effect: the tone of a train whistle appears to increase as the train approaches and reduces as it recedes. The value of the doppler frequency is the same (dependent only upon wavelength and speed). In one case it adds to the real frequency of the whistle note (i.e. when train and listener are both stationary) and in the other it subtracts from it.

There are two less welcome effects, both leading to the same problem. Ideally the filters should have pass-bands which are sharply defined, accepting frequencies within a chosen range and totally rejecting those outside the designed limits. They also should exhibit no sidelobes. Unfortunately each filter falls short of the ideal in both respects. Their pass-bands overlap and their sidelobes encroach upon the neighbouring pass-bands. This leads to filter pollution. We have already seen that a filter designed to separate zero-doppler signals from the rest can also contain wanted signals giving a doppler frequency of the sample rate. This is because the zero doppler filter has an alias at a frequency of the p.r.f. Thus any wanted target generating this doppler frequency will be clouded by any simultaneously present unwanted fixed clutter. The same polluting effect can be present for those filters designed to reject (or filter out) weather clutter signals. Because the pass-bands overlap, weather clutter can appear in the next filter designed to accept signals separated from those caused by clutter. It should be borne in mind always that the object of doppler filtering is to seek more information on 'discontinuities in the atmosphere', many of which can coexist. That is, signals from ships, sea clutter, land clutter, weather, angels, can all be present simultaneously in the same resolution cells. The purpose of signal processing is to seek their differences and present them to the user.

One further difficulty remains to be overcome in doppler filter signal processing. In normal multipulse canceller mti systems, blind speed effects are got round by use of staggered p.r.f. – that is, the sample rate is made to vary. By the device of time varying weights (TVWs) in the higher order cancellers, the adverse effect on cancellation is overcome. In the mtd system, using banks of doppler filters, it is not possible (without massive complication) to employ rapid changes of sample rate. The filter characteristics would introduce losses if the sample rate were not held constant. For the filters to have adequate discrimination there must be at least the same number of samples (at a constant rate) as there are inputs to the summing network of the filter. That is, if there were four samples required per filter, then the interpulse period must be held constant for five successive transmissions (four intervals). In a radar with mtd having perhaps ten effective pulses per beamwidth it would be possible to vary the interpulse period, but only in groups of four. That is, transmissions giving four intervals of t_1 followed by four intervals of t_2.

Only by this 'group stagger' technique can staggered p.r.f. be introduced into mtd. But there is a risk in this. Suppose the target was to be travelling at the blind speed of one of the two groups of p.r.f. in the example above. Half the pulses would appear in the 'zero-doppler' filter output and half in another. If the target was over fixed clutter, unless it exhibited subclutter visibility it would not be visible for half the beamwidth. Moreover, any plot extraction carried out on the target would produce inaccurate azimuth data and cause track wander. Equally, greater reliance must be placed on the radar's target detectability for the remaining group of outputs to be useable.

The consequences of blind speeds have already been pointed out. If they are not dealt with, particularly in air traffic control radars, they can become a positive menace to safety. Figure 9.23(a) is of an L-band radar ppi showing totally unprocessed video. Figure 9.23(b) is of the same radar, but simultaneously showing its mti output with staggered p.r.f. deliberately switched out. Track discontinuities caused by blind speed losses are obvious.

Adaptive signal processor (ASP)

The mti and mtd techniques discussed above are not the only ones in use. There is another (the ASP) which embodies the principle of doppler filtering but implemented in a much simpler and cheaper form. It consists of three simultaneously operated parallel processing channels:

(a) A non-coherent detector
(b) A moving target indicator with zero doppler filter
(c) A moving clutter rejector (auto-doppler) with an automatically variable doppler frequency filter.

The output of the three channels is applied to an OR gate after Temporal Threshold Integration so that any wanted target exhibiting visibility 'in the clear', in fixed or in moving clutter, is made available for display.

The auto-doppler channel has a rejection characteristic or 'notch' of the same form as the zero-doppler mti. The ASP automatically senses the doppler frequency of moving clutter and sets the rejection notch into coincidence in the clutter area. As the clutter moves so does the rejection notch so that all parts of the surveillance area are automatically served.

A photograph of one such processor is shown in fig. 9.24.

Plot extraction

Increasingly, both primary and secondary radars include plot extractors in their signal processing systems. Why? There are three main reasons and the reader will be pleased to know that all are user-directed.

The first concerns the need to transfer target data from the radar sensor to the display area. Sometimes these are separated by many miles or by shorter distances over which it is impossible to use cable. Radar signals consist of pulses of short duration, with rise and fall times of yet shorter duration. A communications channel passing these fast events must have sufficient bandwidth. In a radar system having a pulse duration of 1 μs and rise time of 0.25 μs, a bandwidth of greater than 2 MHz must be available if the target data are to be reproduced without serious degradation. Such a bandwidth requires a transmission link with facilities to send not only the radar signals but also antenna azimuth position data and synchronising pulses.

All these data would allow a radar display to operate in real-time, the data on a given target being renewed each time the antenna beam swept past it. The data renewal rate (antenna turning rate) is usually of the order of seconds – a million times or more greater than the signal event times.

Consider now what basic data the user needs:

- Target presence.
- Target range.
- Target bearing.

Suppose there were to be as many as a hundred aircraft simultaneously in a radar display area of 120 nautical miles radius and the antenna turning rate was 15 r.p.m. There would be $100 \times (4/60) = 6.66$ target reports per second. If the range of each was to be reported to an accuracy of 1/12 nautical mile the range scale would have to contain $12 \times 120 = 1440$ increments. This could be done using a binary word of n bits where $2^n \geqslant 1440$. Thus an eleven-bit word would suffice, ($2^{10} = 1024$, not enough; $2^{11} = 2048$, plenty). If the azimuth was to be reported to an accuracy of 0.1° (one-tenth of a beamwidth) there would need to be 360×10 increments of azimuth; thus a twelve-bit word would be required ($2^{12} = 4096$). Thus all the basic target data in the model above could be sent at a rate of $6.66 \times (11 + 12) = 153.18$ bits per second.

We see from this a massive reduction in the bandwidth required: from 2 MHz down to about 150 Hz. Such small bandwidths are much easier to provide – across land, telephone lines could be (and are) used; across water, ordinary radio channels with speech facilities serve very well.

The second reason for plot extraction is hinted at above. Radar signals are detected in real-time; that is, outputs are developed at the same time as inputs appear (or almost so). They last for as long as, and only as long as, the input is present. Normal ppi displays show the radar target's history by exciting small areas of a phosphorescent screen. This excitation lasts for very short periods resulting in an initial 'flash' of high brightness and decaying rapidly to an afterglow of about 1% of the original brightness in a few milliseconds. In a typical search radar the target exists for about 20 ms at high brightness and reduces to almost extinction in the seconds of time it takes for the antenna to repeat the target illumination. Most users of such radars will know the inconvenience of the need to view ppis in reduced ambient lighting, because the afterglow gives useful track history and its brightness is too low to be seen in normal viewing conditions. A plot extraction system is capable of expressing target position (and in SSR a great deal more data on the target) and retaining the data in digital stores. These can be read and re-read over and over again. Thus it is possible for target data so stored to be read out continually at a rate much faster than the antenna rate. So instead of waiting for seconds between target reports, each report can be repeated (usually about twenty-five times per second) before new position data are available. Such a renewal rate prolongs the brightness of the area of the screen showing the target and no reliance need be placed upon afterglow. The display can therefore be viewed in high ambient lighting conditions.

The third reason is closely allied to the second. It is possible in modern systems to insert data on targets which is derived externally to the radar, and to make this directly readable as alphanumeric characters. Such data are inserted automatically or by operators using keyboards addressing computers which organise the data to be written alongside the relevant targets. In order that the total system can become computer-based, using successive target reports to generate track data, the reports need to be measured and stored. A plot extractor (or in American parlance, a 'target evaluator') does just this.

How does it work? Earlier discussion on target detection gave some clues. The plot extractor is required to emulate viewers of traditional ppi display and, like them, to make decisions upon whether the bright spots constitute a wanted signal or a false alarm. Figure 9.25(a) shows a portion of a ppi with a 'blow-up' of an area taken as example in fig. 9.25(b). In the former, two targets are represented as they are usually displayed. In the blow-up the area has been magnified to show the individual signal returns in a radar having an average of ten pulses per beam width. The target at R_1 is clearly defined, producing an output pulse for every successive transmission throughout the antenna beam's passage across it. The second target at R_2 is less clearly defined. Is it one target whose fluctuating echoing area happens to fall below average for much of the antenna beam's dwell time? Or is it two small targets close together? Or is it one small aircraft with some false detections at the same range? Or are they all false alarms? Traditionally a 'two threshold' form of plot extractor is used and described below. Devices using more of the amplitude data are in being.

The first action or decision to be taken is to decide whether the 'first threshold of detection' is crossed. That is, does the radar output amplitude exceed a level which constitutes an event other than noise. Many signal

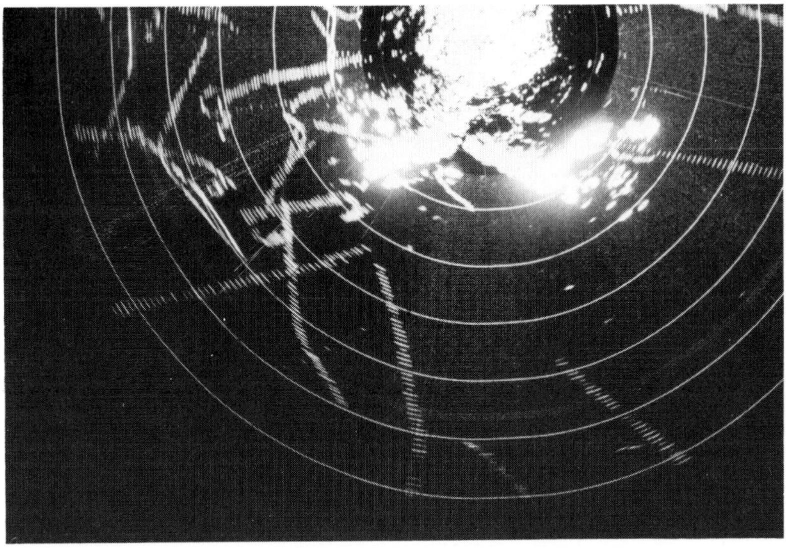

Fig. 9.23(a). Sector of a ppi operating at the same time as in fig. 9.23(b) showing signals with no processing and with staggered p.r.f. deliberately excluded. Comparison with fig. 9.23(b) shows how tracks in the mti processed output are much corrupted.

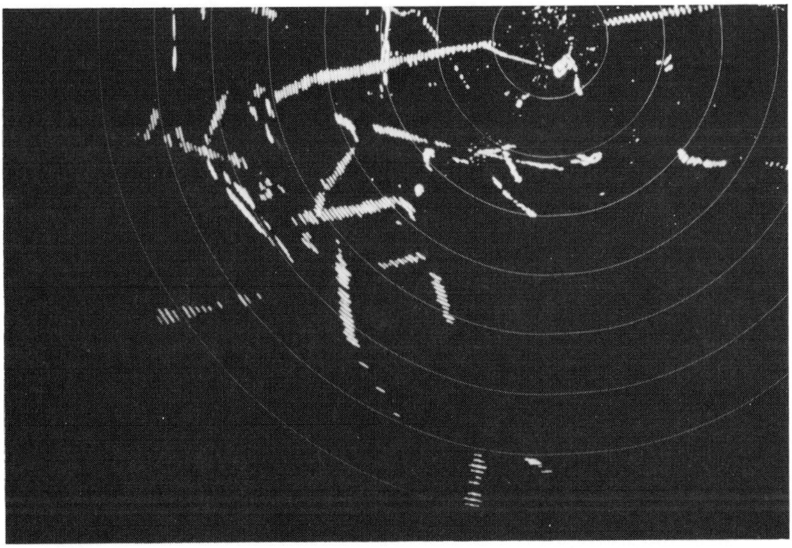

Fig. 9.23(b). Sector of a ppi showing mti processed signals with staggered p.r.f. deliberately excluded and therefore subject to radial blind speeds. (*Photos courtesy of Marconi Radar.*)

Fig. 9.24. A modern adaptive signal processor (ASP). It well illustrates the reduction in size possible by use of micro-electronics. Such a unit would have occupied two large man-size racks in 1960. (*Photo courtesy of Marconi Radar.*)

processors include this 'decision making' circuit but, if not, it has to be incorporated in the plot extractor. These events are looked for in every resolution cell in real-time. Now another decision has to be made – do any of these events form the pattern expected of a set of detections of a target as the antenna beam explores it? This second decision is made by various methods; the most common (and easy to understand) is the 'sliding window' technique.

In the sliding window technique, arrangements are made to store the decision on whether the first threshold has been crossed or not for every range increment of a pulse repetition period. It is usual to provide at least two and there may be as many as four range increments per resolution cell, but the decision is expressed in terms of '1' if a resolution cell contains a first threshold crossing and a '0' if it doesn't. At the next interpulse period the same range resolution increment is examined and its '1' or '0' state is stored with its predecessor. Storage equal to the number of pulses per beamwidth is usually provided. The effect is illustrated in fig. 9.26.

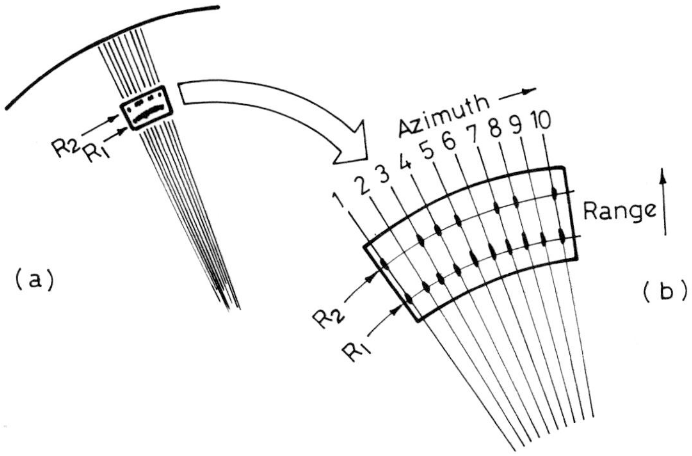

Fig. 9.25(a). Two targets as they might appear on a ppi. That at R_1 is strong, that at R_2 is weak. **(b).** Blown up view of the two targets. The stronger produces detectable signals for each of the 10 successive samples shown; the weaker produces intermittent missed detections.

It can be realised why the technique is called a sliding window. In the example a set of seven cell stores at a given range is swept through the surveillance area and forms a window whose dimensions are one range resolution cell and the azimuth equivalent to seven (in our example) interpulse periods. As the window progresses, i.e. is 'dragged through' the surveillance area by the antenna beam, the window store contents are examined at each new interpulse period and a count of its occupied cells is made. As soon as the count reaches a predetermined value (shall we say three out of seven) a 'plot leading edge' is declared. This is tantamount to saying 'It is so improbable that three contiguous range cells will be occupied by false alarms that there must be a target present'. The counting goes on until another criterion is reached, that is where the number of cells occupied falls to (shall we say) two out of seven. This is equivalent to a decision that even if a leading edge has previously been declared, there are now insufficient detections to declare target presence, i.e. a trailing edge is found.

So we now have decisions on the 'second threshold' criterion: where the plot starts and where it finishes. All this occurs while the antenna position has been noted and a digital expression of the azimuth at which these decisions have been made is also stored. Thus within the plot extractor we have the following data:

- Plot start azimuth (leading edge).
- Plot finish azimuth (trailing edge).
- Plot range (resolution cell position).
- Plot presence (leading edge followed by trailing edge).

There are other operations performed within the extractor. One is concerned with keeping the false alarm rate constant. It usually senses the radar system's thermal noise level and makes sure by automatic means that the first threshold level is maintained to give a fixed probability of false alarm events. That is, if the noise level should increase or decrease the threshold level must follow it automatically. Another operation is to calculate the plot's

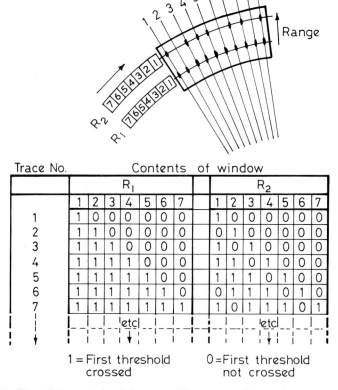

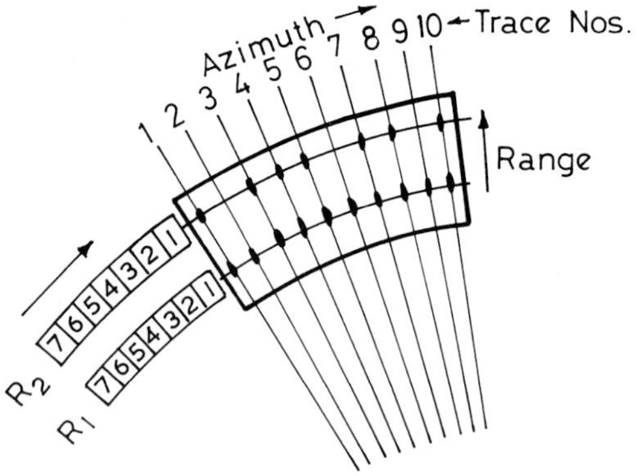

Trace No. — Contents of window

Trace No.	R1							R2						
	1	2	3	4	5	6	7	1	2	3	4	5	6	7
1	1	0	0	0	0	0	0	1	0	0	0	0	0	0
2	1	1	0	0	0	0	0	0	1	0	0	0	0	0
3	1	1	1	0	0	0	0	1	0	1	0	0	0	0
4	1	1	1	1	0	0	0	1	1	0	1	0	0	0
5	1	1	1	1	1	0	0	1	1	1	0	1	0	0
6	1	1	1	1	1	1	0	0	1	1	1	0	1	0
7	1	1	1	1	1	1	1	1	0	1	1	1	0	1
↓				etc							etc			

1 = First threshold crossed   0 = First threshold not crossed

Fig. 9.26. The 'sliding window' detector. The results of detection on two targets are seen tabulated. As the window sweeps across the targets the cells within it contain 1s or 0s dependent upon signal (or noise) level. Various algorithms can be constructed to determine whether the window contents represent 'target detection' or not.

true azimuth by adding the azimuths of the leading and trailing edge of individual plots, halving the result and accounting for the small azimuth difference created by the inequality of the leading and trailing edge criteria. The plot extractor must also store all detected plot data as range/azimuth digital words, ready to release them to the outside world. Often other associated data are assembled in the plot word to express the signal conditions under which it was detected, e.g. was clutter present, what (in an mtd) filter numbers detected the plot? These subsidiary data can be used to advantage in any subsequent target tracking routines performed.

In the example above, a hundred targets were taken as a maximum load. Commonly specifications call for two hundred plus as a maximum capacity. It is unusual for targets to be evenly disposed throughout the surveillance area; commonly they occur in one or two sectors. In order to prevent the extractor overloading and to smooth the output plot transmission rate, the final digital words are stored and read out at regular intervals. If no plots are in store then a set of digits is artificially inserted to form an 'idle code'. Transmission rates are organised so that no plot remains in store for longer than the equivalent of 120° of antenna rotation.

The digital output words are used as modulation of audio tones using devices called 'modems'. The term is a compound word made from 'modulator–demodulator'. Sometimes the digital word is signalled by using one tone for a '1' and a different tone for a '0'. Many systems use a single tone but modulate its phase to mark transitions from '1' to '0'. Naturally a pair of modems are used, coupled by a transmission line with audio bandwidth and commonly with high noise immunity to reduce signalling errors. Having established a link from radar to display area it is often used in the reverse direction to effect remote control of the radar by similar signalling techniques. The forward and reverse signals are separated by frequency or other modulation-separating means.

The manner in which the display uses its digital plot inputs is described in part 4.

Part 3

Secondary Surveillance Radar

10
Basic concepts

10.1 Why?

A pioneer of the SSR technique, Mr K.E. Harris once observed that it is often more rewarding (neglecting the rules of English grammar) to start an investigation not by asking questions in the form '*what* is the bandwidth?' but '*why* is the bandwidth'. I shall take his advice: why is SSR?

The answer lies way back in the history of radar in which the original memorandum of 27 February 1935 from Sir Robert Watson-Watt to the U.K. Air Ministry was a benchmark. This answered the question whether or not it would be possible to destroy operators of enemy vehicles or to stop their vehicles from working by means of a ray of radio energy: the so-called 'death ray'. The answer at that time was 'no'. But it was also pointed out that even if it was possible, first it would be necessary to know if the ray's target was an enemy or a friend.

From this realisation grew the Second World War technique of IFF – 'identification, friend or foe'. Its development has led to the present and projected important future SSR – secondary surveillance radar.

What is 'secondary' about SSR? The answer is immediately obvious from the basic system block diagram in fig. 10.1. The system is one wherein a transmission at frequency F_1 'interrogates' one or a number of co-operative (and therefore friendly) partners. The interrogations are received by these and their replies transmitted at a different frequency, F_2. So we have a second pair of transmitter/receivers in the system. Replies are received by the interrogator, its associated receiver being given the name 'responser'.

Because the spectrum of frequencies of the interrogating and reply signal channels in SSR are separated by more than their respective bandwidths, any reflections from objects generated by the interrogating signals are excluded from the received reply channel. As a consequence there are no clutter signals present as there are in primary radar systems.

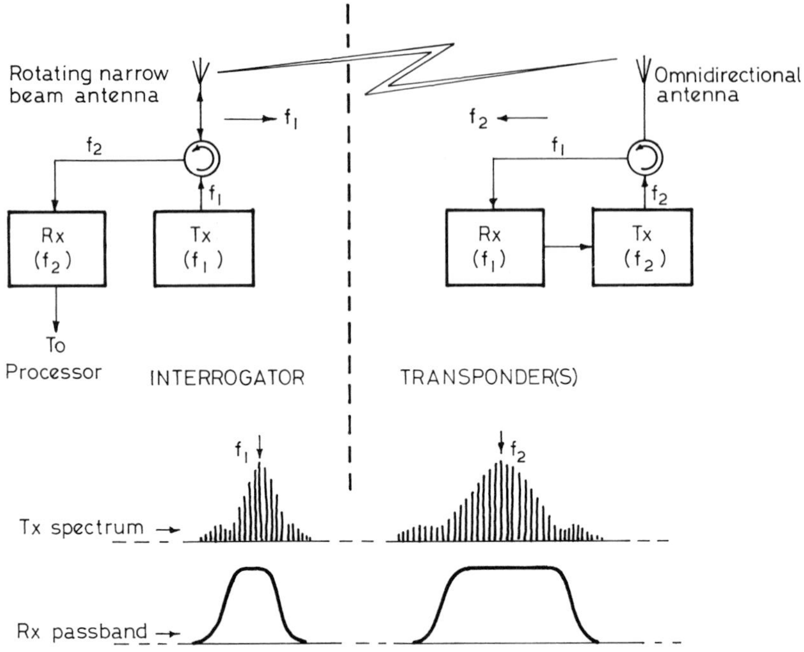

Fig. 10.1. Basic interrogator/transponder principle: Passbands of ground and airborne receivers are narrow enough to exclude reflections and breakthrough from their own transmissions.

10.2 The concept in practice

The concept is one which requires all 'friends' to carry a transponder. In a defence scenario those which answer interrogations in the correct fashion will be identified as friends, those who don't will be classed as foes, and be treated as such! It is clear that the interrogator needs to ask two vital questions: 'where are you' and 'who are you'. The first is answered by use of the radar technique; both interrogator and transponders use pulse transmissions. The round-trip time from interrogation to receipt of reply gives the range (as in primary radar). The bearing or azimuth to the transponder is given by carrying interrogations and replies via a rotating antenna beam which is narrow in azimuth and wider in elevation, just as in primary radar.

The question 'who are you' is answered by making replies take the form of groups of pulses which can be position coded. That is, presence or absence of all or any of the pulses in a group can convey 2^n different answers where n is the number of pulses possible in the group. A practical form of the SSR system is shown in fig. 10.2.

The interrogator transmits a pair of pulses at 1030 MHz. Each pulse has the same duration, shape and amplitude. Their spacing is chosen from a set of

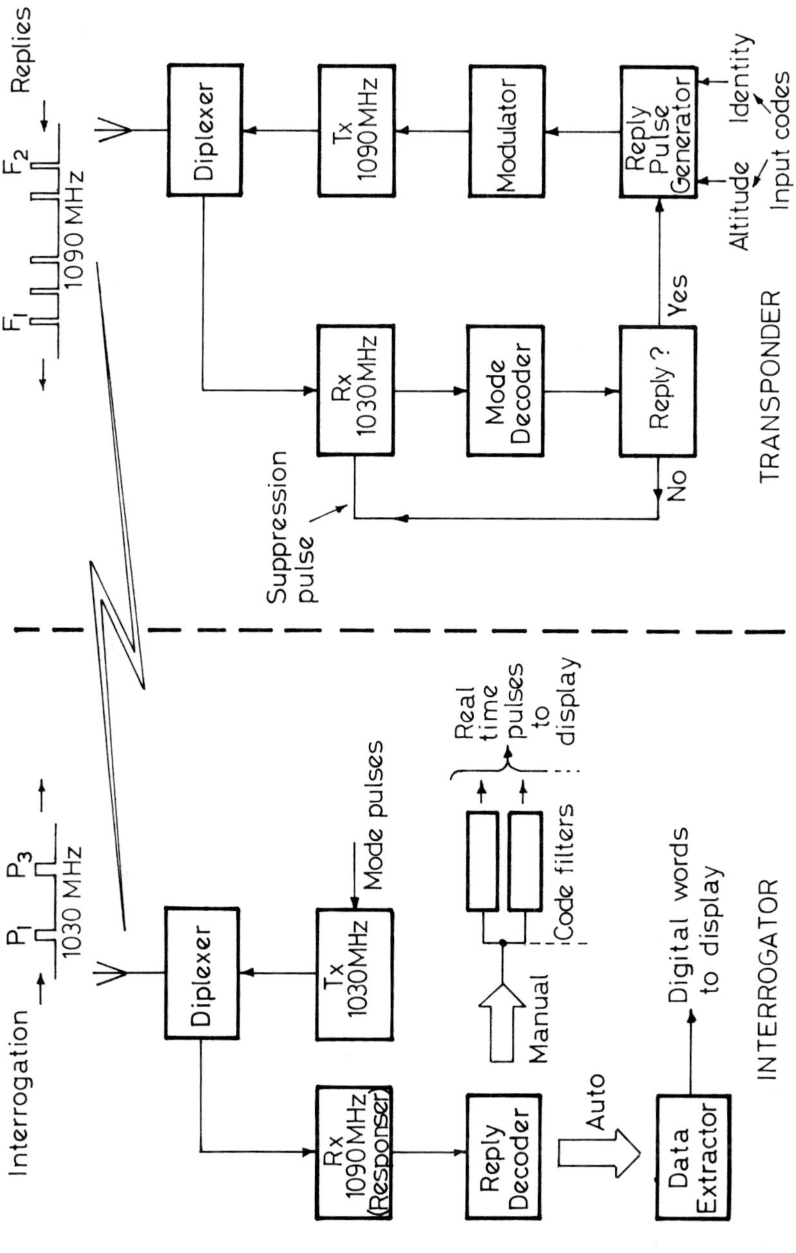

Fig. 10.2. Simple block schematic of SSR in practice; decoding of replies may be by manual or automatic means.

standard agreed intervals. Thus pulse position coding distinguishes various questions or 'modes of interrogations' as they are called and shown in fig. 10.2. By international agreement the context of the modes is as listed in table 10.1. The interrogation pulse pair (P_1 and P_3) are accompanied by a control pulse P_2 whose purpose will become clear later. Transponders within receiving range detect the pulse pair, sense the pair's spacing and make reply according to the context of the interrogation. That is, if a transponder senses an 8 μs spacing the automatic reply will be a code conveying the transponding vehicle's identity. In air traffic control systems, the aircraft altitude is of vital importance to controllers and so an interrogation pulse pair spacing of 21 μs results in replies giving aircraft height automatically in coded form. Mode 3 (military) and mode A (civil) are common to both classes of user, as is mode C. This is of great value in air traffic control since both classes use the same airspace. For the same reason both use the same altitude coding system.

Table 10.1. The use of interrogation modes

Mode	$P_1 - P_3$ Spacing	User	Context
1	3 μs	Military	Secure
2	5 μs	Military	Secure
3/A	8 μs	Military/Civil	Identity (common)
B	17 μs	Civil	Identity (civil only)
C	21 μs	Military/Civil	Altitude
D	25 μs	Civil	Unassigned (for system expansion)

Replies are made to interrogation when certain criteria are met. They take the form of pulses radiated at 1090 MHz which are of 0.45 μs duration, having uniform amplitude and spacing of 1.45 μs (or multiples thereof). By international agreement the reply train of pulses can consist of any, all or none of twelve 'information' pulses contained within two always-present 'bracket' or 'framing' pulses designated F_1 and F_2. The reply format allows space for a thirteenth pulse which has not yet been used. The reply pulses have agreed designations as shown in fig. 10.3. They are in four groups, rather confusingly called A, B, C and D. Each group has three pulses identified by subscript numbers 1, 2 and 4 so that an octal system of code designation is possible, the sum of the subscripts reflecting the octal value required. For example an identity code value of 7213 would require the reply to contain pulses F_1, A_1, A_2, A_4, B_2, C_1, D_1, D_2 and F_2.

There are 4096 different codes possible. This can be appreciated from the following. Each of the three possible pulses in each of the four groups A, B, C and D can be selected to be present or absent; i.e. they represent binary

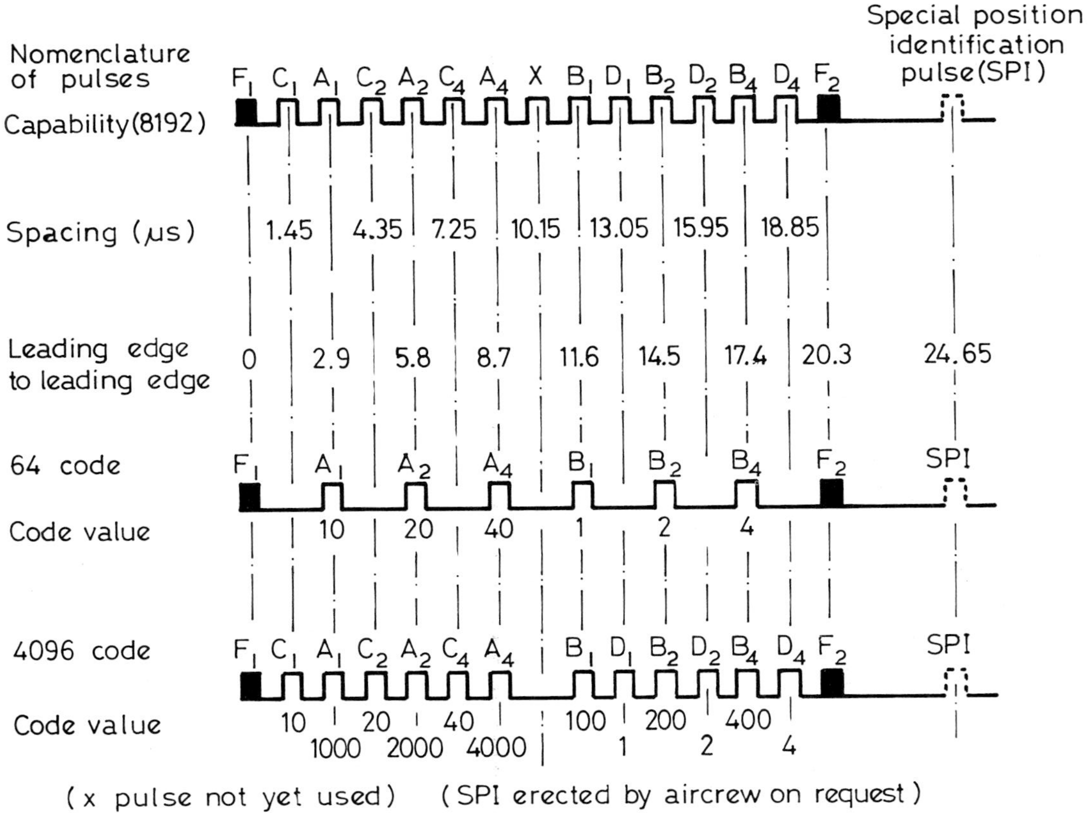

Fig. 10.3. SSR reply formats as specified by international agreement. The two 'framing' or 'bracket' pulses, F_1 and F_2, are always present. Octal coding is used.

conditions '1' or '0'. Thus in total they can form a $4 \times 3 = 12$ bit binary word and $2^{12} = 4096$.

Replies are radiated by an omni-directional antenna so that transponders can be addressed by interrogators at any azimuth in a surveillance system, irrespective of target aspect. When received by the interrogator's responser, the reply pulses are envelope-detected and because the reply format is standardised, the decoding circuits of the interrogator's processing system can detect which pulses are present and hence derive the reply data requested by the interrogator.

10.3 Mode interlacing

The interrogation repetition rates and beamwidths are of the same order as in primary radar. Thus a target illumination time of some 30 ms is obtained, during which about fifteen to twenty interrogations of a given transponder can be made. This permits different modes to be used during the passage of the beam across the transponder. Suppose the interrogator was continually to radiate a pattern of modes A, C, A, C, ..., etc. In a typical illumination time of a given airborne transponder, its identity and altitude reply codes would be repeated perhaps ten times. Thus data of very high integrity can be obtained on all transponders within interrogation range at every revolution of the interrogator's antenna. Of course, positional data are automatically given as well. The number of modes practicably useable in an interlace pattern depends upon the number of interrogations per beamwidth. It is usual in SSR to ensure that there are no less than three replies per mode received. This in turn means demanding at least four because reply probability is never 100%. Thus for a triple mode interlace pattern at least twelve interrogations per beamwidth must be possible. What that beamwidth is and why some interrogations don't result in a reply will be seen later.

10.4 Transponder controls

The cockpit controls of a modern transponder are illustrated in fig. 10.4. In civil aircraft the transponders incorporate a mode selector control and replies are made automatically only to the selected modes of interrogation. It is common to find one of the selector positions labelled 'A/C'. This results in replies to either mode A or C, again automatically. The identity codes in response to modes A, B or D are set in by four front panel switches, one for each of the code groups A, B, C and D. Each switch has eight positions, 0 to 7, so that the 4096 codes are made up from $8 \times 8 \times 8 \times 8$ possible different combinations. The altitude coding for airborne transponders is done automatically by coupling a digital encoder to an altimeter-type

Fig. 10.4. Illustrating the compact nature of SSR transponder controls. The control panel is 35 cm by 15 cm. (*Photo courtesy of Cossor Electronics.*)

transducer. As the transducer senses a pressure change as a result of altitude change, so the code automatically changes in sympathy.

In military transponders replies are made automatically to whatever mode of interrogation has been received. The meanings of codes in response to modes 1 and 2 are kept secure for obvious defence reasons. The mode 3 codes, which are identical to civil mode A, together with altitude reporting, are both controlled in the manner described above.

There are four other important controls on the transponder. The first is a pilot operated switch which inserts another reply pulse after F_2. It is called the *special position identification* (SPI) pulse and is used as a means of ratifying identity upon request by ground control. The second is an 'emergency' switch, for use when the driver dies or in other dire situations. When this switch is selected the transponder issues a preset code of value 7700 when mode A is used. The third is for selection if there is a radio telephone communications failure between the vehicle carrying the transponder and its control centre. When this switch is selected the transponder issues the code 7600. The fourth switch is selected if there is 'unwarranted interference' with the vehicle, such as 'skyjacking' in aircraft. The preset code is 7500 in this case. These three codes, 7700, 7600 and 7500, have been internationally agreed, so their receipt is universally understood.

How the various equipments work, their size and constitution will emerge from the following sections.

10.5 The use of SSR

The main users of the SSR technique are those concerned with air traffic control. There are other users, in the maritime field, who employ a similar system to identify marker buoys (RACONs). In the same way, aviators use another such system to identify beacons on or near airfields (the DME system which is like SSR in reverse). In both of these cases the moving vehicles issue the interrogation and the fixed transponder, by means of a unique and advertised identity code, gives its location – a great help in navigation. If the interrogator has a narrow rotating beam, the transponder position can be determined by range and azimuth data. If the interrogator only has range measuring facilities (as in the DME system) then identification of two fixed transponders and their range will suffice to obtain position data, provided ambiguity can be resolved by other means.

The ideas of SSR mode interlacing, codification of identity and altitude given above are brought together in fig. 10.5. This shows, as an example, the data available at the ground interrogator's responser output for subsequent decoding. Two aircraft are shown: one civil and one military are required to respond a triple interlace of the modes 1, 3/A, C. The military craft will respond to any valid interrogation mode automatically. The civil craft is shown having its transponder set to reply to modes A and C only. The replies

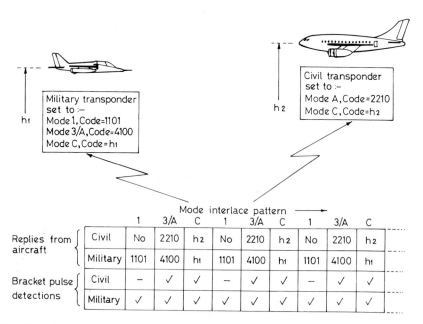

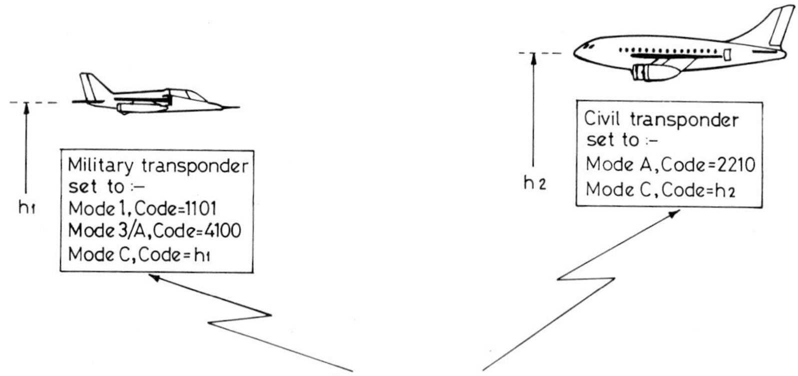

Mode interlace pattern →

		1	3/A	C	1	3/A	C	1	3/A	C
Replies from aircraft	Civil	No	2210	h2	No	2210	h2	No	2210	h2
	Military	1101	4100	h1	1101	4100	h1	1101	4100	h1
Bracket pulse detections	Civil	–	✓	✓	–	✓	✓	–	✓	✓
	Military	✓	✓	✓	✓	✓	✓	✓	✓	✓

Fig. 10.5. An example of reply sequences from two aircraft in response to a triple-mode interface pattern.

made in response to each interrogation are tabulated for each class of aircraft. Note that the civil craft makes no reply to mode 1 interrogations. The net effect is that out of the nine opportunities to detect the presence of a target, the military craft takes them all and nine potential replies are available. In the case of the civil craft only six potential replies of the nine are possible. If neither craft was equipped (or selected) to respond to mode C in the given pattern, then the number of potential replies would go down to six and three in the military and civil craft respectively. In military air traffic control systems it can be appreciated that perhaps three interlaced modes are insufficient, for they would need modes 1, 2, 3/A *and* C in order to get full available SSR and IFF data on a mix of civil and military traffic. This problem is commonly solved by having two different mode interlace patterns which are selected automatically at alternate antenna revolutions or upon demand.

So far, mode interlace patterns have been shown with an even distribution of single modes. It is not necessary to spread the mode repetitions evenly throughout the beamwidth. The weighting and grouping of modes in the interlace pattern are dictated by the decoding and data processing characteristics as will be seen later.

10.5.1 Concentrating upon its deployment in air traffic control, how is the system used?

When pilots are filing their flight plans to go from, shall we say Munich to London, the agreed flight parameters (route, height, speed, time of arrival and departure, etc.) will contain a specific identity code to be used. The flight will cross a number of flight information region (FIR) boundaries. With any luck, all the control centres who need to handle this flight will, through the Eurocontrol Agency, have foreknowledge of the craft's identity code. It should be unique among all the other craft in the air during the intended flight. Thus, as the aircraft progresses along its journey, any control centre which knows the identity code will be able to follow its flight through the FIR for which they are responsible and to hand it over to its neighbouring FIR. Identification of craft entering the FIR should not require any radio-telephone request because the unique identity is previously advertised. This is one of the prime reasons for the use of SSR: the reduction of both aircrew work load and occupancy of overloaded airwaves. If primary radar data only were available on aircraft in controlled airspace, identification doubts must be resolved by asking particular aircraft to execute 'identification turns' – a waste of fuel, time and airwaves. Interrogators with a range of 200 nautical miles are common, thus control agencies using SSR facilities can obtain aircraft data throughout their entire FIR with very few ground equipments. The data available are exactly that needed for air traffic control:

- Position

- Identity
- Altitude

These data are made available to users in a number of ways, as will be described in the following sections.

10.5.2 Real-time decoding

Every reply elicited from a transponder contains data expressed as any or all of twelve information pulses within two ever-present bracket pulses (F_1/F_2). These are always spaced 20.3 μs apart. The interrogation pair received is analysed by the transponder and if a reply is to be made it begins 3 ± 0.5 μs after the receipt of the last of the pulses in the interrogation pair (P_3). Range can therefore be measured to an accuracy of 1/12.36 nautical miles on individual aircraft from a given interrogator. The decoder contains circuits which delay any input pulse by 20.3 μs and looks for a companion pulse in the undelayed input after this delay. Any F_1 input should be accompanied by an F_2 pulse 20.3 μs after it. If coincidence is found, then the system knows to high probability it has a valid response from a transponder irrespective of code and this event is output from the decoder in the form of a pulse of about 1 μs duration. If the transponder continues to reply (as one has every right to expect throughout the interrogation beam's dwell time on target), such coincidences will continue to be found. The output pulses will therefore appear like single detections of a target in primary radar.

These output pulses, signifying 'presence' of a target, are thus able to be sent to a ppi display in virtually real-time and show up as a small bright arc, as do primary radar outputs in a real-time display. The arc lasts as long as the target illumination time. So we have indication of the presence of a responding target given as range and azimuth. How are the coded data within these detected bracket pulses called out? In a real-time decoding system one way is as follows. You reach for a device called a 'light pen'. It looks just like a writing pen, except that it reads! In the head of the pen is a photosensitive cell with extremely quick reaction time. The head of the pen is directed to the area where next the detected target is expected to appear on the following antenna revolution. As the antenna sweeps across the target the coincidence of F_1/F_2 will create a new output pulse for display; the first of a series of a beamwidth history. As soon the output pulse is displayed, the ppi screen will emit light at high intensity for the duration of the displayed pulse. The light is detected by the photo-sensitive cell which almost instantaneously generates a trigger pulse from which a gating waveform is developed ready to catch the next repetition of the reply from the transponder. The gate duration is arranged to straddle the wanted reply thus isolating it from other replies in the same interrogation period. Because the system repetition time is known, the gate will be present at the same time (and only for the time) the next reply

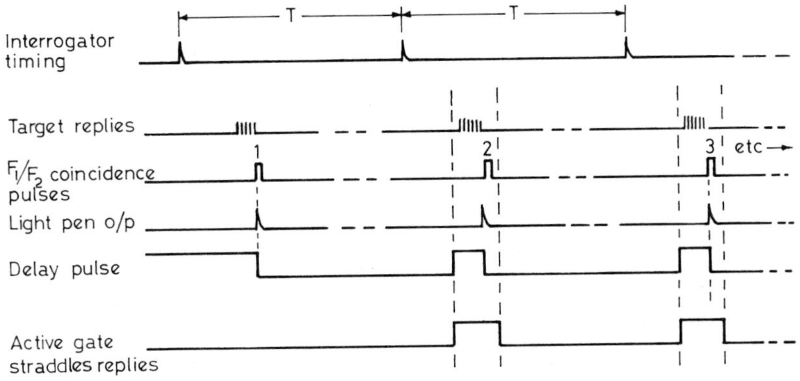

Fig. 10.6. Showing the generation of a real-time 'active decoding' gate. It is organised to isolate individual replies. The light pen is activated by responses displayed when F_1/F_2 'presence' pulses occur.

from the same transponder enters the system. The technique is illustrated in fig. 10.6.

So the light pen captures only the reply pulses associated with one transponder. Circuits are arranged so that during the gating time each possible reply pulse position of the twelve is examined. If a pulse is present at any of them it is directed to summing networks separated into the information pulse groups A, B, C and D. Summation of all the subscript values of the pulses in each group is made and the result shown on indicators, together with the mode to which the replies were made. This process is known as 'active decoding in real-time'.

Having detected a code by active means, the operator can now set up a corresponding 'passive' code filter. It consists of five selector switches. One sets the mode in which the code replies, the other four are the counterparts of those used in the transponder to set the code in the first place. That is, the sum of the subscripts of the information pulses in their groups A, B, C and D. If the code sensed by active decoding was mode A, 2235 then the 'passive' filter switches would be set to correspond. All replies received would automatically be offered to this filter. Naturally only those replies with that code content would pass through it. The process is equivalent to matching a set of pegs into a set of holes. If the pegs fit, an output is given. Subsequently every time this code was received, its passage through the filter would be made to generate a separate output pulse timed so that it occurred after that which indicated 'presence' of a target. Thus a reply detected by the filter would be indicated as two arcs, one slightly behind the other (usually about three displayed pulse durations). By this means individual targets can be isolated from the whole group. This process is known as 'passive decoding in real time'. An example is given in fig. 10.7.

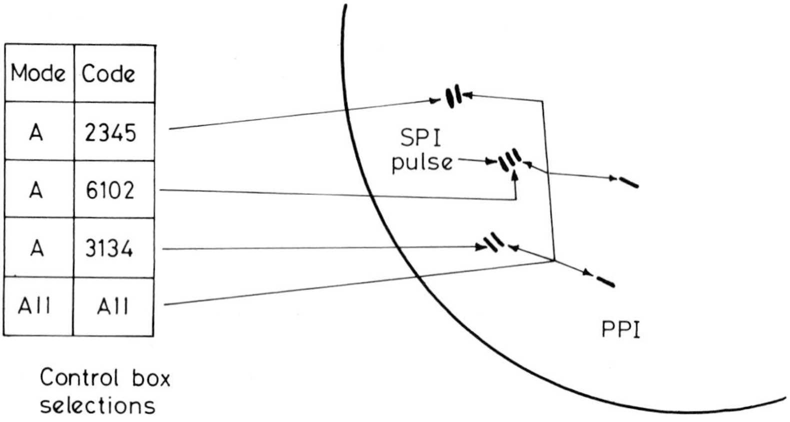

Fig. 10.7. A form of displaying the results of 'real-time' decoded replies. Any reply to any mode of interrogation is registered as a 'single bar'. Replies to selected codes add a second bar. An SPI pulse, if present, adds a third bar. Emergency replies are usually displayed as an elongated bar.

The decoding processes described above indicate active decoding leading to passive decoding. Such a sequence must be followed if code values are not previously known to the operator. However, in most systems the codes to be used *are* known through the control organisation's pre-planning process. Take the example already cited, an aircraft flying from Munich to London. The craft will have a planned estimated time of arrival (ETA) at the London FIR boundary. Some little time before ETA, the radar controller would set up a code filter ready to detect the planned and advertised code. When the craft was within range of the system's interrogator its presence would be automatically indicated as a 'double bar' as described above.

It is also possible to use the filter technique for altitude codes. In this, two sets of switches are used to set upper and lower height limits for acceptance of altitude codes. Only those altitude reports falling within these limits are allowed into the display system. They result also in the 'double bar' presentation. The second bar indicating the target lies between the set limits. 'Active/passive' SSR decoding systems appear to give radar controllers all the tools of their trade.

10.5.3 Automatic decoding and data extraction

Obviously, operating the manual real-time decoding system requires a great deal of 'knob twiddling' on the part of the user, who also has to exercise quite a bit of real-time logic as well, because the display representation (one, two or three bar 'blips') has such limited repertoire. To deduce unique identity requires deselection of all other switches which could cause similar outputs to be displayed, and another full antenna revolution must be executed before reaction to new settings is seen, or active decoding must be used.

The way out of this inconvenience and inefficiency is to use automatic decoding and data extraction techniques. There are numerous different systems in being, and all use methods similar to that illustrated in fig. 10.8. Their operation results in target data being displayed in non-real-time, but in a form which allows the user to read, directly from the screen, all the relevant data. The data comprise position, identity, altitude (if available) and previous history of the target position.

Essentially, the operations performed in the equipment are as follows. Replies to interrogation are fed to a decoder as in the real-time system described. They are examined in the decoder to ensure they match the internationally agreed format and are then declared as 'valid' code trains. These are put to a 'partial plot' former. Here only those replies in response to one's own interrogations are stored, others are rejected. This is done by the process of range correlation and only those replies having the same range are used as potential wanted plots.

Partial plots are put to a 'valid plot' former. Here the history of potential wanted targets across the antenna beamwidth is examined, now using the appropriate azimuth data. During this process the target azimuth is calculated and logic checks performed which raise the level of confidence to be given to its outputs, which take the form of digital expressions of the target's range, azimuth, altitude, code (and the mode used to stimulate it). These digital words are then assembled into a defined format for transmission as a serial digital message; the format definition allows the data receiver to understand what the train of '1's and '0's means.

At the far receiving end of the transmission line (it may be metres or hundreds of kilometres distant) each digital message goes through a 'sorting' process. The range and azimuth words of a given plot are put to a display 'position' generator and are usually converted from their R, θ form to values of X and Y deflection voltages, so that the target position can be properly registered on the display as a recognisable symbol. The code data associated with the plot are put to a 'character generator' so that the waveforms necessary to write the data in alphanumeric form can be reproduced alongside the plot position symbol.

Associated with the 'data sorting' is a storage process by which each valid plot position is remembered for a preset time. As new plots enter the system, their predecessors are output in the form of 'historic' plots; usually these are given a small dot as a symbol. Thus a craft on a straight track would be registered, as illustrated in fig. 10.9, by a recognisable symbol for the latest plot report followed by up to about five dots representing the last position reports. The net effect is to give the following data, up-dated at every antenna revolution:

- Position (square, circle, diamond, etc., symbol).
- Past positions (dots).
- Mode and code (letter plus four digits; in the example the mode was A and the code 2201).
- Altitude (three digits; in atc parlance this represents target flight level in hundreds of feet; in the example the target was at 15 500 ft).

As in the case of primary radar plot extractors, the data on all targets are renewed once per antenna revolution. There is thus a whole antenna revolution time in which to process and send new inputs. This is usually of the order of four seconds or more. However, if all this time was used, the data displayed would be stale and limits are usually placed on input/output delay representing the time to execute about 120° rotation of the antenna.

10.6 We all agree (that is, most of us)

Reference has been made a number of times in this description to 'international agreement'. The need for it is obvious; it would be most inconvenient to find one set of SSR standards in North America, another in Africa and yet another in Europe. If this were so, one aircraft designed for international deployment would need three types of transponder!

To overcome this difficulty, almost immediately after the United Nations Organisation (U.N.O.) was formed, the International Civil Aviation Organisation (I.C.A.O.) was brought into being and operated under the U.N.O. umbrella. I.C.A.O. has over 160 member states. One of the major tasks of I.C.A.O. is to lay down internationally agreed standards of performance for aviation navigational aids and communication systems. SSR performance characteristics, through long years of discussion, debate, development and decision, find expression in annex 10 to the Chicago Convention on Civil Aviation (reference 10).

Because of the SSR system's effectiveness in civil and military air traffic control and the need for commonality in large civil and military communities, N.A.T.O. has its own set of SSR standards laid down in STANAG 4193. But these are almost indistinguishable from the I.C.A.O. annex 10 standards and are entirely compatible with them. Inescapably the one-time Warsaw Pact community set standards which were different. However, both the USSR and China became members of I.C.A.O. and through their subscription to *I.C.A.O. Specifications in Civil Aviation* made (and still make) their contribution to the maintenance of civil aviation safety. Many of the Warsaw Pact countries are striving to join the European Aviation Community, ECAC (the European Conference on Civil Aviation).

There are two other important agencies devoted to maintaining agreement on SSR standards; both work closely with I.C.A.O. They are the Radio Technical Committee on Aviation (R.T.C.A.) and the European Committee on Civil Aviation Electronics (Eurocae).

R.T.C.A. is served almost entirely by prominent members of the aviation electronics industry in the U.S.A. They discuss and decide upon deadly dull but vitally important matters such as a common set of dimensions and fixings for airborne transponders, their plug and socket types and even connector pin numbering. By such measures various manufacturers can be assured that their products will interface smoothly with others.

Eurocae is the European equivalent to R.T.C.A. but to date has shown itself more concerned with system parameters and agreement upon performance characteristics – the important precursor to finer detail.

10.7 Practical equipment

To give perspective to the descriptions above, the following text describes and illustrates some practical aspects of equipment in use today. In later sections, examples of the latest developments will be given.

Although the scene is rapidly changing, small antennae predominate. These are commonly about 3.5 m long, 0.5 m high and 0.75 m deep. They are of light weight and can be mounted on existing primary radar reflectors with little difficulty, or operated independently. As a result of their physical size such antennae have a horizontal beamwidth of some four to five degrees and a gain of about 20 dB (100 times), but their small vertical aperture leads to a very wide beam in the elevation plane of 45 to 60 degrees. Many of these types of antennae are made to be about 7 m long, thus halving their horizontal beamwidth and doubling their gain to about 23 dB (200 times). The choice is dependent upon the operational need.

Connection to the interrogator/responser is made by coaxial rotating joints (see part 1), integrated with that of the primary radar if the antennae are co-mounted. The interrogator output pulses are of modest power, not usually greater than 1 kW peak, although most units can develop 2 kW peak output. Mean power demands are extremely low compared to primary radar transmitters, being of the order of only a few watts. This makes for much greater reliability because of the reduced electric stress in the equipment.

Replies to interrogations are received by the interrogator/responser and passed to the decoder. In a manual decoding system the operator's controls determine which of the decoded output signals are to be displayed. There are many forms of control unit. Some manufacturers have been clever enough, using miniature components, to put the decoder itself into the operator's position; however, most systems use a central decoder to serve up to six display positions.

In automatic decoding and plot extraction systems, the extractor is usually put close to the decoder. In many modern equipments they are totally integrated to ease the technical problem of interchanging their fast signals which are at low voltage levels. As illustrated in fig. 10.8, the distance between the extractor and display equipment can be very great. They

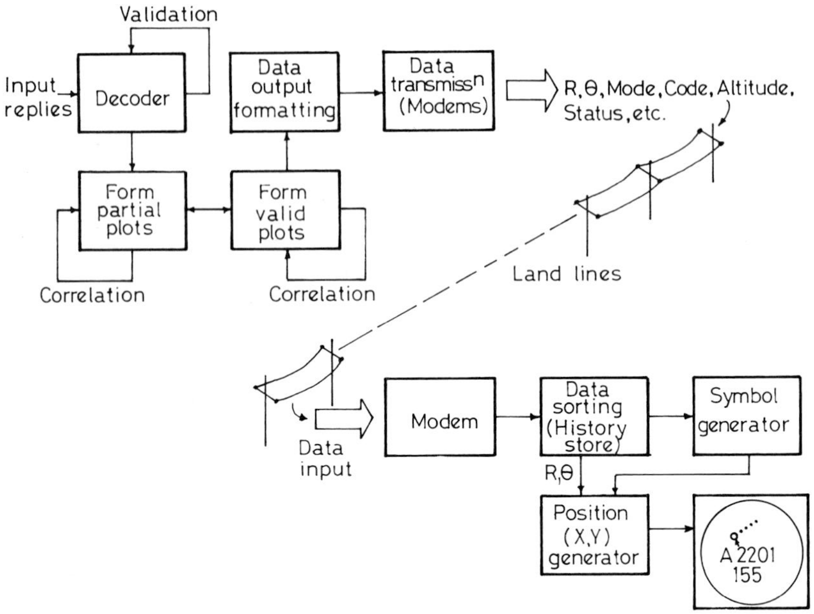

Fig. 10.8. Representation of an automatic decoding and plot extraction system. Important functions which contribute to the transmitted data integrity are indicated.

communicate in the same manner as described in part 2, chapter 9, by use of modems.

Data are selected for display by use of a keyboard and a marker symbol whose position on the display is under the operator's control. The control device is commonly a 'rolling ball', the marker moving about the screen in sympathy with the ball rotation. A typical example of its use would be where the operator wished to remove alphanumeric data on all targets except one. He would first use the appropriate keys (perhaps 'delete labels' and 'execute'); then he would move the marker to coincide with the position symbol of the target of interest and press the 'display label' key. The available data on this target would then appear near its position symbol and a small line would be written, to associate the two. The operator can then repeat this process as he desired.

As in the case of primary radar plot extractors the data can be played out of the digital stores at a much faster rate than the antenna rotation rate. This permits the displayed data to be refreshed many times per second, leading to its legibility in very high ambient lighting conditions.

10.8 SSR range calculation

There are significant differences between primary and secondary radar systems in regard to range performance. In primary radar the magnitude of the returned signal depends upon the size and character of the reflecting surface of the target. In SSR this dependency disappears; it is replaced by the characteristics of the transponder which can be likened to an amplifier carried by the target. Since SSR transponder characteristics are governed by internationally agreed specifications which set minimum input signal levels required to produce a reply and a minimum power output, all targets behave much the same to the interrogator and responser – the 'tiny tripper' looks just like the jumbo jet.

Equally the minimum sensitivity of the responser's receiver is specified, and users are asked to radiate the lowest practical value of 'effective radiated power' (ERP). ERP is the product of the antenna gain and the power across its terminals; if an antenna had gain of 20 dB (100 times) and 100 W (20 dBW) applied to it, the ERP would be 20 + 20 = 40 dBW. This is engineer's shorthand for saying 'the effective radiated power would be 10 000 times one watt, i.e. +40 dB relative to 1 W or 10 kW'. An important point to note is that an ERP of 40 dBW could equally be provided by an antenna of 30 dB gain with only 10 W (10 dBW) as its input. The I.C.A.O. specification at one time limited the interrogator's ERP to 52.5 dBW. Although no longer a limit, and because it is traditionally associated with the provision of 200 nautical miles' range, this ERP value will be used hereafter as a point of reference.

The introduction of a transponder into the system breaks the range considerations into two distinct parts. These are commonly called the 'up link' and 'down link' ranges. The 'up link' range is that to which the transponder can go before it ceases to reply to 90% of the interrogations. The 'down link' range is that to which the transponder can be taken before its reply pulses cease to be detected at useful levels.

In passing it is interesting to note that the SSR system's overall performance is usually couched in terms of 'data integrity' rather than 'probability of detection'. The phrase 'round-trip reliability' (rtr) will often be encountered. This attempts to express the probability that a useable reply will be detected as a result of issuing an interrogation; thus a required rtr of 81% would mean a 90% probability of a reply resulting from any interrogation and a 90% probability of detecting useable replies ($0.9 \times 0.9 = 0.81$). Many specifications call for an rtr of 90%. With the I.C.A.O. specification set at the level of 90% reply probability, this implies 100% probability of reply detection – an impossibly high level since it would mean absolutely perfect performance on the part of the interrogator's receiver and decoder. However, because the transponder issues very high power output (in the radar sense), very high levels of detectability are achievable from long but practical interrogation ranges.

The SSR performance in terms of maximum range is seen to be the lesser of two values, the up link and the down link ranges.

10.8.1 The up link range

The transponder requires a minimum signal power level at its input before it can reply, call this S_t. It will be that power intercepted by the transponder's antenna, of gain G_t. Using the same derivation of power density at the transponder as was used in the primary radar case (see part 2, chapter 7), we see that

$$P_{den} = \frac{P_i\ G_i}{4\pi R^2}\ \text{W/m}^2 \qquad (10.1)$$

where P_i = peak interrogator power at its antenna
G_i = interrogator antenna gain

The transponder antenna is immersed in this density and the power gathered (P_t) will be determined by the antenna's equivalent area, A_t.

$$\text{Thus } P_t = P_{(den)}.\ A_t \qquad (10.2)$$

We have already seen (part 2, chapter 7) that antenna area and its gain are related as follows:

$$G = \frac{Af4\pi}{\lambda^2}$$

Thus assuming a totally efficient transponder antenna of gain G_t, ($f = 1$) we have:

$$A_t = \frac{G_t\lambda^2}{4\pi} \qquad (10.4)$$

Combining (10.1), (10.2) and (10.4) leads to:

$$P_t = \frac{P_i\ G_i}{4\pi R^2} \times \frac{G_t\lambda^2{}_i}{4\pi} \qquad (10.5)$$

where λi = interrogation wavelength.

$$\text{Thus } P_t = \frac{P_i\ G_i\ G_t\ \lambda^2{}_i}{(4\pi)^2\ R^2} \qquad (10.6)$$

The received power must be at a certain minimum specified level to produce transponder replies. This has been defined as S_t (W). So this sets a maximum value to the range, beyond which S_t falls to too low a level. Calling this range $R_{i(max)}$, we can rewrite equation (10.6) as:

$$S_t = \frac{P_i\ G_i\ G_t\ \lambda^2{}_i}{(4\pi)^2 R^2{}_{i(max)}} \qquad (10.7)$$

Re-arranging to express maximum achievable interrogation range gives:

$$R^2{}_{i(max)} = \frac{P_i\ G_i\ G_t\ \lambda^2{}_i}{(4\pi)^2\ S_t} \qquad (10.8)$$

Practical values

Equation (10.8) is a theoretical form of the up link situation. In practice there are a number of loss factors to be accounted for. We have for instance written the interrogator output in terms of its ERP. In most practical cases the interrogator unit's output is attenuated before reaching the antenna by cables, connectors, measuring instruments and rotating joints. This attenuation represents loss. Another similar accountable loss is that between the transponder unit and its antenna. Yet another is the atmospheric loss in carrying the interrogation to the transponder. This is a function of range, wavelength and elevation angle of the up link path. Call these losses respectively L_i, L_t and L_a. Incorporating them into equation (10.8) gives a practical version as follows:

$$R_i^2\,(\max) = \frac{P_i\, G_i\, G_t\, \lambda^2{}_i}{(4\pi)^2\, S_t L_i L_t L_a} \tag{10.9}$$

10.8.2 The down link range

The down link range equation is the exact reverse of the up link case and so we can write the maximum achievable range from which useable replies can be received as:

$$R^2{}_{r(\max)} = \frac{P_t\, G_t\, G_i \lambda^2{}_r}{(4\pi)^2\, S_r\, L_i\, L_t\, L_a} \tag{10.10}$$

where P_t = peak transponder output power in watts
G_t = transponder antenna gain
G_i = interrogator antenna gain in receive mode
λ_r = wavelength of transponder output
S_r = minimum useable reply signal power at the responser antenna
L_i = losses between responser and interrogator antenna in receive mode
L_t = losses between transponder and its antenna in transmit mode
L_a = atmospheric attenuation

10.8.3 The case of I.C.A.O. specified systems

SSR is well regulated by the I.C.A.O. Annex 10 specification which determines performance of both the interrogator and transponder elements.

Foremost among the parameters is the interrogator ERP which, if 52.5 dBW implies that an antenna whose gain is 23 dB (200 times) must have 52.5–23 dBW applied to it, i.e. 29.5 dBW. This equals $100 \times 8.913 = 891.3$ W (from $20 + 9.5$ dBW). The ERP is given by compounding some of the factors in equation (10.9) as follows:

$$\text{ERP} = \frac{P_i\,G_i}{L_i} \tag{10.11}$$

Thus it is possible to write

$$R^2_{i(max)} = \frac{\text{ERP} \times G_t\lambda^2_i}{(4\pi)^2 S_t L_t L_a} \tag{10.12}$$

Signal level required at the transponder to stimulate a reply, S_t

The I.C.A.O. specification requires that both pulses in the interrogation pair received at the transponder antenna are '. . . nominally 71 dB below 1 milliwatt with limits between 69 dB and 77 dB below 1 milliwatt'. In power terms these represent:

$$\begin{aligned} -71\ \text{dBmW} &= -101\ \text{dBW} = 0.8 \times 10^{-10}\ \text{W} \\ -69\ \text{dBmW} &= -99\ \text{dBW} = 1.26 \times 10^{-10}\ \text{W} \\ -77\ \text{dBmW} &= -107\ \text{dBW} = 0.2 \times 10^{-10}\ \text{W} \end{aligned}$$

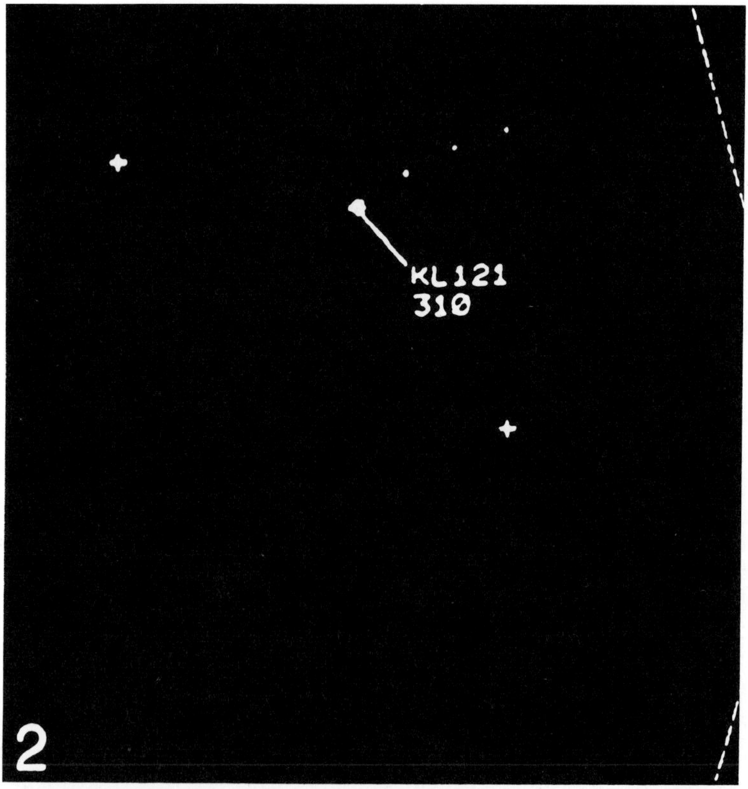

Fig. 10.9. Automatically decoded and plot extracted data on an aircraft. Flight number (KL121) and altitude (31 000 ft) are derived from SSR reply codes. Dots indicate past plot position. (*Photo courtesy of Marconi Radar*)

Transponder antenna gain, G_t

The radiation pattern of the transponder antenna according to the I.C.A.O. specification, is to be '. . . essentially omni-directional in the horizontal plane'. In aircraft, the antenna is usually mounted on the craft's underside so that it is not obscured by the airframe when flying. Sometimes two antennae are used, particularly in military aircraft. They are placed above and below the airframe to allow the interrogator's access during steep banking manoeuvres. The transponder is switched to each in turn for periods of about 25 ms (i.e. a 40 Hz switching rate). This can have unfortunate effects on performance, as will be seen later.

In some civil aircraft the dual antenna arrangement is different, each antenna being served by a separate transponder. Decision circuits select whichever channel has the higher signal level, thus avoiding the disadvantages of antenna time sharing.

Although the antennae have measurable gain of modest proportion, typically 3–6 dB, their placement on the airframe results in the gain varying about these average values at different aspect angles. For this reason it is common in calculations to assume the average gain is zero. The variation in gain is the result of energy from direct and reflected paths combining simultaneously, producing a 'scalloping' effect to the radiation pattern by the same mechanism as described in part 1, chapter 3. The reflecting surfaces in this case are various parts of the airframe such as engine nacelles, leading and trailing wing edges, etc.

Up link losses, L_i, L_t and L_a

The first term (L_i) is covered by its integration into the ERP; it is a loss which is overcome either by increasing the interrogator unit's output or the provision of more antenna gain.

The transponder loss factor, L_t, is dependent upon the type and quality of installation in the vehicle. In aircraft it is typically 3 dB, resulting in reduction of the signal input to the transponder by a half.

The atmospheric loss, L_a, is significant at ranges in excess of 60 nautical miles, growing to values of about 1 dB for very long range systems. Actual values may be found from tables (see reference 6).

10.8.4 Practical cases

The graphs in fig. 10.10 have been drawn to show the maximum interrogation ranges achievable by an interrogator producing half the reference ERP of 52.5 dBW. The starting point has been calculated to show the signal power available at 1 nautical mile by substituting values into equation (10.7) assuming the transponder antenna gain is unity, but including L_i and L_t, each of 3 dB. The rate of power fall-off on given straight lines from the power source is an inverse square law, so S_t is proportional to $1/R^2$. In decibel notation this represents a decrement of 6 dB for each doubling of the range. Thus at 2 nautical miles the value of S_t has fallen by 6 dB relative to that at 1 nautical mile and at 4 nautical miles, by another 6 dB and so on. This decrease can be continued until S_t has fallen to the level required to trigger the transponder. This is represented by three values, corresponding to the minimum, maximum and nominal levels quoted in the I.C.A.O. specification. It is interesting to note that the difference in maximum interrogation range between minimum and maximum performing transponders is 8 dB, representing a range ratio of $8/2 = 4$ dB; 2.5 to 1. In practical terms this means an interrogator reaching a maximum sensitivity transponder at 250 nautical miles will only see a minimum sensitivity transponder to 100 nautical miles at the same elevation angle.

It should be borne in mind that the curves in fig. 10.10 do not include any atmospheric loss which in effect will make the true decremental line dip away from the straight line illustrated. As stated above, a value of 1 dB at 200 nautical miles is a practical value. It will reduce the range by a factor of 0.89 so 200 nautical miles become 178 nautical miles: a significant reduction.

The same technique to express the down link case has been applied to produce fig. 10.11. The interrogator antenna gain for receiving is taken to be 26 dB, with losses as for the up link. A few interesting points of note need to be made.

Transponder output

The I.C.A.O. specification permits the peak power applied to the transponder's antenna to be 27 dBW or 500 W, but it must not fall below a quarter of this (i.e. 6 dB less than 27 dBW). General aviation aircraft which fly at low altitudes are permitted a lower output peak power limit of 70 W (18.5 dBW). Once again we see that a similar ratio of range performance is permitted, being the equivalent of 6 dB or 2:1 in range. Fortunately, the usual situation is one wherein the up link range is the limiting case in system performance and thus successful interrogations generally result in successful reply detections.

Responser input signal level

As in the primary radar case, reply signals have the receiver's noise added to them before amplification and detection. So we have the same need to define probability of detection. But in the case of SSR the target characteristics are very different from those in primary radar. The mechanisms which produce target fluctuations of various types in primary radar are altogether different in SSR. We have seen that the 'equivalent echoing area' of a primary radar target changes rapidly from one antenna sweep to the next or from one

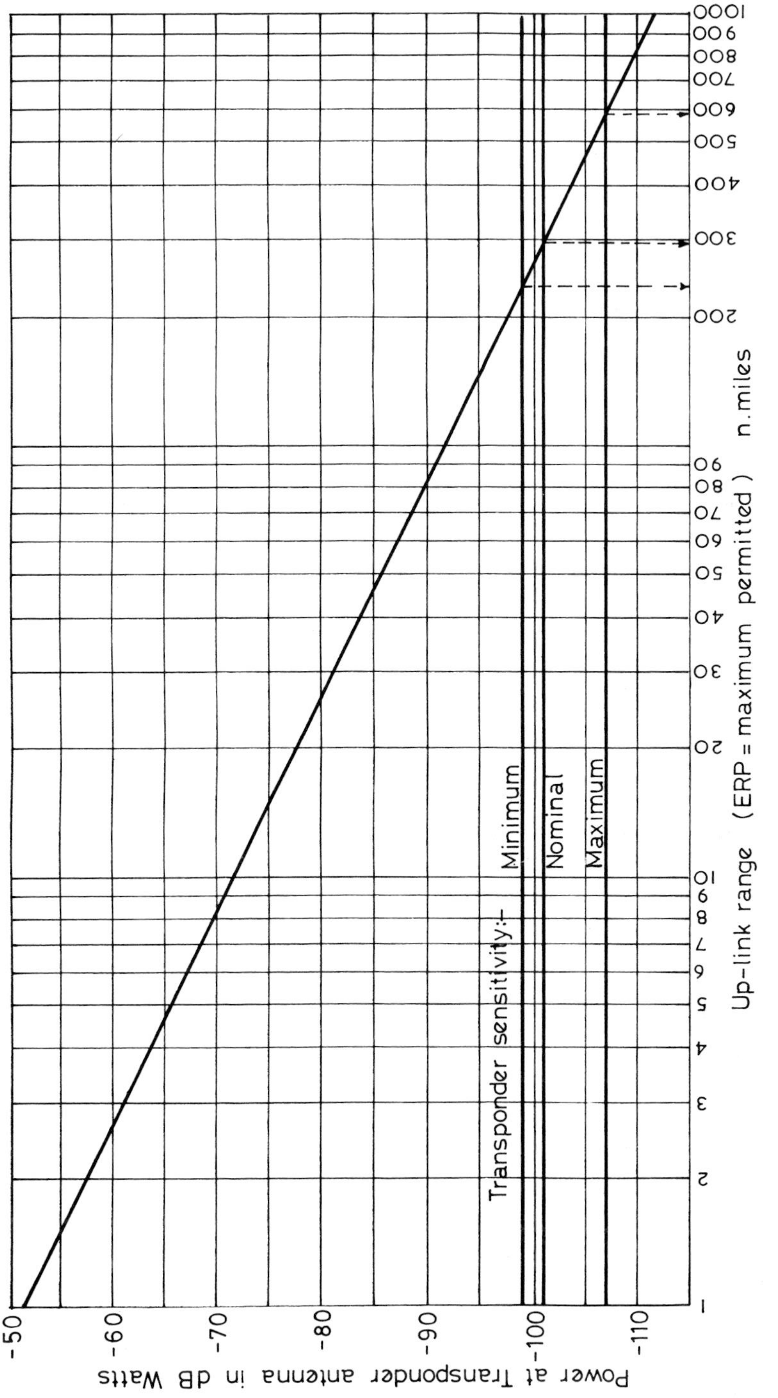

Fig. 10.10. Nomograph showing peak interrogation range limits for an ERP of 48½ dBW (half the permitted maximum). Note the effect of transponder sensitivity.

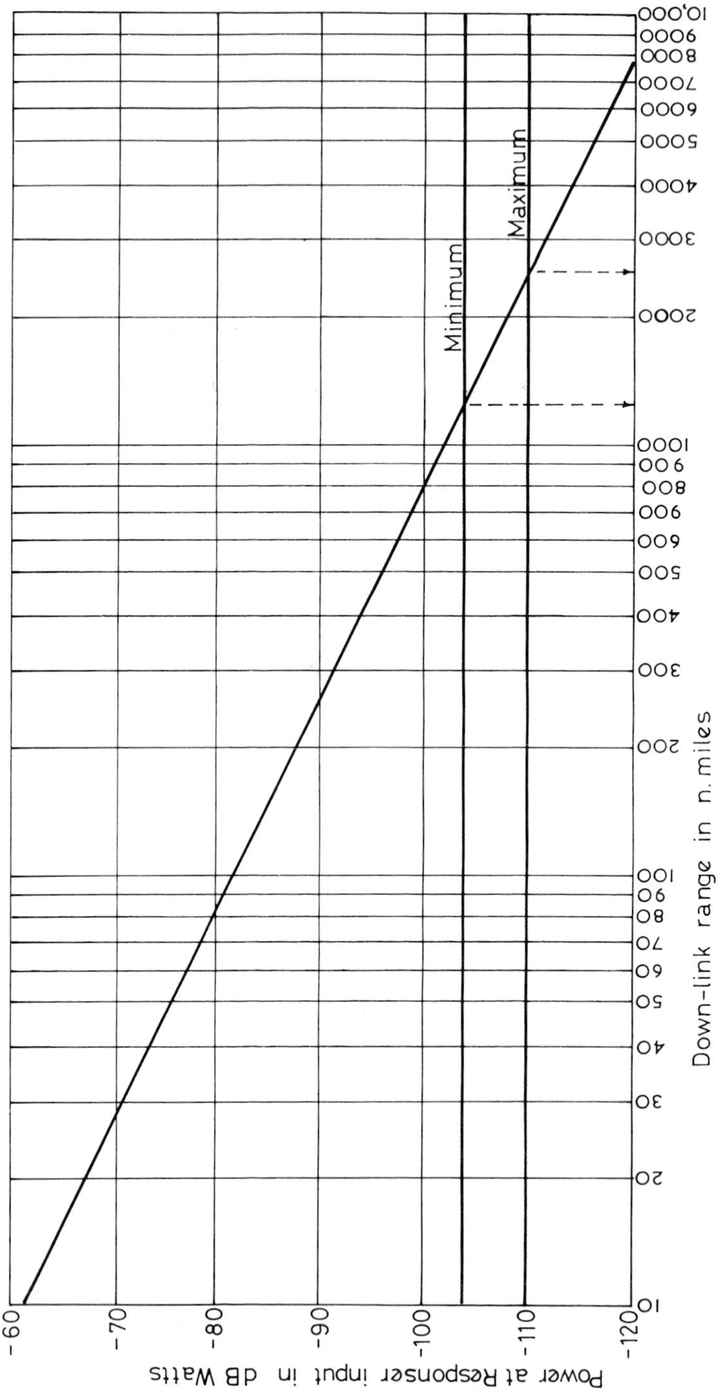

Fig. 10.11. Nomograph showing reply range when the ground antenna has a gain of 20 dB and subsequent losses are 3 dB. Responser sensitivity = 85 dBmW, signal to noise ratio = 13 dB. Maximum range is for for transponder output of 500 W, minimum is 6 dB less. System range is usually set by a much smaller up-link range. Thus at this value, signal to noise ratio at the responser increases.

transmision to the next. Such rapid changes are rare in SSR. The rate of change of the transponder antenna's gain with aspect is very slow, hence it remains fairly constant for successive interrogator antenna revolutions. Following this theory, SSR targets are generally considered to be 'non-fluctuating'.

An average reply would consist of two framing pulses and six information pulses. The system is designed to use the high degree of correlation between successive replies on a given mode. This allows the 'Single pulse' false alarm probability to be set higher than usual in primary radar.

For this reason a figure of 10^{-4} probability of false alarm for a single pulse is usually taken. To achieve a detection rate of 90% success for a code train of eight pulses requires that the detection probability for each pulse is higher. This notion can be expressed as:

$$(P_d)^n = P_c \tag{10.13}$$

where P_d = probability of detection for a single pulse
n = number of pulses in the train
P_c = probability of detecting all of the n pulses.

From (10.13) we have:

$$P_d = \sqrt[n]{P_c} \tag{10.14}$$

and for the values assumed above:

$$P_d = \sqrt[8]{0.9} = 98.7\%$$

L.V. Blake (reference 11) gives in fig. 4 of his work on radar maximum range calculation a value of 13.2 dB signal to noise ratio for a P_d of 99% at 10^{-4} probability of false alarm.

System performance

The range calculations addressed above are maximum values. They are not useable in practice because philosophically they represent the range achievable at the interrogator antenna's gain peak. In a system with a rotating antenna this point would have too short a dwell time on target to stimulate and receive a reply. As does primary radar, SSR relies upon receiving repeated signals from a given target; thus we must seek points on the antenna's beam shape between which the gain is equal to or greater than that producing the required signal level at the transponder. The points either side of the beam's peak, representing a beamwidth, will take a calculable time to pass across the target. This time must be at least that required to generate and receive the necessary number of repetitions of replies.

Suppose the calculated beamwidth necessary to generate the required replies was 2°. The practical (as opposed to peak) interrogation range is obtainable by reference to the interrogator antenna's horizontal polar diagram. The gain reduction from peak to the 2° beamwidth will be found from this. Suppose it was 2 dB. From the up link equation (10.8) we see that the R_i^2 max value would have to reduce by 2/2 dB = 1 dB or 0.79. Similar calculations can be made for the down link case.

11
Problems and solutions

11.1 Introduction

The reader could be forgiven for believing from the foregoing that SSR is the ideal system for control of traffic – of all kinds. Because it is a radar with a communications capability it gives target position and identity data and its code capability can be extended to give other important information for control purposes. In the case of aircraft, altitude data are available. There is another mode as yet unused in the system and an extra code bit (the thirteenth '*X*' pulse) available to double the system reporting capability. These could both be used to signal, shall we say, a vehicle's heading and speed, its intentions, etc. Using a decoder of mode sequences, it would be possible to stimulate other reply data which are slow to change but nevertheless important to the control organisation – sea state in vessel control springs to mind.

Alas, no system is ideal. SSR is no exception; it is beset by problems both technical and operational but, as with all problems, they have solutions. This chapter is devoted to describing what the problems are and how they are solved, for both have direct impact upon the user. The application of the solutions has proved so successful that SSR is being used more and more by air traffic authorities, particularly for long range control purposes in preference to primary radars. Calculating costs on a 'pound per bit per second' basis gives SSR a tremendous advantage.

A dual channel primary radar plus a dual channel height-finding radar could currently cost about £5 million pounds. A modern SSR, with no clutter problems to overcome and with much more accurate position and height data would cost about £1.5 million pounds. Although these figures are much dependent upon specific requirements they nevertheless represent a typical ratio of more than 3 to 1 in favour of SSR.

The communications capability is assuming great importance and the future extensions planned for SSR focus attention upon realising its potential as a data link. This topic is separately addressed in chapter 12.

11.2 'Fruit' and 'defruiting'

11.2.1 'Fruit'

SSR uses two set frequencies: 1030 MHz for the interrogations on the up link and 1090 MHz for transponder replies on the down link. As presently organised the interrogations are broadcast by the antenna beam's continual rotation and issue of repeated modes. Imagine that you were with your friends Joe and Nikki (I'm not too sure about Nikki!) at a large party and the lights went out. If you acted like an SSR in locating your friends you would ask repeatedly (probably shout) 'everybody, tell me where you are and what your name is'. When you get around to hearing 'Joe and I'm at the bar' and 'Nikki and I've found another friend here at the buffet table' you would have achieved your aim. But at what cost? Everyone who was co-operative would have answered you, many of them at the same time. Furthermore there would be others who wished to locate their friends and all would join you in a babble of confused question and answer. The confusion results from asking all the people for information all the time, resulting in your getting lots of information you didn't really want.

The same is true in SSR. Groups of interrogators in a service area will all be continually seeking data on targets of interest to them and because the transponders have, perforce, omni-directional capability, replies made to one interrogator can be and are received by others who didn't call for reply. The unwanted replies are called 'fruit' to distinguish their special nature from cross-channel interference experienced in primary radar. It is special because fruit is always at the same frequency (within the I.C.A.O. tolerance of ± 3 MHz), and within each interrogator's receiver pass-band. It was quickly realised that one way out of the confusion caused by fruit was to give each interrogator its own signature by way of a specific repetition rate different from any of its neighbours likely to be a fruit generator. By this means, every station's decoder and display system would have its own sample rate. Any replies not sharing this would become decorrelated in time (and hence range) and would not appear in a given interrogator system grouped at a constant range within an interrogation beamwidth. A good example of the effects together with the nature of a beamwidth's reply history is seen in fig. 11.1. The picture is of an expanded ppi display showing the envelope detected output of a responser with no processing. Two groups of reply trains are prominent. They are at a range of about 120 nautical miles from the interrogator. The display centre is well to the left of the page.

The individual code pulses in the repeated reply trains can be seen. The first and last of these of course are the always present F_1/F_2 bracket pulses. The beamwidth contains twenty replies. The broad lines are indicating air lanes (red one airway in this case). They are 10 nautical miles apart. The other signals are fruit. Their structure in range clearly indicates that they are

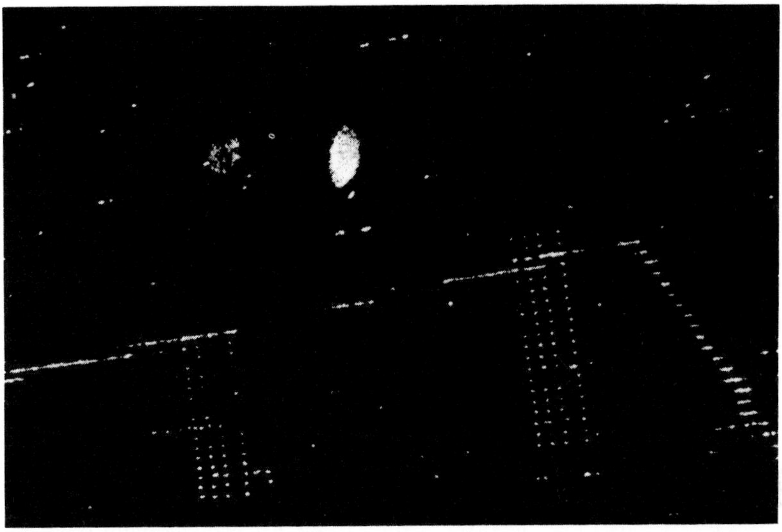

Fig. 11.1. Unprocessed SSR replies from two aircraft at about 180 nautical miles range. Bracket and interspersed code pulses can be clearly seen. The target on the right has made 20 replies. Fruit replies (uncorrelating in range) are evident. Broad lines drawn by electronic mapping represent an airway, 10 nautical miles wide. (*Photo courtesy of Marconi Radar.*)

SSR replies. Their decorrelation in range can also be seen. Because transponders have such powerful output, fruit replies can be received from great range via even the lowest of sidelobes of the interrogator's antenna, thus their nuisance value is exaggerated. Why is it called fruit? In early systems not fitted with 'defruiters' or means to restrict replies to interrogator's main beams, the unwanted fruit inputs would be displayed as long crescent shaped arcs emanating from the display centre. They looked like rudimentary bunches of bananas, hence the name 'fruit'.

Why, apart from the sheer nuisance value created by the look of the display, is fruit a problem? Quite simply because it corrupts wanted data. The corruption takes numerous forms. In fig. 11.2 twelve replies out of a beamwidth's history are shown. Six are those which the interrogator has asked for and carry their bracket pulses labelled F_1, F_2. They have three pulses in their coded reply. The other six, labelled f_1/f_2 are fruit replies received at a different interrogation rate caused by another interrogator. The wanted replies are at a fixed range from the 'home' interrogator, those from the 'foreign' station will appear to shift in range because of the difference between the two stations' repetition rates. I have deliberately chosen the difference to be the equivalent of one reply time per repetition period. At repetition period number one, all seems well because the two replies are well separated. At repetition period number two an unfortunate circumstance occurs; the spacing between F_2 and f_1 becomes equal to that for a standard

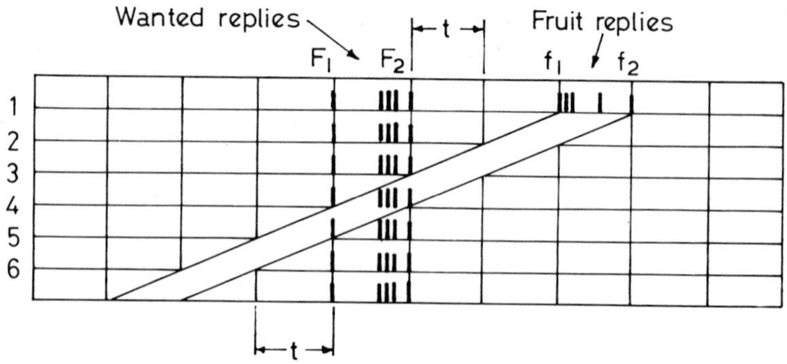

Fig. 11.2. Fruit replies processing through a series of 6 wanted replies. When the time of arrival of fruit is within t of wanted replies, corruption and/or ambiguity results. 1 to 6 represents successive interrogation periods.

reply, i.e. 20.3 μs. The decoder, which looks for such spacing, will now see three examples, only one of which it requires. At repetition period number three F_2 and f_1 become coincident and there may well be other instances of code pulses in the two contiguous code trains that are 20.3 μs apart. At the next repetition, the two code trains are overlapped. This can produce total confusion about the code content. Periods five and six are replicas of previous circumstances.

The fruit replies have been shown conveniently (or rather, inconveniently) separated from their wanted neighbours. Obviously these circumstances are dependent upon chance because there is no designed correlation between the wanted and fruit target ranges or the home and foreign interrogation times. But they can and do occur, not always corrupting to the degree illustrated, but corrupting nevertheless.

11.2.2 Defruiting

The method of overcoming fruit is almost obvious from the above description of why it occurs. Wanted replies appear in the 'home' station's system at a time directly proportional to the range of the target. Unwanted fruit does not. All that needs to be done is to store all the responser output pulses from one interrogation period to the next and look for range correlation. If found, the signals can be rightfully passed to the decoder for further processing; if not then they can be rejected as fruit.

Early (but still used) defruiters do just this. Finely structured digital stores having 0.5 μs storage cells are addressed by a 'clock' in synchronism with the interrogator. A store is provided for each mode of interrogation so that comparisons of code content in the range dimension for repetitions of a given mode can be made.

Various criteria of correlation can be preset so that the system will for instance output replies only when any two out of a consecutive four trials for each cell location have been successfully made. The simplest way of providing such a system is to use the stored data as an input for an acceptance gate. The input arriving from the next interrogation period will then be allowed through only if it matches the gate input condition in time and it becomes both the decoder input and the next gate input signal. It is thus obvious that in such a system, valid data are denied to the decoder, for if on the first two trials (interrogation periods) the same code data appear at the acceptance gate (one from store and the other 'current' signals) only the second current data bits enter the decoder. The first set is lost. If the criterion is set at a level which requires any three out of four consecutive trials, the same reasoning leads to loss of the first two correlating replies (it needs to wait until the first two repetitions are in store).

In general we can say then that a defruiter which has criteria requiring any 'n' out of 'm' consecutive replies then the decoder loses $n-1$ inputs. This can be very significant. Suppose an SSR had a system beamwidth which gave twelve replies and it was a triple mode interlace. This represents a requirement for four replies per mode. If these were first put to a defruiter with a criterion of 'any three out of four consecutive trials on a given mode', from the above we see that $3-1$ wanted reply pulses are lost. So the wanted data into the decoder are halved and only two examples of the wanted data on a given mode are gathered. This is the bare minimum required for any kind of certainty that the data are valid. Such conditions would pertain even if no fruit was present.

The above description relates to what one might call real-time defruiters and is the first processing of the SSR data after detection. We have seen in part 2 how primary radar target plot extractors work by seeking correlation of signals in time and hence range. Most modern SSR systems incorporating plot extractors work in this way and, since their operation is based in range correlation, the action of defruiting is automatically a by-product of SSR plot extractors. It is very rare to find an SSR plot extractor system preceded by a separate defruiter because of the potential data loss incurred.

The calculation of 'fruit densities' which a system is required to combat is a very difficult matter. So difficult that most specifications echo the highest value that their writers have seen in previous attempts. Commonly a figure of '. . . fruit at the rate of 20 000 per second' is used to describe the environment. No distribution figures are quoted nor any statement about reply content. I commend future users to ask their advisers to explain their thinking – because if it is lax they could be over-specifying your requirement.

11.3 Sidelobe interrogations and their suppression

The phenomenon of sidelobe interrogations and ways to suppress them were among the major system factors studied, discussed, politicised and resolved by I.C.A.O. members over a very long period from about 1950 to the mid 1960s. At the end of this time a choice between two suppression methods was made. Only this long-fought and hard-won method (the 'three pulse system') will be described below. Those wishing to understand both methods are referred to a paper by the author published in 1961 (reference 12).

11.3.1 Sidelobe interrogation

The polar diagrams of an interrogator system can be looked upon as lines showing equal and constant signal strengths in space. They represent the bounds within which interrogations and replies can be expected to achieve a stated probability of detection. Consider the case of an interrogator whose horizontal polar diagram is as shown in fig. 11.3. It represents the antenna up link situation plotted as gain versus azimuth for a full 360°. Taking the above definition of this diagram it is possible to assign a range at the peak of the pattern to which interrogation can be made at the required probability level (usually 90%). If a transponder of maximum sensitivity were to be under the influence of this interrogator, we see from fig. 10.10 and the basis of its calculation, that the maximum range obtained is 570 nautical miles. Since the ERP in the case cited is only half the reference value of 52.5 dBW, this range can extend to a further value of $570 \times \sqrt{2} = 806$ nautical miles. The $\sqrt{2}$ factor follows from the relationship in equation (10.8).

$$R^2 = K\, G_i \qquad (11.1)$$

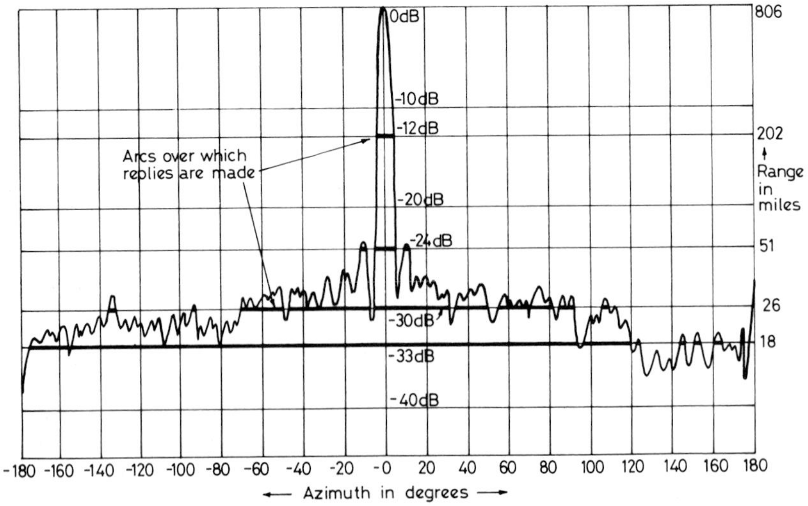

Fig. 11.3. Showing the azimuth extent of replies which would result without interrogation sidelobe suppression (ISLS) as range reduces.

Using this value (806 nautical miles) as the range at the beam's peak, the azimuthal arcs over which interrogations can be made at various reducing ranges are indicated in fig. 11.3. The supposition here is that no attempts are made to restrain interrogation or consequent replies. Although the polar diagram is notional, it represents many practical cases in use today.

All seems well from maximum range to about 50 nautical miles if we neglect the implications of actual beamwidth in terms of target resolution. At −24 dB, interrogations caused by the first sidelobes are seen. At −30 dB these sidelobe interrogations extend significantly, over almost 180°. At 18 nautical miles (−33 dB) there is hardly an azimuth arc over which no interrogations would result. A number of unfortunate effects would follow from this, were nothing to be done about it. One obvious effect is the massive increase in unwanted replies which represent fruit to neighbouring interrogation stations. If the wanted number of replies was generated in 2° beamwidth then at a range of 26 nautical miles the interrogator would be creating 90 times the required number.

Transponders in aircraft have to occupy as small a space as possible. For a given technology there is a limit set to their power output capability, and its transmitter stages protected from overload. Currently SSR transponders are allowed to have 500 W peak power and output fifteen pulses of 0.45 μs duration. The I.C.A.O. specification, recognising this power limitation, calls for reduction in transponder receiver sensitivity to limit the output rate to 1200 replies per second. Thus if transponders are in an area where they are bombarded with interrogations at this level, their reduced sensitivity denies access to remote interrogators. That is, the I.C.A.O. specification calls for the transponders to 'put cotton wool in their ears'. This is known as the 'capture' effect: the nearby interrogators, raising the interrogation rate above 1200 per second, capture the transponders away from distant stations. The rate of desensitivity is given in I.C.A.O. annex 10 specification (reference 10).

Another more obvious effect, particularly in real-time SSR decoding and display systems is the extension of aircraft responses in azimuth. In the examples shown in fig. 11.3, aircraft at 18 nautical miles would produce responses at this range which would be almost a continual circle upon the display – the so-called 'ring-around' effect. Under this condition azimuth data are entirely lost, or at best can only be guessed at.

What is done to prevent these drawbacks? In system design terms, the I.C.A.O. specification resolves most of the difficulties. It requires interrogations to be made at no more than 450 per second.. It recommends reduction of ERP, but above all it requires all interrogators to have interrogation sidelobe suppression (ISLS) facilities. Note that a distinction is drawn between interrogation (I) and receiver (R) sidelobe suppression, the latter being italicised as RSLS. These and other acronyms have caused great confusion in the past through specification and authorities' lax use of terms. Particularly is this true of another technique (to be described later) called improved interrogation sidelobe suppression (IISLS or I^2SLS). Many people construe the 'I' of ISLS itself to mean 'improved' sidelobe suppression. Sometimes one can be forgiven for believing such confusion is deliberate.

11.3.2 Interrogation sidelobe, suppression (ISLS)

In principle the current ISLS system is very simple. The interrogator uses not one but two radiation patterns and introduces another pulse into its transmissions. This extra pulse is positioned between the interrogation pulse pair (P_1 and P_3) and is called, not unnaturally, P_2, the Control pulse; it always occurs 2 μs after P_1 of any mode. It is radiated by the second of the two antenna patterns referred to above. The action brought about to inhibit interrogation by sidelobes is best understood by reference to fig. 11.4. The directional polar diagram radiates the interrogation pair, P_1 and P_3. Their spacing and amplitude is sensed within the transponder; the spacing defines the mode of interrogation. The other

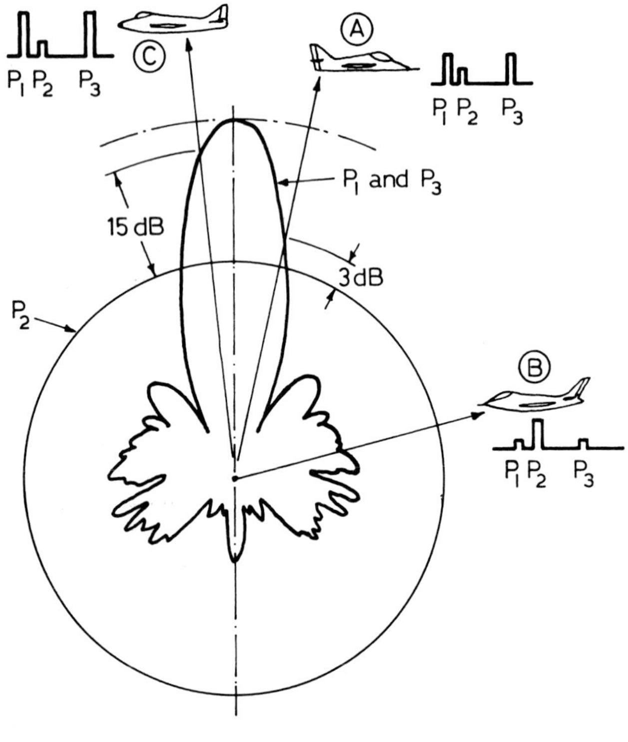

Fig. 11.4. Reply and suppression situations in SSR with ISLS.

pattern, shown as omni-directional, radiates P_2, the control pulse. Its radiation in any given direction is set to exceed that of the interrogator's sidelobe levels. It is usual for all three pulses to be of equal amplitude as fed to the two antenna patterns. Three aircraft A, B and C are indicated. Aircraft A will receive pulses P_1 and P_3 of equal amplitude, it will receive P_2 at a lower level. Circuits within the transponder measure and record the amplitude of P_1 and match it against the P_2 pulse (always 2 μs later) amplitude. The transponder carries out real time logic, again specified by I.C.A.O., as follows:

(a) $P_1 < P_2$ = must Suppress (no reply)

(b) $P_1 \geqslant P_2$
and
(c) $P_3 = P_1 \pm 1$ dB
} = may Reply

(d) $P_1 \geqslant P_2 + 9$ dB
and
(e) $P_3 = P_1 \pm 1$ dB
} = must Reply

In the case of aircraft A, the logic of (b) and (c) applies and the transponder *may* reply. Aircraft B's position results in signals satisfying the logic of condition (a) and its transponder suppresses for a time between 25 and 45 μs after receipt of P_2. Aircraft C is well in the main interrogator beam and logic conditions (d) and (e) apply; therefore its transponder *must* reply. Replies are made within 3 μs of receipt of P_3.

11.3.3 Antenna patterns

The two radiation patterns, one directional and the other omni-directional shown in fig. 11.4 should ideally share a common origin. Early systems, some still in use, produce the control pattern by mounting an omni-directional antenna on a mast near the rotating directional antenna. Later systems have the omni-directional antenna mounted on top of the rotating directional array so that they at least share a common point in azimuth.

By far the most effective system is to produce a control pattern which is not omni-directional but follows the general shape of the directional antenna's sidelobe structure and is generated by the self-same array, producing coincidence of phase centres of two patterns in both azimuth and elevation planes. The necessity for this will become apparent later. This desirable circumstance is the result of using what is called the 'sum and difference' antenna technique. It is readily understood by reference to figs. 11.5 and 11.6.

In fig. 11.5(a) the phase of energy applied across a linear array of aperture n to ℓ is shown to be uniform. The various elements in the array have an amplitude distribution which is weighted to be a maximum at the centre, tapering symmetrically across the aperture (n to ℓ) as shown in fig. 11.5(b). The notion of a linear array and the effect of tapering the power distribution

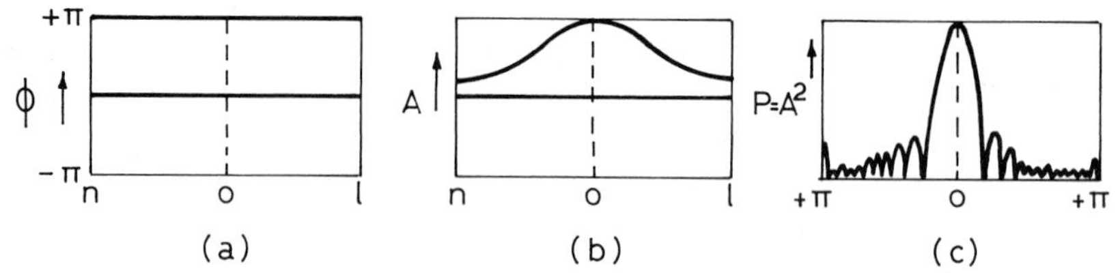

Fig. 11.5. Generation of a sum (Σ) pattern. Phase distribution (a) is uniform. Amplitude (b) is tapered, resulting in radiation pattern (c).

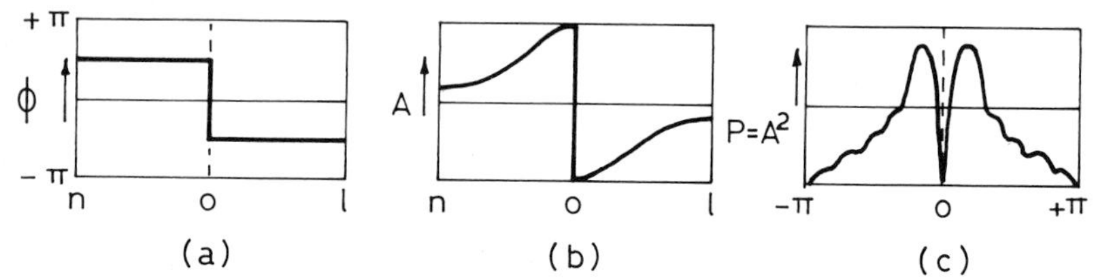

Fig. 11.6. Generation of a difference (Δ) pattern. Phase distribution (a) is uniform with reversal at mid-point. Amplitude (b) is tapered but phase reversal reverses its vectorial sense at mid-point. Radiation pattern (c) has deep null on boresight.

has already been introduced in part 1, chapter 3. The far-field radiation pattern produced by this arrangement is shown in fig. 11.5(c). It has the classical main beam, sidelobes and a back lobe (at $\pm\ \pi$ away from the boresight, 0^0). It has been drawn in terms of power gain and so is proportional to A^2. It is called the 'sum' pattern because it is produced by the vector addition principle already described in part 1, under the section discribing how beams are formed. It usually bears the Greek symbol Σ used in mathematics for a summation process.

In fig. 11.6(a) we see that the phase of energy across the aperture n to ℓ has a 180° difference introduced at its centre point. This is achieved in practice by making the path length of energy fed to the left and right elements in the array differ by half a wavelength. To effect this the power fed to the array is first split into two equal halves, the path length difference is introduced into one half and then each half is distributed to individual elements in the ratio demanded by the desired taper. The result of this, drawn vectorially, is shown in fig. 11.6(b). The phase reversal is clear at boresight. Using the same principles of vector addition illustrated in part 1, chapter 3, the effect of this phase reversal is to produce a deep null on boresight. This can be readily appreciated, for in the far field a point on the line at 90° to the array centre will receive exactly the same amplitude of energy from symmetrically disposed pairs of elements, except that their phase is 180° apart. The resultant is of course zero. The overall far-field pattern is shown in fig. 11.6(c).

Another way of understanding the effect is to consider the array as being made from two antennae (call them a and b); one of aperture n to ℓ and the other of 0 to 1. They are contiguous mirror images of one another and in the same plane. The far field vector combination with both antennae fed with the same phase produces the 'sum' pattern because they are vectorially producing $(+a) + (+b) = a+b$. With both antennae fed 180° out of phase the vector combining in the far field produces the 'difference' (Δ) pattern from $(+a) + (-b) = a-b$.

That both patterns share the same physical phase centre as seen in the far-field is obvious, since they are both generated by the same elements. The patterns of a typical antenna are shown in polar co-ordinate form in fig. 11.7. They are drawn to illustrate the field strengths produced in a practical SSR system when the powers of P_1, P_2 and P_3 are all of equal amplitude into the array. Also shown are the array configurations to produce the sum and difference patterns. To produce the sum pattern the antenna has equal phase across the array for both P_1 and P_3. Arrangements are made by using a very fast acting switch of the PIN diode type to connect the phase shifter (∅) into one arm of the equally dividing power splitter (s). This is done by a switching waveform shown at (a). It produces the difference pattern but only for the time needed to radiate P_2 and then reverts to the sum pattern ready to accept P_3 and subsequent replies. Other methods can be used, but that illustrated has the advantage of requiring only

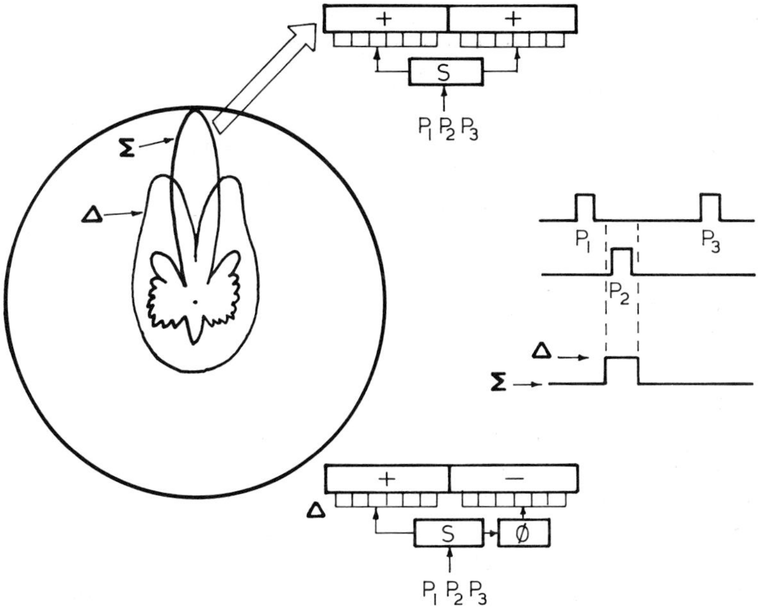

Fig. 11.7. Sum pattern produced when P_1 and P_3 are present since all elements of array are fed in phase. Difference pattern produced by introducing phase shift (∅) during the presence of P_2.

one rotating joint channel. The switching waveform can either be derived from envelope detection of pulses after they emerge from the rotating joint or by passing a switching waveform to the power splitter switch by slip rings which produce rotating contact much more readily.

The Σ and Δ technique produces much more efficient interrogation sidelobe suppression (ISLS) relative to the Σ and omni (Ω) method. Even better is a combination of Δ and Ω patterns such as produced by the antenna shown in fig. 11.8. In this, part of the P_2 power is diverted into an omni antenna mounted centrally above the main array which produces enough radiation to cover the general level of low sidelobes of the sum patern. The difference pattern 'sits on top' of this so that the fall-off of gain of the difference pattern far away from boresight is supplemented by the power radiated by the 'omni' antenna. Such an arrangement produces the patterns shown in fig. 11.9.

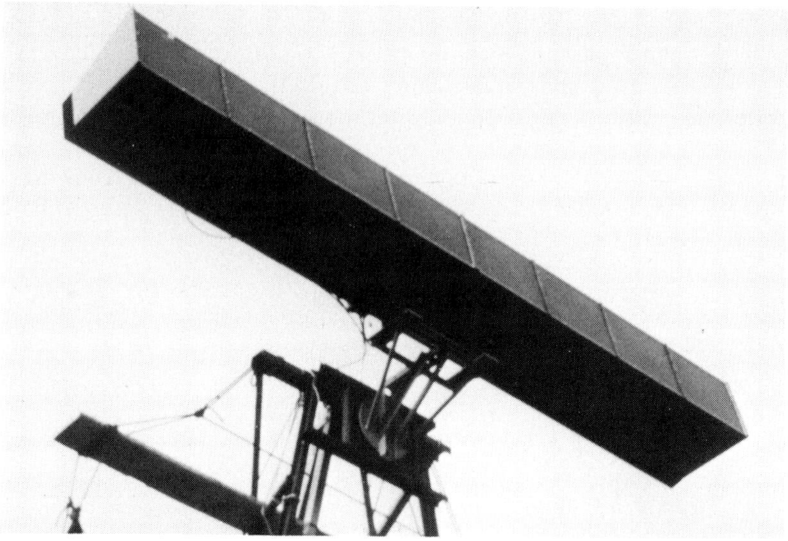

Fig. 11.8. A 14ft (4.3 m) 'hogtrough' antenna mounted on a test tower. A supplementary omni-directional antenna (obscured) is centrally mounted on top of the array. (*Photo courtesy of Radiation Systems Inc.*)

11.3.4 System beamwidth

One of the important improvements of the Σ and Δ system is the effect on interrogation system beamwidth. It is necessary to invoke the concept of 'system' beamwidth because of the action of the logic in transponders governing their 'reply' or 'suppress' conditions.

It will be remembered that the I.C.A.O. specification allows a 9 dB 'grey region' of uncertainty between the imperatives of *must* reply and *must* suppress (see section 11.3.2). Figure 11.10 shows a sum (Σ) and omni (Ω)

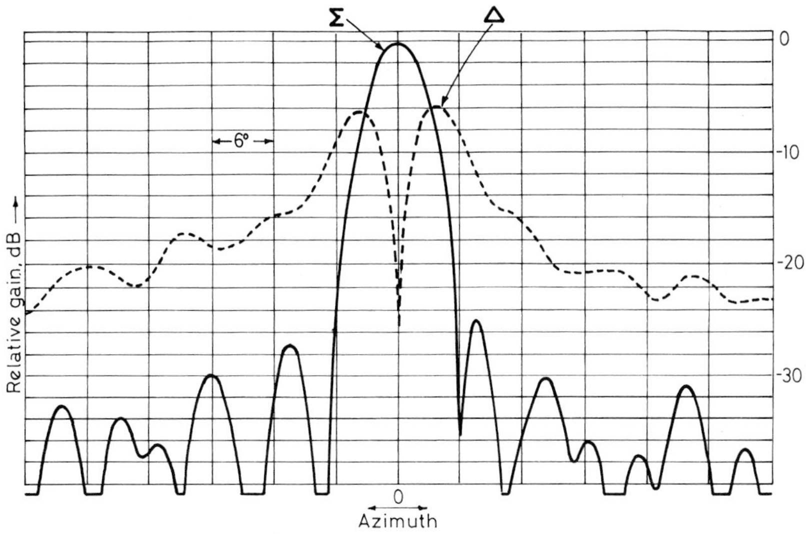

Fig. 11.9. Measured horizontal polar diagrams of the antenna in fig. 11.8.

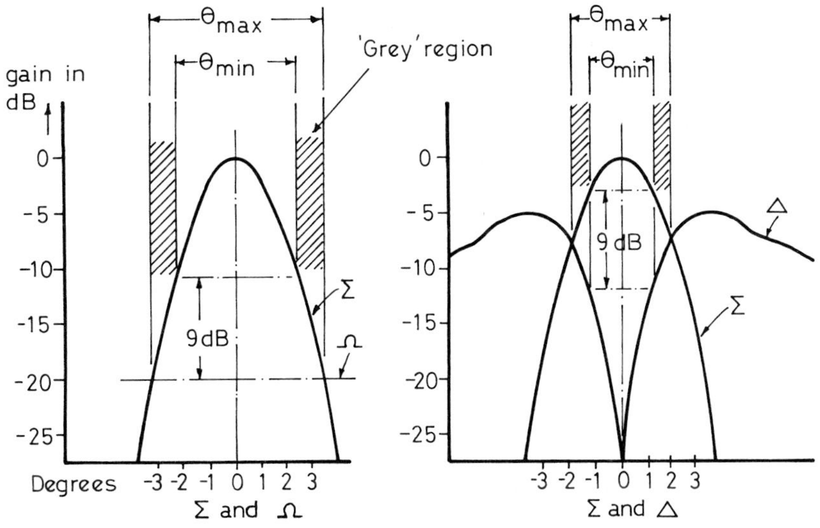

Fig. 11.10. Two ISLS arrangements: that using sum (Σ) and difference (Δ) pattern produces more closely controlled interrogation system beamwidth limits than the sum (Σ) and omni-directional (Ω) arrangement.

pattern pair. The omni-directional pattern used for the control pulse (P_2) is shown at a practical level relative to the interrogation pattern. The logic in the transponder determines that if it receives P_1, P_2 and P_3 it may reply over the arc $\theta_{max.}$ (where $P_1 > P_2$) and it must reply over the arc $\theta_{min.}$ (where P_1 and P_3 are 9 dB or greater in excess of P_2).

Using the same sum pattern and scales, a companion pair of sum (Σ) and difference (Δ) patterns are shown, again in a practical relationship. For this second pair, three significant points are clear:

(a) The radiated power of P_2 is greater in the Σ beam region than is the case for the Σ and Ω pair. This means the sidelobe suppression logic can be exercised to much greater range.
(b) The $\theta_{max.}$ beamwidth is narrower, permitting a better azimuth resolution.
(c) The difference between $\theta_{max.}$ and $\theta_{min.}$ is much smaller because the rates of change of gain with azimuth for the Σ and Δ patterns are divergent around their cross-over points.

The transponder logic can be seen to act as a controller of the number of replies per beamwidth in a system with fixed antenna rotation and interrogation rates. In the Σ and Δ system the number of replies is stabilised within much closer azimuth limits than for the Σ and Ω pair. This is of great value in decoder and plot extractor design. Practical measurements on a wide sample of operating transponders confirm this effect. A typical example is shown in fig. 11.11.

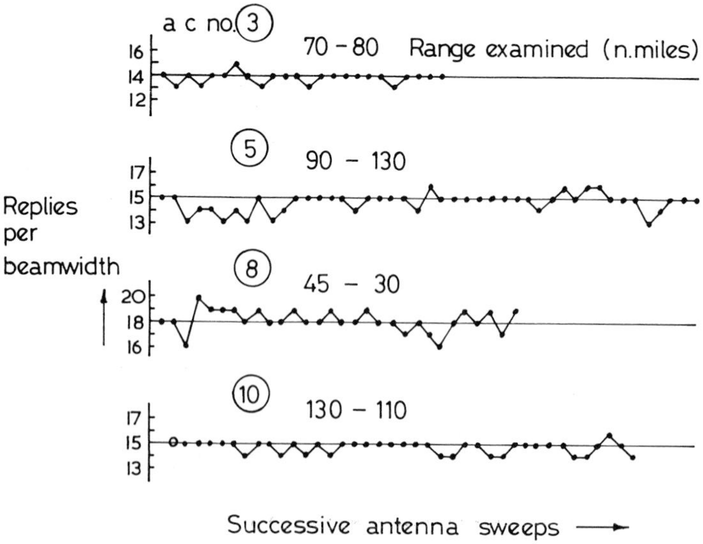

Fig. 11.11. Four examples from a large group of aircraft whose replies per beamwidth in a sum and difference ISLS system were examined. They are typical in exhibiting small variations about a well-defined mean.

11.3.5 Receiver sidelobe suppression

In the previous section the 'system beamwidth' concept applied to the up link regime. There is another concerned with the down link. It is possible to use the ISLS control pattern, not only to govern the arc over which replies are made, but to use it to control the arc over which they can be received. In this, a separate receiver takes in replies gathered by the control pattern. The relative beam shapes give the replies received in the control channel a higher amplitude than in the main beam channel's sidelobe region, much as in the ISLS situation. By comparing the two receiver outputs it is possible to restrict replies into the processing system to only those whose amplitude is greater than in the control channel. This will be across the arc between the antenna pattern's cross-over points. Either side of these the signals from the control channel will be in excess of those in the other. Thus any replies, including fruit, will be accepted only over this restricted arc. As a consequence fruit densities are reduced by a large factor.

It is obvious that such an arrangement governs the down link system beamwidth. Careful system design has to be carried out to ensure the system beamwidths on up and down links are balanced.

11.3.6 'Dead' time

We have now seen how ISLS works and that it has other beneficial effects in that system beamwidth can (within limits) be controlled by varying the levels of P_1 and P_3 relative to P_2. It also greatly reduces unwanted 'fruit' replies. However, once again, benefit is bought at cost. In order to suppress, the transponder must be inhibited from receiving the P_3 pulse following the P_1 which started the suppression action. Since P_3 can be at any one of six positions between 3 and 25 μs after P_1, the transponder receiver must be completely shut off for at least 25 μs after receipt of P_1. Thus if one's own 'home' interrogator creates suppression in a transponder, it cannot reply to interrogations from other 'foreign' stations during the suppression period. Twenty-five microseconds may not sound a very long time, but if we remember that an interrogation period can be of the order of 2200 μs, then one suppression period represents more than 1% of this time. If a transponder is within suppression range of shall we say four stations (quite a common situation) then up to 4% of the transponder's time is taken up by these 'dead' periods.

Equally, when a transponder replies to an interrogation it must have its receiver gated off for at least the effective reply time (25 μs). This dead time is allowed by the I.C.A.O. specification to stretch to 125 μs. Thus whilst responding to interrogations at a rate of 450 per second (2200 μs interrogation period) each reply creates about 6% dead time. It is easy to see now why interrogation and ISLS power, interrogation rate and beamwidth all should be reduced as much as possible (as the I.C.A.O. specification recommends), for their unbridled increase can (and does) seriously reduce the reply probability. Suppose a system designed to operate at 90% round trip reliability had a system beamwidth yielding an average of fifteen replies. If during the passage of the beam across the target its transponder was found 'busy' at the instant the wanted interrogations arrived at the target, no reply would result. Only three such 'busy' occasions would result in the fifteen potential replies reducing to twelve and so the round trip reliability would fall to (12/15) × 100 = 80%. The effects can be more profound as will be seen later.

11.4 Wide elevation beams and their reshaping

11.4.1 How small, how convenient!

In the earliest days of IFF and SSR the quickest and cheapest method of implementation was to use a separate antenna mounted on top of existing primary radar antennae. Because the SSR technique virtually put an amplifier in the target, ground based antennae need not have very much gain and hence could be small in comparison to their primary radar antenna hosts. In turn this meant that the vertical dimension (the most significant for the mechanical designer) could be very small and everyone was satisfied, for quite a long time. The antenna illustrated in fig. 11.8 has a horizontal aperture of 14 ft (4.3 m) and a vertical aperture of 22 inches (0.44 m). Its weight is only 112 lb (about 44 kg). Such antennae are often called 'hogtroughs' for fairly obvious reasons. Mounting this on top of a primary radar antenna is thus no great problem, provided the increased wind resistance can be tolerated by the antenna turning motors.

How wide, how awkward!

We have seen the consequence of narrow apertures. The typical 'hogtrough' SSR antenna has a vertical aperture of about one and a half wavelengths and thus produces a beamwidth of about 45°. If this beam is directed at an angle of about 25° elevation, then the radiated power at zero elevation is about half the peak value. Such power will be reflected from the surrounding ground, which is usually flat, being on or near an airfield or on a ship at sea. In part 1, chapter 3, the effect of this upon the vertical polar diagram has been described. Such large amounts of reflected power produces massive lobes and gaps in the vertical coverage diagram. Even though the vertical polarisation used reduces the reflection coefficient at some angles (see fig. 3.12, chapter 3, part 1) to very low values, at the low elevation angles of consequence there is little to be gained from the reducing effect. Because the SSR antenna is at a height of many wavelengths, the lobes and gaps are numerous and narrow. A typical example is shown in fig. 11.12. The flight path of an aircraft is indicated. It can be seen that it flies through a significant number of lobes and gaps and its track history in the radar system will be discontinuous and thus a great nuisance and source of worry to air traffic controllers. Because the beam is so wide in elevation there is little to be gained by tilting the beam up yet further; the rate of change of gain with elevation is very slow.

Another obvious disadvantage is that the radiated power is not distributed efficiently in the vertical plane. Radar coverage is massively provided where there are never going to be wanted targets.

11.4.3 Lobing mismatch

Another disadvantage of this multiple lobing effect is found in systems where an omni-directional antenna is used for ISLS. As remarked earlier, some of

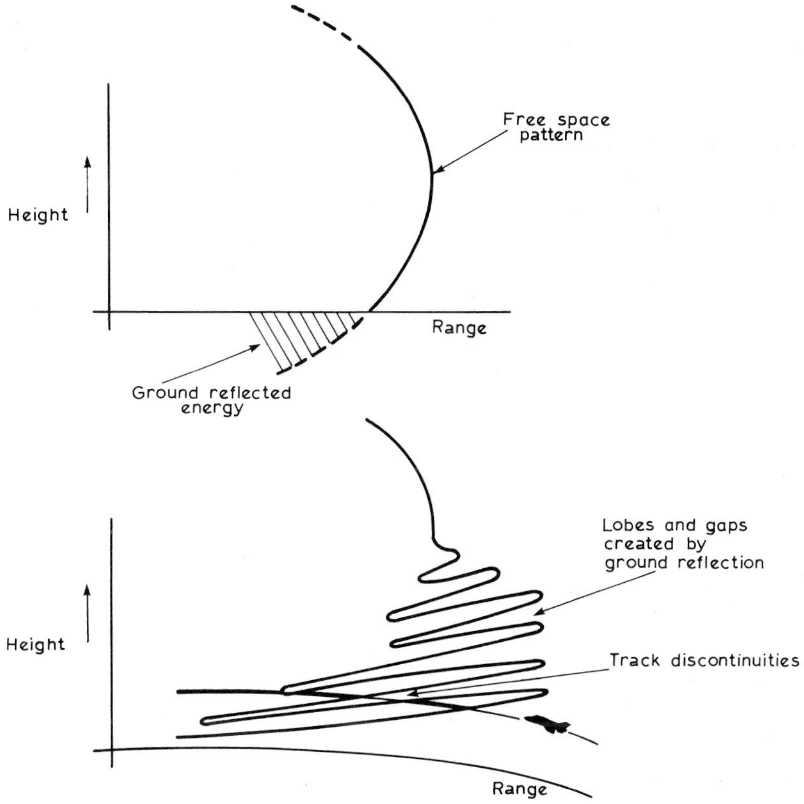

Fig. 11.12. Showing how lobing effects create track discontinuities and wide vertical beamwidth produces inefficient high coverage.

these are mounted on a mast close to the rotating antenna and, to avoid obscuration, the omni antenna overlooks its brother. Thus the two antennae have different phase centres in both azimuth and elevation planes. The ground reflection effects of each will be different and so the lobes and gaps in space will not coincide. This produces inconsistent ISLS performance and further disrupts track continuity. A very good and readable paper has been written on this topic and will be found in reference 13.

11.4.4 Reshaping the beam

To overcome the problems discussed above it is necessary to change the antenna design in radical ways. What is required is an antenna which:

(a) Fits the vertical coverage requirements better.
(b) Reduces power directed at negative angles of elevation to reduce reflections.
(c) Preserves a common phase centre for all its radiation patterns in both elevation and azimuth planes.

A new generation of antennae has emerged to satisfy these requirements – the large vertical aperture (LVA) antennae.

As the name implies, the vertical dimension of the array is increased to something approaching six wavelengths (1.65 m). The narrowing of the elevation beamwidth compared to a 'hogtrough' of one and a half wavelengths is obvious from the relationship given in part 1, chapter 3, and from this reduces it by a factor of four. But design engineers have been more cunning than to rely upon this simple relationship. The physical aperture of 1.6 m allows a large number of elements (about ten) in the vertical plane to have their amplitude and phase distributions to be tailored in a very carefully controlled manner. Using this new degree of design freedom, a technique has been developed which puts the antenna efficiency per square metre of area at a very high level of about 85% compared to the generality of 60%. This technique of array synthesis is called 'SYNFF', an acronym for 'synthesis of far field' patterns.

One such design, now in service with the United Kingdom's Civil Aviation Authority (C.A.A.) has produced an elevation gain pattern shown in fig. 11.13 which also shows the equivalent elevation pattern of the much-used 'hogtrough' antenna, for comparison purposes. It has been drawn to a common base of ERP which discounts the extra gain of the LVA. The important features are:

(a) The LVA distributes its power in the vertical plane much more efficiently, being amenable to design incorporating the 'cosecant squared' principles described in part 1, chapter 3.
(b) The looked-for reduction of ground directed energy is present to a marked degree; it provides 'fast bottom cut-off' characteristics.

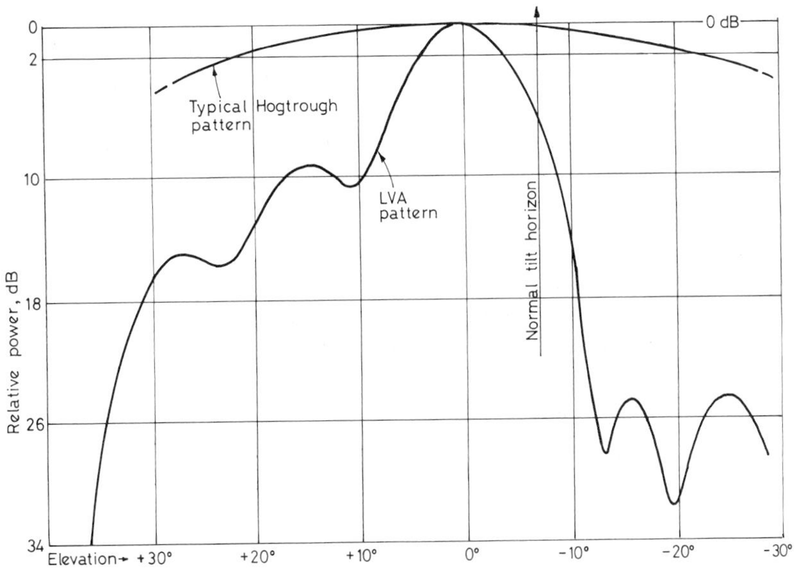

Fig. 11.13. Gain versus elevation angle patterns showing the advantages of LVAs (large vertical apertures). They restrict radiation to required areas much more efficiently. The patterns are normalised to a common ERP. (*Data courtesy of Marconi Radar.*)

(c) Vertical sidelobes are kept to very low levels.
(d) All the patterns required in a modern SSR system are produced from one array of elements which share a common phase centre for both azimuth and elevation planes in the far field. Thus their relationships in space are unaffected by ground reflection effects.

A photograph of the LVA is shown in fig. 11.14 mounted upon an ASR. Its area of 1.65 m by 9 m gives 29 dB gain which, by use of the SYNFF technique, is nearly twice that of traditional designs for the same aperture.

11.4.5 LVA Operation

Discounting the finesse with which power distributions are executed, all LVAs consist of a row of columns of dipoles. The vertical columns are usually identical mechanically and electrically and are equispaced in the same plane. The vertical columns are fed with energy from an azimuth distribution network which tapers their excitation symmetrically across the row of columns.

To show the advantages of the LVA in relation to the effects listed above concerning vertical cover and reduction of lobing, the diagrams in fig. 11.15 are presented.

Fig. 11.14. A modern high gain LVA open plan array antenna, now in service by the UK Civil Aviation Authority, co-mounted on an ASR. (*Photo courtesy of Marconi Radar*).

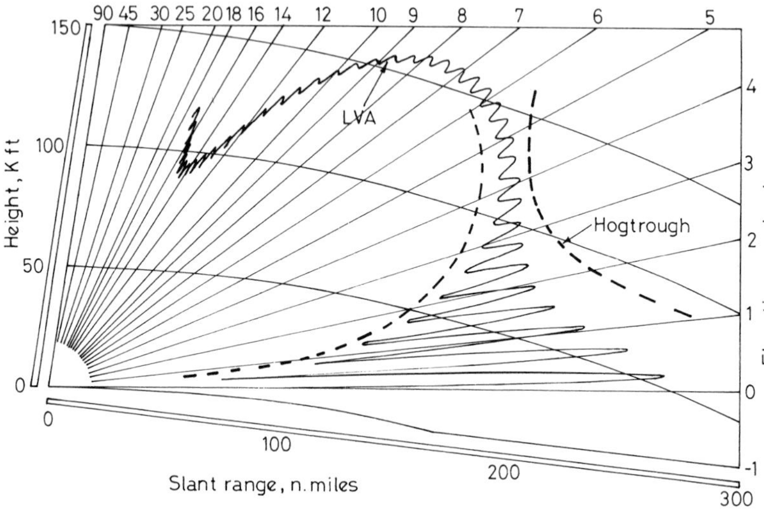

Fig. 11.15. Vertical polar diagram of an LVA with ground reflections causing lobing. The increase expected from a 'hogtrough' antenna adjusted to meet the same cover requirement is shown as dotted lines indicating the extent of the peaks of nulls and lobes.

11.5 False replies by reflection and their suppression

11.5.1 False reply generation

The SSR system is far more efficient than primary radar because it exercises logic in both the up link and down link paths and incorporates an intervening amplifier – the transponder. We have already seen that this efficiency brings with it certain drawbacks. There is yet another: the generation of false replies. How do these occur?

Imagine a commonly found situation, as shown in fig. 11.16. An interrogator is located in the vicinity of a large vertical reflecting surface such as an airport building, a hanger or the tail fin of a large plane. An aircraft following the indicated track, in the position shown, will be illuminated twice in each revolution of the interrogator's beams. Once at azimuth θ_d, the direct path and once at θ_r, the path via the reflecting surface MN. This surface can be of such proportions as to make it 'transparent' to the aircraft's transponder, i.e. the surface acts as a mirror and does not disturb the relationship between the interrogation beam carrying P_1 and P_3 and the control pattern carrying P_2.

Interrogations made via the reflected path will allow any stimulated replies to enter the interrogator system by the same path. These will be indicated at

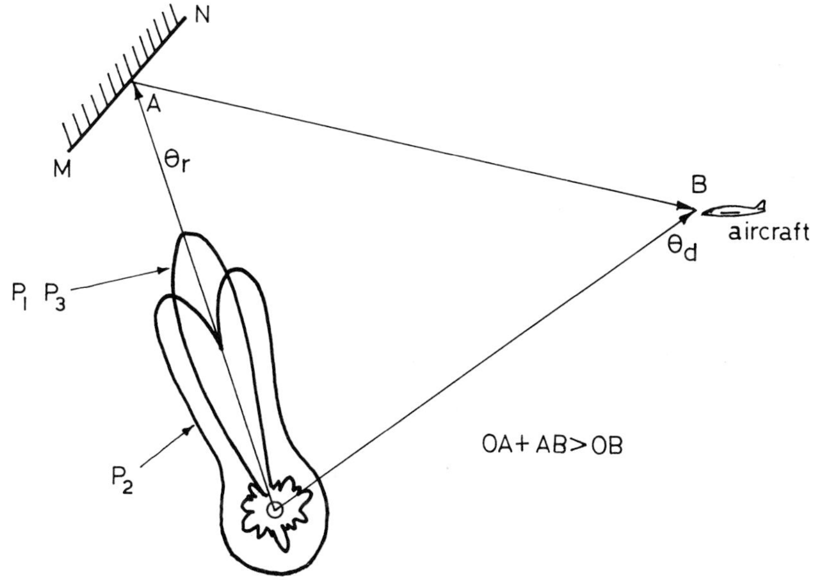

Fig. 11.16. Interrogation reflection geometry: θ_r is the bearing of the antenna boresight; θ_d, that of the aircraft. Interrogations can reach the aircraft by reflection from a vertical surface MN.

the wrong azimuth, for the antenna is pointing at the reflecting surface and not the aircraft. The decoding and display system will thus have two indications of one aircraft at each antenna revolution – one via the proper path and another false indication at a slightly longer range (OB < OAB) and at a totally wrong azimuth. The confusion is compounded since both the true and the false replies will have common identity and altitude codes.

These false replies can only occur if there is a failure to suppress the transponder via the shorter path OB when the geometry of the situation is as drawn. According to the I.C.A.O. rules governing suppression and reply, the aircraft transponder at B should have been suppressed by the arrival of a P_1 and P_2 pulse earlier than an interrogation pair P_1 and P_3 via the longer reflected path. But here an irony of the SSR system comes into play: because of the trouble caused through interrogation by sidelobes, designers have continually and successfully striven to reduce sidelobe levels.

For the ISLS action to take place the transponder *must* receive P_1 at sufficient strength to begin the suppression action. If P_1 is not detected, P_2 cannot be construed as such and suppression will not take place. In the sidelobe regions P_1 power can be as little as −30 dB (1/1000) relative to beam peak and at ranges as little as 5 nautical miles, wanted suppressions can fail to take place. Together with this is the observed ability of reflecting surfaces to stimulate false replies out to ranges of up to 50 nautical miles.

11.5.2 Suppression of reflected interrogations

From the discussion above on the effects of wide elevation beamwidths it will be easy to appreciate that one of them is the tendency to engender false replies by reflection. The wide beam directs more power than necessary at small positive and negative elevation angles. The introduction of the LVA to replace the 'hogtrough' produces remarkable improvement in avoiding reflected interrogations. The LVA directs far less power into these elevation angles. To illustrate this, figs 11.17(a) and (b) have been reproduced from a report on an evaluation of different LVAs by the U.K.'s C.A.A. In this, a site was chosen especially for its propensity to generate reflections. An aircraft was flown orbitally around the interrogator station at about 15 nautical miles radius and at 8000 ft. Figure 11.17(a) shows results achieved by use of a 'hogtrough' antenna. Actual true plots are indicated by dots. These are accompanied by others marked as crosses and indicate plots detected from false replies received by reflections from local features. Special code correlating algorithms in the plot detector and a knowledge of the reflection geometries allowed the false plots to be identified. Nearly 60% of true plots were accompanied by false plots.

The test was repeated using the same aircraft, following the same track and using the same Interrogator ERP, but with a Marconi Radar high efficiency LVA. The false plots created fell dramatically to about 3.5%.

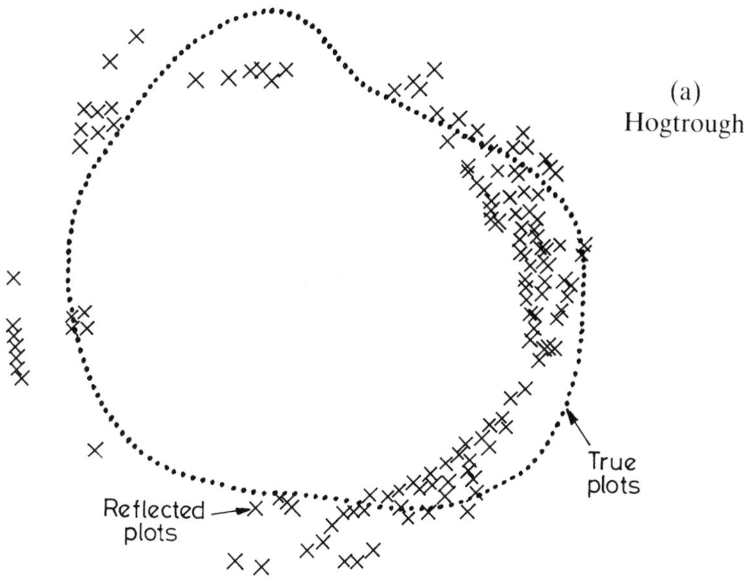

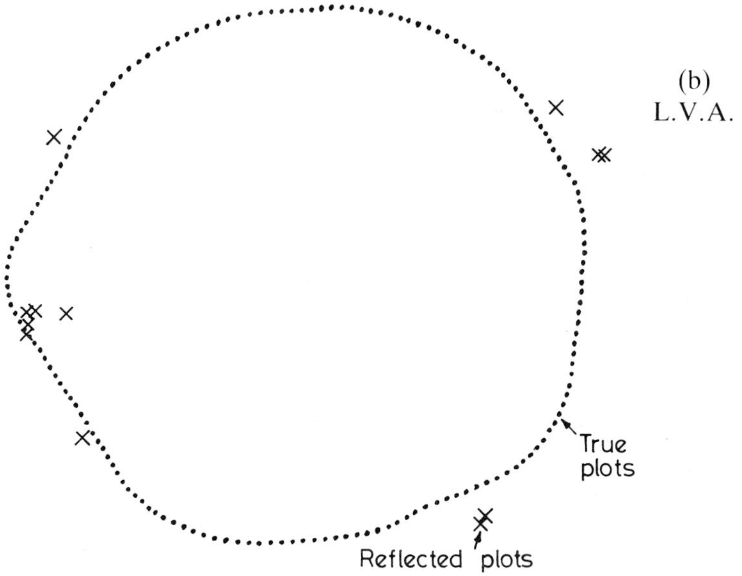

Fig. 11.17(a) and (b). Results of comparison between 'hogtrough' and LVA antennae showing the vast improvement by LVAs in protection from reflected interrogation effects. False replies reduce from 60% to 3½%. Flight and site conditions were kept constant. The aircraft flew an orbit of 15 nautical miles radius at 8000 ft height. (*Reproduced by kind permission of UK CAA.*)

11.5.3 Improved interrogation sidelobe suppression (IISLS)

The reduction of reflected interrogations achievable by use of an LVA still leaves a significant residue which could create troublesome, even dangerous, control situations. Their further reduction is achieved by the IISLS technique. Various methods are used and share one common aim: to increase the signal strength of P_1 in sidelobe regions to a level where ISLS can take place. This is not as easy as it sounds because if the P_1 level is increased it must not exceed existing levels of P_2, neither must the accompanying P_3 be allowed to increase or any failure of suppression could result in unwanted reply.

One method of implementation of the author's devising is described in reference 14. Figure 11.18 has been reproduced from this to show the measured efficacy of the system in the field. Out to about 25 nautical miles the improvement factor is more than 10 to 1 with much higher levels at closer range. The LVA improvement factor is about 17 to 1. Thus the expected overall improvement factor is 170 to 1. What will this mean for the user? If we posite an approach controller handling thirty aircraft in a two-hour period, a reasonable estimate of the number of radar plots he will experience is 3000. If 20% of these result in false reports he would have to resolve 600 ambiguities. If an LVA and IISLS were fitted, these ambiguities would fall to a likely value of 600/170 = 3.5, a much more tolerable value. The point of this numeric estimate is to show that no matter how clever engineers get, there is no 100% guaranteed solution to any problem.

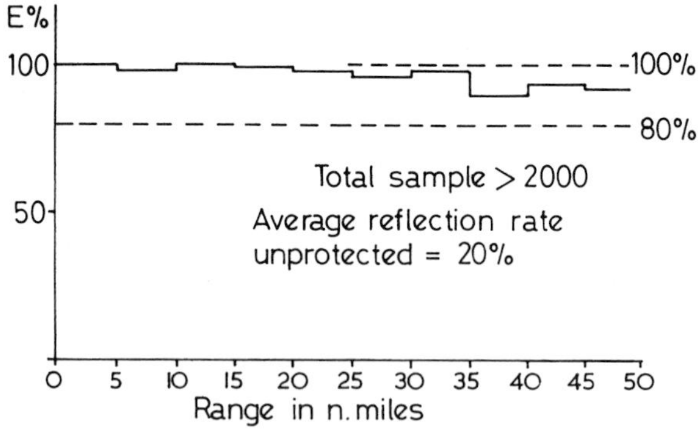

Fig. 11.18. Measured results of a practical ISLS system. Efficiency (E%) expresses the degree to which the system was protected from false replies stimulated by reflected interrogations.

There will always be a shortfall from the ideal; judgement has therefore to be exercised as to what is an acceptable residue. The principle of 'risk accountancy' is the only rational method of assessing this level of acceptability, for we speak here of a residue which could create potentially dangerous situations. The author commends readers to study Lord Rothschild's Dimbleby Lecture on the subject (reference 15).

11.5.4 Reflection suppression by plot processing

The two methods described above are properly aimed at preventing the *generation* of false replies by reflection. The small residue of false replies will enter the SSR decoding and plot processing system. It is possible to suppress most of these using signal and plot processing logic which acts upon characteristics distinguishing the real from the false. Some of these are:

(a) The range of the false reply will always be greater than that of the true.
(b) The code content of both will be the same.
(c) The azimuth of the two will be different.
(d) The track behaviour of the two will be mirror images about the point of symmetry of the reflection geometry (largely knowable for fixed reflecting surfaces).

Using such data and premises, algorithms within the processing can identify the false replies and suppress them. A number of papers on the subject have been written; a very comprehensive one is found in reference 16.

11.6 Poor azimuth data and their improvement

In section 10.5.3 the process of automatic plot extraction was touched upon. It is relatively easy to see how the range on a particular target and the code content of its replies can be handled because these do not change in value during the target's dwell time within the radar beam. Azimuth is a different matter. Some interrogators have effective system beamwidths in excess of 3° with a variability from perhaps 2° minimum up to 5° maximum. In a real-time display system azimuth determination is fairly easy even with the exampled variability; the operator can judge fairly accurately the centre of the response's arc. But in a plot extractor this 'centre of gravity' representing the target's true azimuth must be calculated. It is the uncertainties associated with this calculation that leads to poor azimuth data under certain input conditions. How these arise follows from an understanding of the calculation method.

11.6.1 Automatic azimuth detection

Although there are variations between equipments the following principle, used in many modern systems, illustrates how they are subject to error. The techniques of the 'sliding window' described in part 2 is used. It will be remembered that this seeks to determine the leading and trailing edges of the series of target responses within the beamwidth. If the logic behind the plot formation is satisfied, the plot is declared valid. In general this logic is as follows:

(a) Do signals cross the 'first threshold'?
(b) If yes, at what azimuth do they correlate in range for '*n*' successive samples (repetitions)?
(c) Does '*n*' increase successively?
(d) At what azimuth are there '*m*' failures to satisfy (a) and (b)?
(e) Calculate true target azimuth from values of (b) and (d).

For an immaculate and uninterrupted 'run length' (a series of replies correlating in range) the logic will produce an azimuth value whose accuracy depends upon the number of bits expressing the full 360° antenna rotation. This is usually twelve or thirteen, giving incremental values of 360/4096 or 360/8192 degrees – 0.088 and 0.044 degrees respectively. However, the radar sample rate, or pulse repetition frequency, is usually lower being typically 250 per second with a limit set by I.C.A.O. of 450. If it is the latter it will probably be accompanied by an antenna rotation rate of 15 r.p.m. and thus only 450 × 4 = 1800 interrogations per revolution will be available. The azimuth reporting system is not usually synchronised with the prf but this leads to only one azimuth increment as an error. This is entirely tolerable in most systems.

The inaccuracy occurs when the 'run length' is broken or the count of '*n*' or '*m*' is disturbed. Remembering the 'dead times' introduced into the SSR system we see immediately how such disturbance can be created. Add to this any failures to cross the first detection threshold plus the small but finite probability that false alarms enter the detection logic, and the full realisation is before us.

Examples are given in fig. 11.19 of the azimuth jitter or track wander created in SSR plot extraction. It is obvious that 'across track' errors are greater, for a given azimuth error, as target range increases.

11.6.2 Improving azimuth accuracy by monopulse technique

Modern SSR systems markedly improve their azimuth accuracy by incorporating a monopulse 'direction finding' technique. In this, each and every pulse entering the system has the azimuth of its source measured to high accuracy; hence the term 'monopulse' – only one pulse is needed to determine the azimuth. There are two types of monopulse system and each needs at least two antenna patterns to be simultaneously operated in the down link path. The two patterns have separate receivers. Any signal entering the antenna patterns will be passed simultaneously to the two receiver outputs. Comparison of these allows the azimuth to be determined.

The two methods are very similar and differ largely in the way output comparisons are made. They are called the amplitude comparison and the

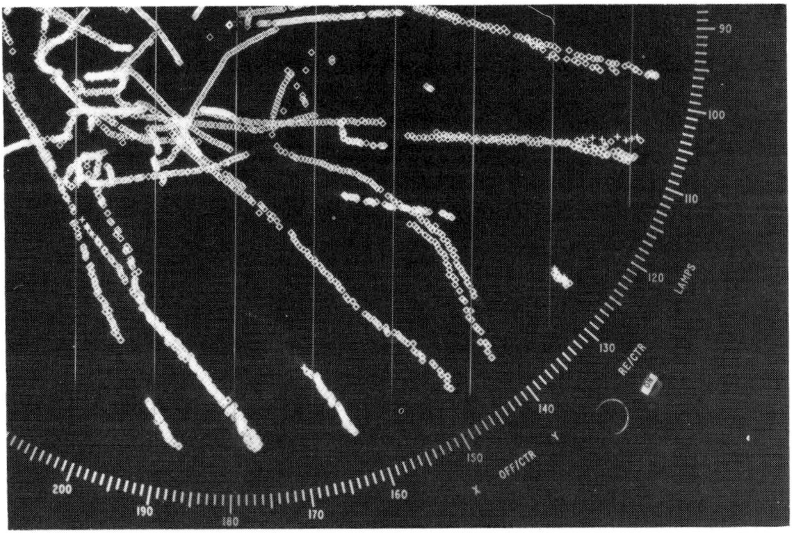

Fig. 11.19. Long term exposure of ppi picture showing SSR plots using 'sliding window' azimuth detection. Azimuth jitter is obvious. Display radius is 120 n. miles. (*Photo courtesy of Marconi Radar.*)

phase comparison methods. It is common to find that both use an antenna of the 'sum and difference' type described earlier. The two patterns must be present together and so the switching arrangement illustrated in fig. 11.7 is by-passed for the down link path. The antenna thus has two ports (Σ and Δ), each feeding its own receiver.

11.6.3 Amplitude comparison monopulse

The most simple form of the monopulse technique, chosen for ease of understanding, is illustrated in fig. 11.20. Sum and difference patterns, simultaneously present, each feed separate receivers designed to have balanced performance over the wide dynamic range of input signal amplitude. Recalling the earlier description of interrogation system beamwidth, it will be appreciated that replies can be elicited from transponders at angles either side of the interrogating antenna's boresight. Any reply entering the system at the instant its source is on the antenna boresight will present a plane wave phase front parallel to the antenna array. Replies given at angles away from boresight will arrive as plane waves not parallel to the antenna array. This 'out of parallel' angle represents (by geometry) the 'off-boresight angle' (OBA) which the system has to measure and express.

In fig. 11.20 a reply pulse is indicated as arriving at an angle 'α' away from the boresight angle, θ_{bs}. The resultant signals in the sum and difference channels will have different amplitudes. Their difference is indicated as '*h*'.

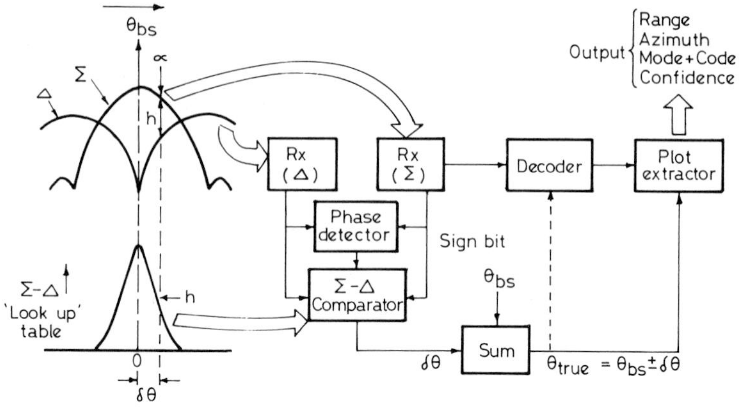

Fig. 11.20. An amplitude sensing monopulse system. Angle of arrival of reply (∝) allows off-boresight azimuth (δθ) to be derived by fore-knowledge of antenna beamshapes. Their amplitude differences versus azimuth are expressed in a 'look-up' table.

The shapes of the sum and difference patterns are known by design (and usually by measurement at individual installations). It is therefore possible to express the 'angle of arrival' of pulses relative to boresight as a function of *difference* in amplitude between the outputs of the two balanced sum and difference channels.

Such a hypothetical law is illustrated. It is sometimes referred to as a 'look-up' table. It shows the expected peak on boresight where the difference is greatest, falling to zero at the antenna's sum and difference pattern cross-over points. Somewhere in its range the value of '*h*' will be found and an output from the $\Sigma - \Delta$ comparator will be given (usually in a binary coded word of six to eight bits) to express the off-boresight angle. This is designated as $\delta\theta$. It will be noted that the law expressing the difference between Σ and Δ outputs is symmetrical about the boresight and thus creates ambiguity as to whether $\delta\theta$ was left or right of θ_{bs}. This ambiguity is resolved by detecting the phase of the Σ relative to Δ outputs.

From the earlier description of how sum and difference patterns are produced it will be realised that at either side of boresight the difference signals undergo a 180° phase shift. Thus on one side of boresight the sum and difference signals will be of the same phase polarity. On the other they will differ by 180°. The phase detector thus provides a 'left or right' sign bit into the $\delta\theta$ expression. In order to determine the true azimuth of the source of the input signals, it is necessary to account for the position of the boresight at the time $\delta\theta$ is measured; it must be noted that such measurements are taken a number of times during the presence of each reply pulse in the system – typically three to eight samples in 0.45 μs. The boresight position is continually reported by digital encoders of thirteen- or fourteen-bit discretion, dividing a revolution into increments of 2.6 or 1.3 minutes of arc. This 'running azimuth' is fed to a summing network which also accepts the value of $\delta\theta$ plus its 'left or right' indication. Thus the summed output represents the true azimuth of the source of the input signal:

$$\theta_{true} = \theta_{bs} \pm \delta\theta$$

The sum channel also provides input to the decoder/plot extractor chain and so each pulse of every reply can be assigned corresponding data representing the true azimuth from which it came. This gives monopulse systems great power in signal processing, as will be seen later.

11.6.4 Phase comparison monopulse

The aim of phase comparison monopulse is precisely the same as described above: to obtain and report true azimuth data on every pulse entering its antenna. A typical arrangement is shown in fig. 11.21. Again, sum and difference antennae are used. The difference between amplitude and phase measuring monopulse systems lies in the manner in which the signals from the antenna are processed. In phase measuring systems, the signals from the two separate Σ and Δ channels are put through a limiting process to reduce all input signals to a common level, slightly above that of system noise. The limited outputs of both channels are then compared in phase by a phase detector whose output will be a dc voltage proportional to the phase difference between the detector's inputs (see part 1, chapter 6).

The shape of the characteristic (exampled in fig. 11.21 as 'e') will, as in the

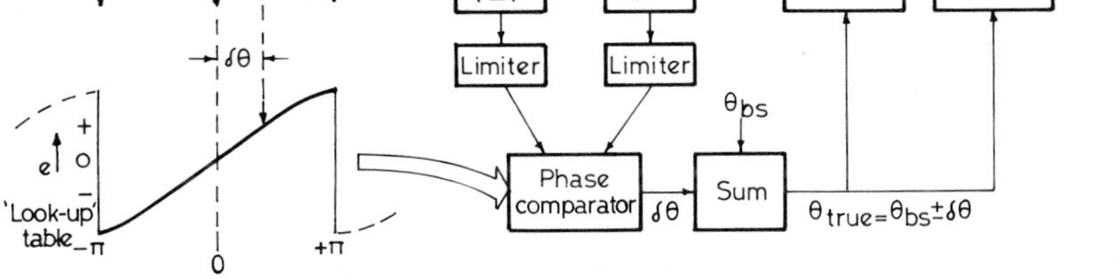

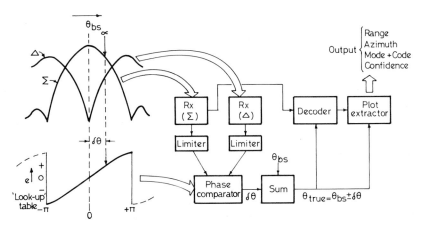

Fig. 11.21. A phase measuring monopulse system. The 'look-up' table is now a function of phase differences between antenna sum and difference outputs versus azimuth.

amplitude comparison method, be dependent upon the actual antenna used. Again it is necessary to create from this foreknowledge, a 'look-up' table to relate the phase detected output to OBA. There is no need to provide a separate phase detector to resolve the 'left or right' ambiguity since this is given automatically by the phase detector already in the system. As before, the OBA value is converted to a digital word and summed with the running azimuth value θ_{bs} concurrent with it. From thence the system is directly equivalent to the amplitude comparison method already described.

11.6.5 Summary

A few points are noteworthy. The amplitude comparison method illustrated is usually engineered upon slightly different lines to improve the angle over which OBA values can be measured. As illustrated it is obvious that the OBA measuring aperture is limited to the angles between the sum and difference pattern cross-over points; beyond these, further ambiguities would result. The system is improved by seeking the *ratio* between the amplitudes of the sum and difference outputs. This is accomplished by the use of logarithmic receivers and the mathematical device based in the familiar logarithmic relationship:

$$\frac{A}{B} = \text{Antilog}\,(\log A - \log B)$$

By giving the sum and difference channels logarithmic characteristics their output difference ($\Sigma - \Delta$) expresses their input ratio. Because the rates of change of the two antenna patterns are monotonic (i.e. have no polarity reversals) either side of boresight up to the roots of the sum pattern, it is possible to use nearly the whole of the sum pattern beamwidth for OBA measurement. The limit is set by the reducing signal to noise ratios away from boresight and the reducing sensitivity of OBA to the sum and difference ratio near the flattening top of the difference pattern. There is also a region about boresight, where the difference pattern's null is very deep, whereas distant targets' OBA values cannot be ascribed with certainty to be left or right of boresight – again because of signal to noise ratio limitation, coupled with the finite depth of the difference pattern's null point. From the description of how such patterns are generated it will be realised that the null depth increases as the rate with which the phase difference is effectively created at boresight: the faster the rate of change producing the phase reversal, the deeper the null. A given irreducible noise level obviously sets a limit to detectability of the phase of sum and difference signals and hence to the ability to say whether the OBA is left or right of boresight.

In the phase measuring system illustrated it is possible, as with the amplitude sensing method, to improve detection of OBA. By using two phase detectors the illustrated OBA characteristic is formed, giving improved detectability across the whole of the sum pattern beamwidth. Again the limitation is imposed by the reducing signal to noise ratios far away from boresight. It is also interesting to note that the characteristic shown goes on repeating itself away from boresight as the main beam of the sum channel is replaced by successive sidelobes. The consequent ambiguities have to be resolved. This is accomplished by using the Receiver Sidelobe Suppression (RSLS) technique described earlier. The RSLS system restricts input to the phase comparators to the main sum beam region. A further distinction between the two monopulse methods is that the dual phase measuring system is much less sensitive to differential phase errors caused by differential temperature drift.

11.6.6 How the monopulse technique improves azimuth accuracy

Having now some insight into the workings of both the sliding window and monopulse direction finding techniques, it remains to discuss how the latter improves azimuth accuracy. It is almost self-evident.

Take as example an aircraft giving twenty replies in one beamwidth, with each reply containing eight pulses. The sliding window azimuth detector will produce only two azimuth measurements, θ (leading edge) and θ (trailing edge) from which the true azimuth is calculated. Both of these values are subject to error if the '*n* out of *m*' and 'not *n* out of *m*' logic fails in some way. In monopulse azimuth detection, for the example taken, there are at least 8 × 20 azimuth measurements after averaging samples within each reply pulse. Thus there is 80 times the data in monopulse. Further, in monopulse systems, range correlation can be effected by seeking an '*n* out of *N*' criterion where *N* is the total number of replies in the beamwidth. There is no requirement for the '*n*' replies to be contiguous in an early group within the beamwidth: if '*n*' was given a value of three, the azimuth detection could be made if the second, seventh and ninth replies in the group of twenty correlated in range.

An indication of the power of the monopulse azimuth detection relative to sliding window is given in fig. 11.22. It is calculated from a theoretical basis but with practical values. We have seen that the measuring discretion of a typical sliding window detector is about twelve minutes of arc. Monopulse systems commonly have an azimuth measuring aperture of about 4°. This is capable of expression by an eight-bit word, so each increment is 4/256 degrees or just under one minute of arc. Systems in operation have azimuth rms errors of only about 3 minutes of arc.

11.7 The loss of data integrity by garbling and its recovery

11.7.1 Garbling

As its name implies, garbling occurs when two replies enter the system close enough in time for their pulses to interleave or overlap. There are two distinct

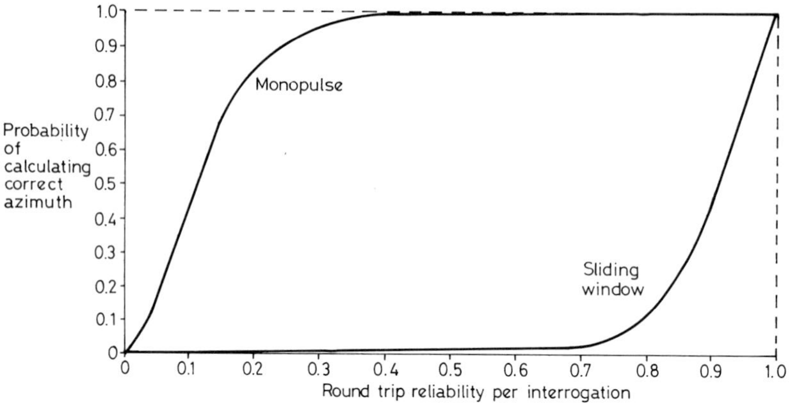

Fig. 11.22. Illustrating the superior ability of monopulse systems to produce azimuth data from few replies. (*Reproduced by kind permission of Marconi Radar.*)

types of garble. The first is classed as 'asynchronous' where the garble is created by fruit replies interfering with those generated by one's own interrogations. These are usually transitory and lead to short term data corruption. The second type is the 'synchronous' garble. In this, two or more transponders are stimulated by the interrogator. If their range is the same to within about 2 nautical miles then pulses of replies from individual transponders become intermixed. Note the important parameter is range. In an atc situation, a number of aircraft at greatly differing heights on one airway can create garble even though their *ground* range is very different – the radar measures slant range. In such a situation it is possible for garble conditions to exist for quite long times if the garbling aircraft are on the same track with similar speeds. For example two craft travelling in the same direction at 300 and 360 knots respectively would have a potential garble time of roughly four minutes. A perfectly feasible garble situation is shown in fig. 11.23. The two range/azimuth segments represent the area occupied by two aircrafts' replies as the interrogator beam sweeps past them. The beamwidth is assumed to be about 3° and the craft at ranges r_1 and r_2. The shaded area represents that wherein garble can occur; a significant part of the interrogation time for the two aircraft.

11.7.2 Consequences of garbling

Restricting consideration to the most serious form, synchronous garbling, there are a number of consequences, dependent upon the precise time relationship between the garbled pulses. There is a subset of the group of synchronous garbling. One member has pulses of one reply occurring in the time between those of the other in such a manner that leading and trailing

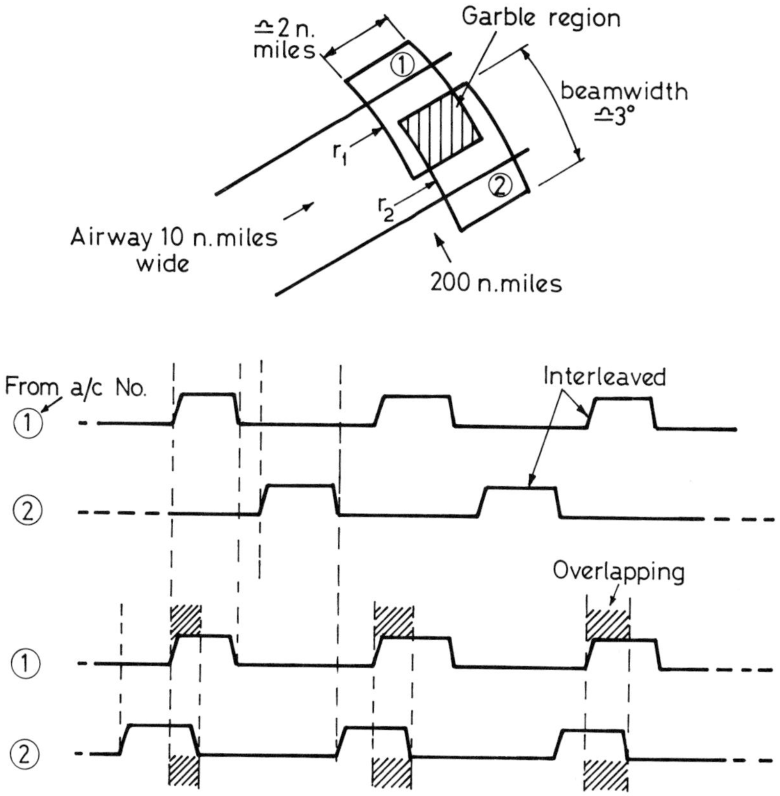

Fig. 11.23. Synchronous garble situation. Two types are possible: interleaved or overlapped reply pulses.

edges of all pulses are distinct and separated, this is called 'garble by interleaved pulses'. The other member has the pulses of one reply so disposed that they overlap (partially or completely) so that there is masking of leading or trailing edges of some pulses, this is 'garble by overlapped pulses'. Both of these conditions are shown in fig. 11.23.

One of the most confusing consequences of garbling follows when the two reply trains are so disposed that they have their pulses spaced by an integral multiple of 1.45 μs (the specified I.C.A.O. spacing). This circumstance can, and does, create 'phantom' replies and code ambiguity.

In non-monopulse systems which rely purely upon pulse timing to detect and validate codes, there are no data to resolve these confusions. As a result, data integrity is lost. In such systems the 'presence' of a reply is signalled by sensing the two framing pulses F_1 and F_2, always 20.3 μs apart. Every such conjunction signals the 'presence' of an individual transponder. As shown in fig. 11.24, two garbled replies can be so disposed as to generate three sets of 20.3 μs spacings, so two transponders look like three! The spurious third man is a 'phantom'.

By a similar mechanism two synchronously garbled replies can create ambiguous codes. Replies can be considered as digital messages consisting of '1's and '0's. If one reply has a '0' replaced by a '1' of the garbling reply, the code becomes corrupt. If the code was signalling aircraft altitude then the consequences could be dire; indeed fig. 11.25 shows how two garbled replies can produce two totally spurious codes by this ambiguity mechanism.

The consequences of garbling by overlapping are clear to see. Most non-monopulse systems operate in digital mode as soon as can be arranged. The reply signals have their amplitude limited before they are input to the decoder. Thus, envelope detection of overlapped pulses will produce composite pulses of non-standard duration. Unless the decoder is fitted with circuits which detect and operate upon the spacing between leading *and* trailing pulse edges, data can easily be lost. In fig. 11.23 we see that the leading edges of reply number one and the trailing edges of reply number two are mutually 'masked'. If the decoder had a 'dual edge' detector it would be able to infer with high probability that if valid leading edges are present then they must be accompanied by trailing edges. The decoder would then

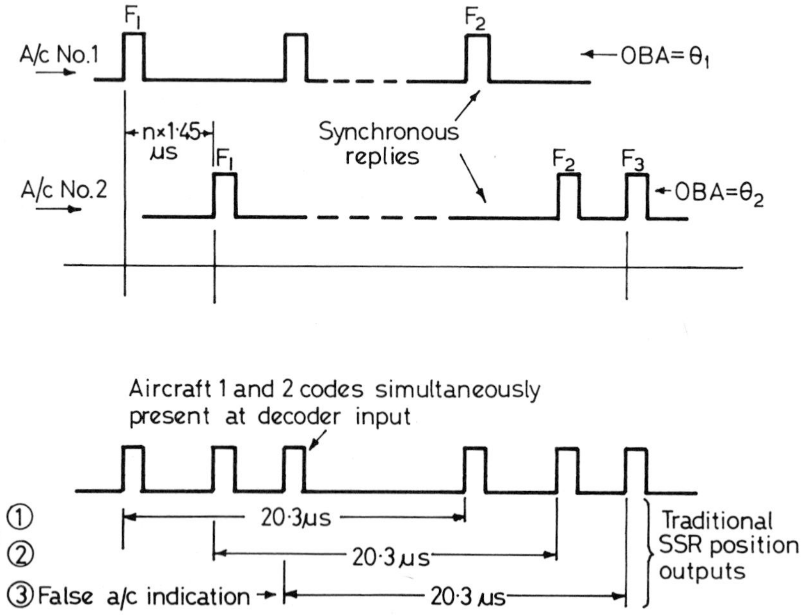

Fig. 11.24. Two garbled replies can generate a 'phantom' aircraft. The SPI (F_3) of aircraft number 2 can appear falsely as F_2 of a non-existent aircraft.

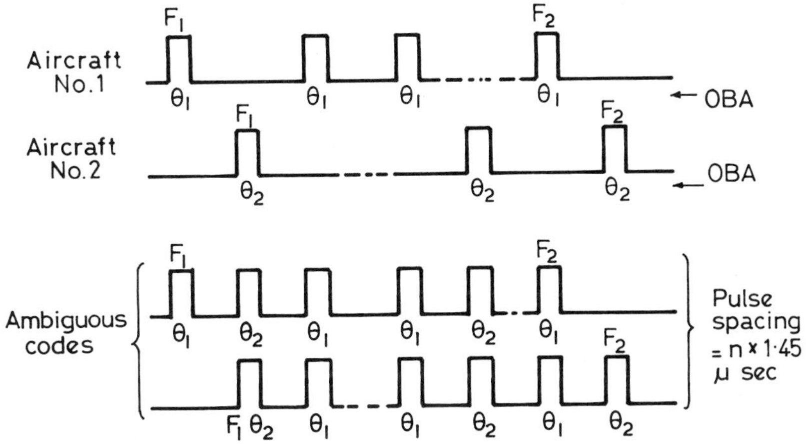

Fig. 11.25. Two garbled replies can produce ambiguous codes. Non-monopulse systems have no OBA values.

manufacture separated and 'repaired' reply pulses. Decoders without such facilities would lose the data.

11.7.3 Monopulse to the rescue!

We have seen that monopulse systems can recover azimuth data to very high accuracy, even in highly polluted signal conditions. What is not so readily appreciated is their power to improve data integrity by their ability to measure azimuth of each and every received pulse. Such azimuth data, by suitable processing, can be kept in association with every pulse. The OBA values of reply pulses from a given transponder will be the same (± the small measuring error). Those from a garbling transponder will be a different but also constant value. Using these enhanced data in fast logic circuits the system can reject phantoms, resolve ambiguous codes and, by taking OBA samples throughout individual pulses, improve the integrity of data in overlapped pulses. The diagrams in figs. 11.26 and 11.27 illustrate the principles employable in monopulse systems to effect these highly desirable data integrity improvements which are not available to non-monopulse systems.

In fig. 11.26 the two aircraft replies arrive simultaneously in the decoder. Because a traditional decoder is only sensitive to pulse position it detects that there are three aircraft present (three conjunctions of pulses, 20.3 μs apart). Association of OBAs with individual pulses immediately resolves the confusion by categorising the SPI with aircraft number two's reply. A different merged input pulse train is shown in fig. 11.27. The code content of aircraft number one has two 'interlopers'; that of aircraft number two, three unwanted pulses. The monopulse system will associate every pulse with the azimuth from which it emanated. Thus more elegant processing can assign

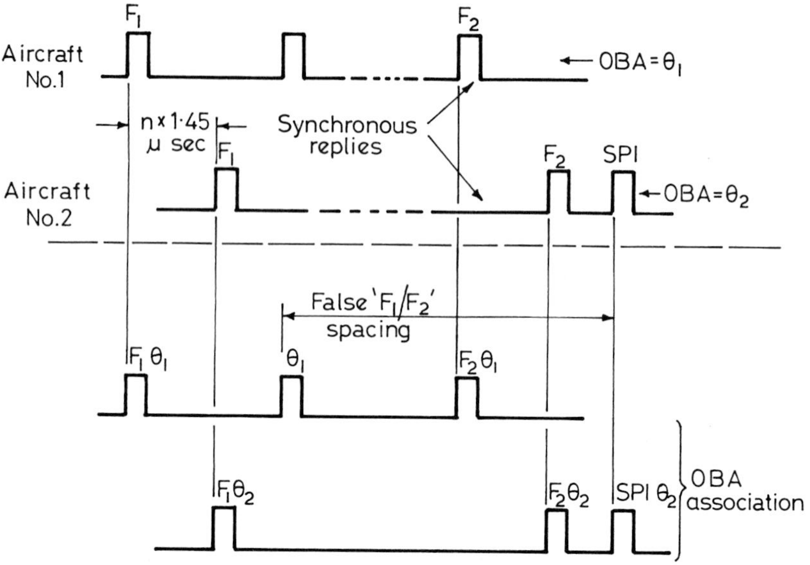

Fig. 11.26. 'Phantom' aircraft can be rejected in a monopulse system by associating OBA values of individual pulses in the decoder/plot extractor processing.

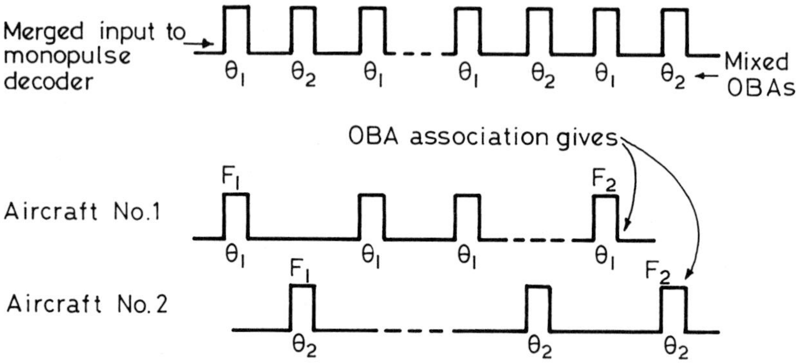

Fig. 11.27. OBA association in monopulse system processors can resolve code ambiguities.

pulses (which are validly disposed according to I.C.A.O. format) to the proper code train. This separation is possible provided the two aircraft are at azimuths differing by the discrimination of the monopulse direction-finding system – typically 6 minutes of arc, equivalent to 1/3 nautical miles (620 m) at 200 nautical miles (372 km) range.

12
The future

12.1 Introduction

The cost-effectiveness of SSR systems and their value to traffic control agencies led to accelerated useage from the mid-1970s onwards. The rate of implementation is still growing. As a consequence the SSR's many problems outlined above became more widely known and means were devised to overcome them. A summary of these will be found in reference 17. They are essentially 'fixes'; that is they may be and in some cases have been, successfully applied to already existing equipments. However, as long ago as 1970 it was realised that an evolutionary (but nevertheless pretty radical) change to the SSR system itself would solve all the known problems (see reference 18). In the United States the idea was translated into the DABS (direct addressed beacon system) and the original U.K. idea, grown from an early seed (reference 19) developed into a hardy plant as the Adsel (addressed selectable) system.

These two almost identical systems (Adsel and DABS) have now been fused into one fast-growing tree which by extensive experimental trials have shown that it not only overcomes the SSR system problems but can greatly expand the data capacity on both up and down links. Its data formats and other basic parameters have already been agreed and ratified by I.C.A.O. in 1983 as the mode 'S' system (reference 20). Although described under this chapter heading as 'the future', mode S is virtually here and now. The national airspace plan of the United States calls for its universal application in the U.S.A. by 1990. Most prospective users throughout the world are actively anticipating its use by issuing SSR specifications calling for equipment which can be upgraded to mode 'S' standards retrospectively. The system is designed to improve data integrity and volume for the user. But once more we see the technology leading the user. There is already 'user–technocrat' imbalance in the development of mode 'S'; the author knows of no agreement on data presentation, or user facilities, comparable with those in the technical field. Unless this imbalance is redressed, mode 'S' could become another 'solution looking for a problem'. Perhaps what follows will contribute to this redress.

12.2 Mode 'S'

The ability of mode 'S' to overcome SSR's problems is largely the result of avoiding (apart from infrequently gathering new target data) the need to broadcast interrogations. Reverting to the cockail party analogy drawn in chapter 10, you can locate and converse with your lost friend by saying 'Joe – tell me where you are': only Joe should answer. We are now 'selectively addressing' transponders.

The interrogation and reply phases are exactly the same as in SSR, using the same up and down link frequencies. The differences are largely:

- Message format.
- Type of modulation.
- Data integrity checking.

Both interrogator and transponder equipment in mode 'S' are organised to be totally compatible with current SSR standards.

12.3 Operation in a mixed environment

In the near future, the atc arena, typical of others, will contain a mix of standard SSR transponders and those also capable of mode 'S' response. It is therefore necessary to have interrogators which can elicit data from either type. Because mode 'S' is a selective address system, it becomes necessary to know where such mode 'S' transponders are, and more importantly where they will be the next time the interrogator's antenna beam sweeps past them. Matters will have to be so organised that every now and again the interrogator puts out an 'all call' signal. In response to this, all mode 'S' equipped transponders will include their own address code in their reply. It will then be possible, if its track is known, on the next antenna revolution to issue an interrogation of particular transponders at the appropriate time, i.e. when the interrogating beam is directed in the desired direction. In the meanwhile the interrogator will issue normal SSR interrogations in the broadcast manner.

In the fullness of time as more and more transponders are raised to mode 'S' standards, the full benefits of the system will be felt. Until that time (the author suggests it will be decades of years) the up and down link equipments must be of dual standard. Moreover, because the 'broadcast' mode of operating must still be maintained and the intermittent use of mode 'S' interrogations disrupts contiguous replies from normal transponders, the sliding window azimuth detection system will produce more errors. This means that monopulse direction finding and data processing (which *must* be used in mode 'S') also provide protection of data integrity in normal SSR modes.

12.4 Message formats and modulation in mode 'S'

Message formats and types of modulation in mode 'S' are radically different from the current I.C.A.O. modes. The mode 'S' up link format is shown in fig. 12.1(a) and (b). As already mentioned it is necessary to have an interrogation regime wherein new mode 'S' targets are acquired. The 'all call' format is used for this. It is like the normal SSR up link pulse train but with the addition of a fourth pulse (P_4) which can be either of 0.8 or 1.6 μs duration. The former duration is used for obtaining normal mode A or C SSR transponder replies (the position and duration of P_4 doesn't affect their reply capability). Mode 'S' transponders will not reply to this. If P_4 is of 1.6 μs duration they will recognise it and reply with their unique addresses in mode 'S' format. Thus the position and address of all mode 'S' transponders can be downloaded to an 'interrogation scheduler' for subsequent use.

The mode 'S' interrogation proper is shown in fig. 12.1(b). It consists now of four pulses; P_1 and P_2 are of equal amplitude in the main interrogation beam and are therefore construed by normal transponders as a suppression pair. The mode 'S' transponders will then seek the presence of the start of P_6 (P_3 and P_4 are only involved in normal SSR and 'all call' interrogations). It consists of a long pulse of 1030 MHz with much closer frequency tolerances than cited in the current I.C.A.O. annex 10 specification and its duration is either 16.25 or 30.25 μs. The frequency stability is required because the modulations throughout its duration are of the 'differential phase shift keyed' type (DPSK). In this a binary '1' is signalled by changing the phase by 180°.

By dividing the pulse duration into 'chips' of 0.25 μs and sampling the pulse at these intervals, phase reversals sensed between samples mean '1's and absences of reversal, '0's. Such modulation has high immunity from noise and interference. Thus P_6 can form a 56- or 112-bit word, dependent upon the duration chosen. To establish a reference by which the phase reversals can be sensed, the binary message proper doesn't begin until 1.25 μs after the P_6 leading edge. The phase of the interrogation's r.f. during this period is 'stored' within the transponder (in much the same way as the coherent oscillation in a coho-stalo mti system). A phase sensitive detector uses this sustained oscillation as one input and the DPSK signals as the other. Phase reversals detected generate digital outputs of '1's and '0's from the detector for subsequent decoding and reaction within the transponder. As in normal SSR an ISLS pulse, P_5, is transmitted by an interrogator 'control' pattern. This is to obviate any normal SSR transponders' replies should they mistake part of P_6 as a normal P_3.

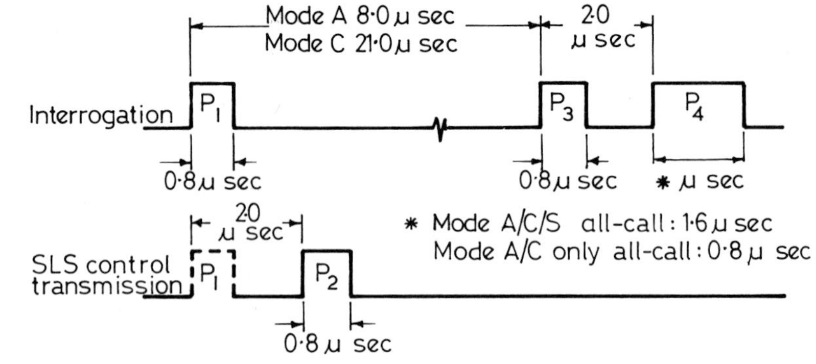

Fig. 12.1(a). Mode 'S' all-call interrogation formats.

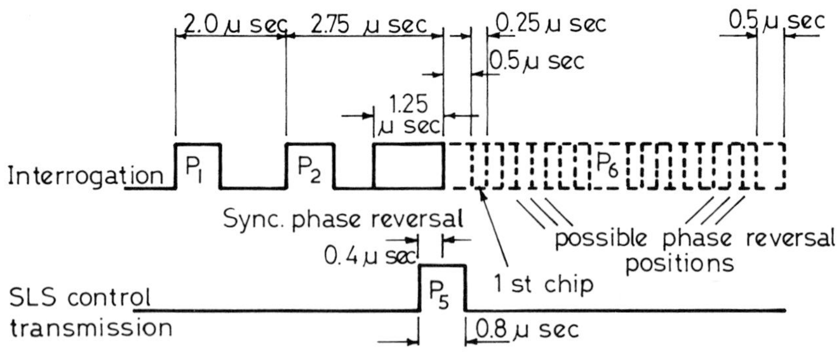

Fig. 12.1(b). Mode 'S' interrogation format P_6 can be 16.25 or 30.5 microseconds duration and uses differential phase shift keying (DPSK) modulation.

The mode S reply format is shown in fig. 12.2. Its increased duration relative to normal SSR is immediately obvious. After a four-pulse preamble, it can express either a 56- or 112-bit binary word using pulse position coding, as does normal SSR, but with greater economy and integrity. Each 1 μs time slot contains a 0.5 μs pulse. This occupies either the first half of the one microsecond period to signal a '1' or the second to signal a '0'. Thus both states are indicated by the *presence* of a pulse; in the I.C.A.O. format, a zero is signalled by the *absence* of a pulse.

12.5 The use of mode 'S' data

It is already obvious that the data capacity of mode 'S' is vastly greater than SSR as it currently stands. When used in its data link role mode 'S' messages can be yet further extended by repetitions of its standard form in what are termed 'extended length messages' (ELMs). Up to sixteen blocks of eighty bits can be strung together giving a total message content of 1280 bits. The formats, protocols, identifying bit patterns, etc., are all described in reference 20. It instances potential uses to which the signalling power of mode 'S' can be put, summarising these in its chapter 5, part of which is reproduced here:

> *Airborne Collision Avoidance*
> 5.2.2 The Mode S data link provides a reliable means for air-to-air collision avoidance systems to co-ordinate their activities. The use of the same aircraft address for surveillance and communications eliminates the possibility of transmitting a data link message to the wrong aircraft.
>
> *ATC Services*
> 5.2.3 Mode S data link can provide a back-up to many ATC Services that are provided today by VHF voice communications. This data link back-up will improve system safety by reducing communications-related errors within the ATC system. Many types of messages are potential candidates for data link back-up and other ATC services. These include:
> a) flight identification
> b) altitude clearance confirmation
> c) take-off clearance confirmation
> d) new communications frequency for sector hand-off
> e) pilot acknowledgement of ATC clearance
> f) transmission to the ground of aircraft flight parameters, and
> g) minimum safe altitude warning.

There are a number of very significant functions listed here. Automating the messages normally transmitted by voice has great impact for the user. Indirect usefulness will result for users of modern computer-based tracking

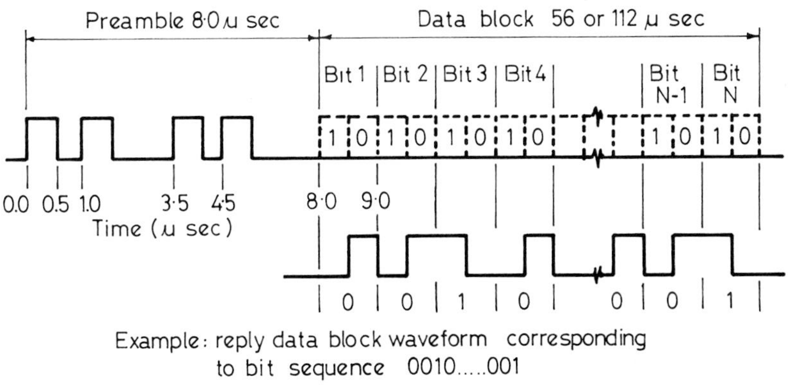

Fig. 12.2. Mode 'S' transponder reply format. Note that a '0' is transmitted as a signal and not as *absence* of a signal.

systems, by down link messages signalling aircraft rate of turn, heading, rate of altitude change, etc. These will all aid tracking algorithms resulting in better turn detection with fewer false turn indications and quicker reaction time. The potential use of the data link in collision avoidance is undeniable, but it is worth citing another small paragraph in the I.C.A.O. advisory circular (reference 20). In the section dealing with communication protocols it says:

> 3.4.2. *A definition of the operational content of messages is not a part of the Mode S system specification* (author's italics) but it is recommended that each data message start with an 8-bit field identifying the content of the message.

It is to be hoped that users increase their dialogue with their planners and technocrats so that their voice can be heard before it is too late and they are saddled upon a monster out of their control.

One last point of note. In order to reduce interrogations in space and to make 'more efficient use' of the SSR's great range, it is suggested in the I.C.A.O. advisory document that those stations capable of interrogating across FIR boundaries become the source of control data for the neighbouring FIRs. Again, it is to be hoped that the user is widely consulted before such arrangements are set in concrete.

12.6 Technical implications of mode 'S'

12.6.1 Transponders

The technical implications in airborne transponder's design will already have been realised in the main. Obviously they will have to cater for the dual standards of mode 'S' and SSR for as long as the latter is part of the I.C.A.O. annex 10 specification. Another implication is concerned with interfacing between the aircraft's (or vehicle's) existing sensor system and providing such data as speed, heading, fuel state, rate of turn, etc., as input to the new transponders – not a quick or cheap trick. Once again the United States has taken a solid stance in insisting that all new transponders to be fitted in airframes operated in the U.S.A. after 1987 shall be of the required dual standard.

12.6.2 Interrogators

The transmitter output stages of current SSR interrogators need have only a very small mean power output capability, of the order of only 2 W typically. This is because peak power of less than 2000 W can be employed at repetition rates of not more than 450 per second for three pulses. The duty cycle is very low, being not more than 0.001 (0.1%). In mode 'S' systems the output pulse duration grows from three pulses of 0.8 μs duration to just over a total of 30 μs. These must be able to be repeated sixteen times with very short spacing between bursts resulting in the need to support duty cycles of about 60% for fractions of a second. As a result either solid state devices or vacuum tubes of the requisite mean power capability have to be used. This means that most interrogators currently installed will need radical modification, possibly even replacement. Most modern designs, of course, have anticipated this need, since the requirement has been known for some time.

Another less obvious implication is in the need to support the dual standard. This results in increased complexity of interrogators and the need to provide the means of 'scheduling' interrogations, to establish their priority, and to sort, format, queue, and route the two sorts of reply data for transmission back to the user who demanded it.

Part 4

Displays

13
The plan position indicator (ppi)

13.1 Display tubes

Most users will be already familiar with the principle of the cathode ray tube (CRT). In essence a CRT is a vacuum tube device which has:

(a) A rich source of free electrons.
(b) Means for directing them as a stream towards a display screen.
(c) Means for turning the stream into a finely focused beam.
(d) Means for deflecting the beam under control across the screen either by electrostatic or magnetic deflection.
(e) Means to convert the beam energy into light upon screen impact.

Those readers who wish to have insight into the design of such tubes are commended to seek reference 4. It is sufficient here to draw attention to salient points affecting the user. The display is made visible by the last item of the list above. The energy conversion is carried out by coating the inside of a transparent screen (usually glass) viewing surface with a phosphorescent layer. The viewer sees the converted energy as light emitted through to the other side of the screen, whose intensity varies as the beam energy; the faster the electrons and the higher their concentration in the beam, the greater the light output.

Thus CRT design centres around generating fine beams with high velocities. The first of a number of tantalising design problems arises from this requirement; for the higher the beam energy the more energy is required to deflect the beam into desired directions. The light output is also a function of the time for which a given small area of the screen is excited by the beam. Various chemical menus for the phosphorescent layer produce different light colours with varying persistency. The energy conversion is not instantaneous, neither is its dissipation. The time taken for this latter gives the output light persistence. Figure 13.1 illustrates the two regimes. The first creates screen fluorescence during the beam's dwell time on a given screen area, the second is the phosphorescent period which can be given various decay laws and durations. Values ranging between microseconds and greater than a second

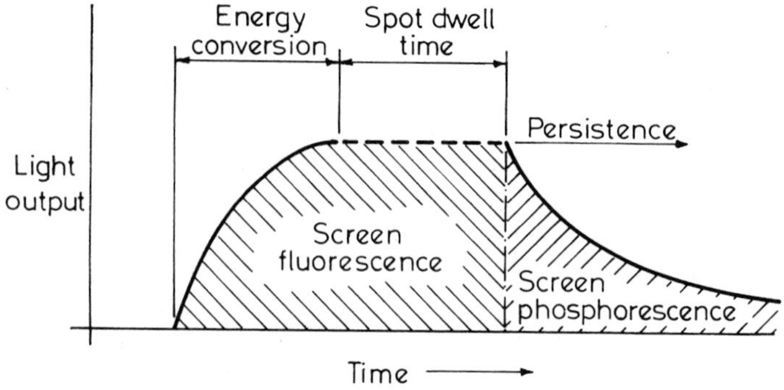

Fig. 13.1. History of ppi tube light output as the beam passes a point on the screen.

can be obtained, dependent upon the display requirement. Persistence time is usually quoted as that taken for the light output to fall to 1% of its peak. Radar display designers usually aim for beam cross-sectional diameters which result in spot sizes of between 0.5 and 0.25 mm. Obviously the smaller the spot size, the greater is the display's resolving power. If we take the average viewing distance as 50 cm and the average viewer's acuity as being one minute of arc, a spot size of 0.5 mm would not limit the viewer's ability to resolve displayed data.

13.2 Representing the range domain

Radar range (that is the slant range distance between the radar and a reflecting object) is proportional to the two-way journey time of transmissions to and from the reflector. This is because the velocity of propagation of electro-magnetic waves is constant. To make a CRT display directly capable of indicating range, it is necessary to deflect its beam from a fixed point (representing the radar's position) at the instant a transmission is made and to govern the deflection so that the beam explores the screen at a constant rate of x cm per unit of time. If the screen was 20 cm across and was required to represent 20 nautical miles of range then because the round trip time is at the rate of 12.36 μs/nautical mile then the beam would be required to be deflected at the rate of:

$$\frac{20 \text{ n. miles} \times 12.36\ \mu\text{s}}{20 \text{ cm}} = \frac{247}{20} = 12.36\ \mu\text{s/cm}$$

If the deflection rate was doubled to 24.72 μ/cm, the display screen would represent 20/2 nautical miles = 10 nautical miles. The designer seeks to match the deflection sensitivity of the CRT tube (V/cm in electrostatic deflection tubes, amps/cm in magnetic deflection systems), the tube dimensions and the required range scales so that the linear representation of range is accurate throughout the display.

The presence of a signal along this scale has to be indicated. In the 'A' scope technique this is accomplished by using the X deflection axis to move the CRT spot across the screen as described above, and using the orthogonal Y deflection axis to indicate signal amplitude. In a typical radar scenario such a display would be as shown in fig. 13.2. Such types of display were among the earliest used in operational radar systems. It clearly shows the history of receiver output with time and hence the range at which signals appear. It is equally clear that for a given beam deflection sensitivity the range scale is alterable by varying the rate of change of v with time.

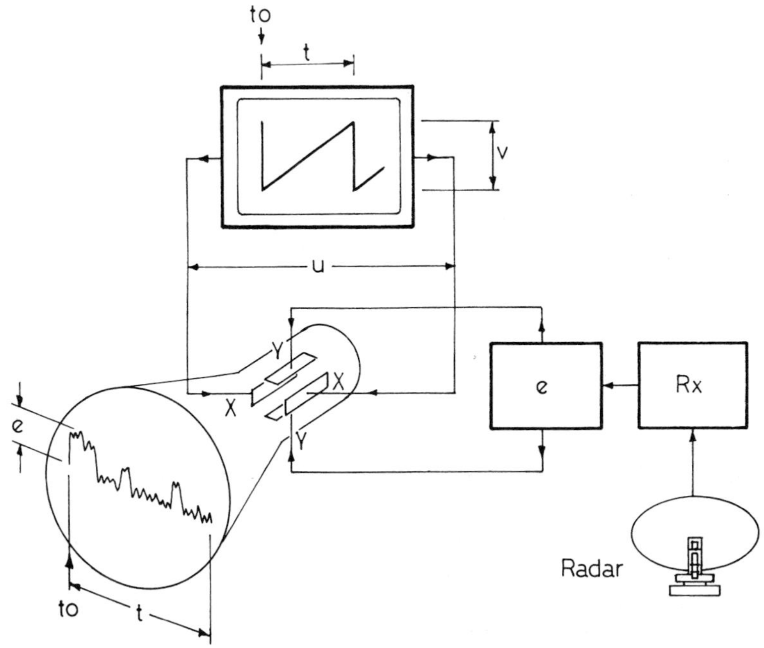

Fig. 13.2. The 'A' scope technique. Electron beam is deflected from left to right from t_0 at a linear rate for duration t by a time base generator of output u applied to the X plates of the crt. Simultaneously the radar output signals (e) are applied to the crt's Y plates. The radar signal's history with time then appears on the tube screen.

13.3 Representing the azimuth domain

13.1.1 Amplitude to brightness

The ppi is probably the best analogue of the radar system that can be devised. True to its name (plan position indicator) it gives a God's eye view of the radar output from overhead the antenna. To do this, the voltages expressing time and amplitude in fig. 13.2 must undergo changes. The trace indicated by t must be preserved but made *to take up a position on the screen which coincides with the antenna beam azimuth.* The amplitude e must be turned through 90° (i.e. to go in and out of the page) and made to express beam energy, to vary the brightness of the area of screen when a signal appears. The latter is easy to do; all that is required is to take e and make it modulate beam intensity. This is done by applying it across the grid and cathode of the CRT (see reference 4). The cathode is the source of the beam energy and the grid (an open wire structure) is interposed between it and the subsequent beam forming elements. If the grid–cathode voltage is varied, the beam energy varies almost instantaneously and in sympathy. Making the trace take up a direction which changes in sympathy with the antenna beam azimuth is more complicated but nevertheless easy to understand.

13.3.2 Azimuth telling

The first thing required is an element which expresses the antenna beam's azimuth. There are many ways of doing this. A common method, still in use, is to couple the antenna driving shaft to a continuously rotated potentiometer with two rotating pick-up arms at right angles to each other. If a constant voltage is applied across the potentiometer the two arms will pick off voltages that are directly proportional to the sine and cosine of the shaft angle. As the antenna rotates the sine and cosine voltages will vary between two equal peaks, one positive and the other negative. At the time of installation, setting up the alignment of the antenna and potentiometer can be arranged so that the sine voltage is at a minimum and the cosine a maximum when the antenna is pointing north (or any other desired reference). Subsequently, when the antenna rotates the beam's azimuth relative to the reference will automatically be expressed by the potentiometer's two outputs, $v \sin \theta$ and $v \cos \theta$ where θ is the beam's direction.

13.3.3 Producing θ on the tube

In fig. 13.2 the linear 'sawtooth' waveform has a constant rate of change of amplitude with time. One commonly used method of generating such a waveform is to use an integrating circuit whose voltage output increases in proportion to the time for which its constant input voltage is present. Its action is illustrated in fig. 13.3(a). A constant amplitude input, e, continues for a time, nt. During a small increment of time (t) the integrator gathers energy represented by $A = et$. It produces an output voltage $v = ket$ where k is a constant amplifying factor. As the area under the e curve grows, so the output voltage, V, grows in proportion. Thus after a period of $2t$,$V = k2et$. Because the input amplitude is held constant, the output will rise linearly with time

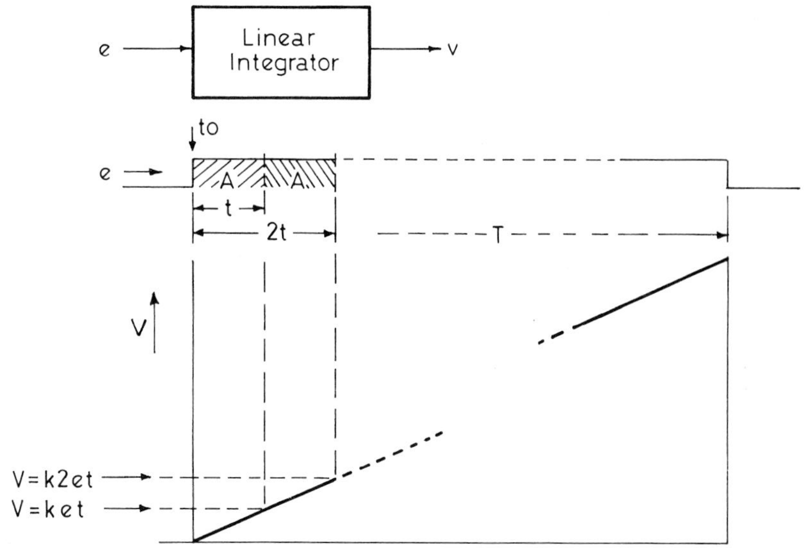

Fig. 13.3(a). A linear integrator. An input, e, is given at t_0 and exists for controlled period, T. As time increases, the circuit action of the integrator gives output V which is the sum of equal small increments A+A+A+A+ ... etc.

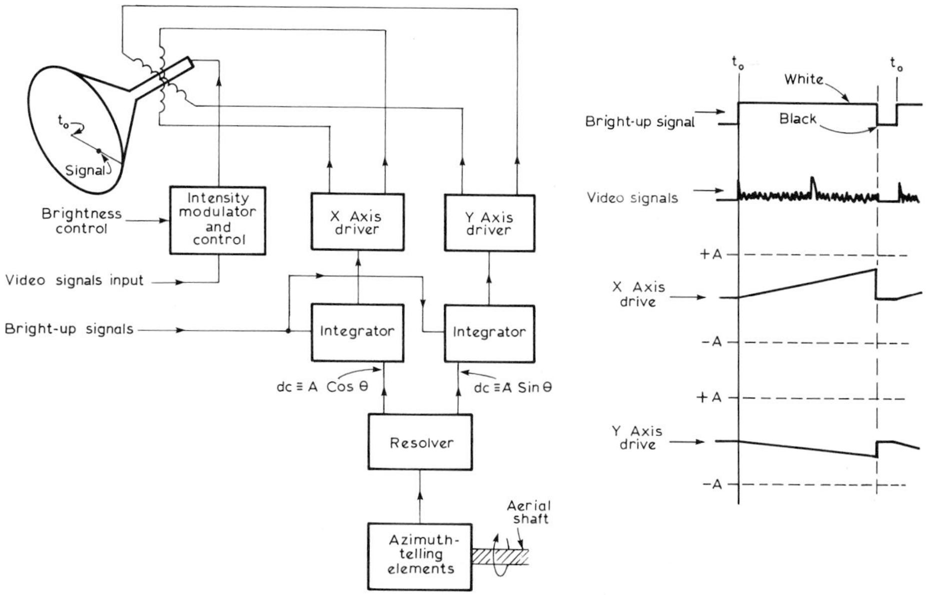

Fig. 13.3(b). How rotating ppi traces are made. (*Reproduced by kind permission of Butterworths.*)

until it grows to V = kT. To produce a ppi trace this principle is used as follows, illustrated in fig. 13.3(b).

Two integrator circuits are allowed to run from the transmitter firing time (t_0) for as long as required by the display to represent a given range. That is, $n \times 12.36$ μs, where n is the number of nautical miles to be represented. They receive dc input voltages proportional to $v \cos \theta$ and $v \sin \theta$. These change as the antenna rotates but at such a slow rate they can be considered as constant during each integration period, which lasts at most 4 ms. The integrators will generate linear sawtooth waveforms as described above. They change their amplitude and polarity in sympathy with the antenna's rotation as the resolved dc outputs are caused to change. The CRT beam will be deflected across the tube's face at the dictates of the two simultaneously present deflection outputs. In the example shown, the trace will be from the tube's centre at an angle of 105°. At the end of the integration period the beam will have been deflected to the tube's edge. The integrators are then returned to their quiescent state ready for the next integration period. These are generally synchronised on a one-to-one basis with the radar's repetition rate producing a ppi trace for each radar period.

The signals from the receiver are displayed by using their amplitude to govern the beam's intensity. To effect this, the grid and cathode of the tube have their relative voltage set so that in the absence of any radar signal input, the beam energy is just at the point of producing fluorescence of the screen. Any increase in grid–cathode voltage increases beam energy and hence screen excitation. This low level is controlled by a 'bright-up' waveform to which signals are effectively added.

Thus, as illustrated, a signal at range r will cause the beam energy to increase at the correctly scaled place along the tube's trace. Signals from ships or aircraft will be present at a constant range for a number of radar repetition intervals as they continue to be illuminated by the radar beam. As a consequence the returned signals will be drawn as an arc upon the screen, its subtended angle will be dependent upon the effective width of the antenna beam. The arc is actually a series of brightened spots side by side. If they are absolutely contiguous or overlapping they merge into the single short arc.

Thus the ppi models almost exactly the radar principle and its performance:

- The spot on the screen moves from a fixed point, representing the radar's position.
- The spot's rate of travel models the passage of the radar transmission through space.
- The radial dimension of the screen's excited area models the range extension of the reflecting object in space, complete with any 'shadowing' it causes.
- The direction of the trace models the antenna boresight and its movement.
- The azimuth dimension of the screen's excited area models the reflector's beam occupancy.

14
Non-real time operation

14.1 Introduction

The description so far has been restricted to real-time operation of the ppi. By real-time is meant that there is a continual and unbroken time relationship between the system input and its output reaction. For example, the trace representing range is drawn by the spot moving at a constant rate, the trace rotation continually follows that of the antenna, the signals to be displayed occur in time ordered sequence directly proportional to their measured range. As a consequence, isolated signals generate data for display only during the antenna's dwell time.

Consider a radar with beamwidth of 1.5°, rotation rate of 15 r.p.m., pulse duration of 1 μs and p.r.f. of 600 Hz. If its display tube of 40 cm diameter had its range scale set to represent 100 nautical miles radius an isolated target would be present as ten samples, each lasting for only 1 μs. This represents a display range dimension of only 1/1236 of the tube diameter. The ten samples would be spread across 1.5°, or 1/240 of the full 360°. In linear measure, the signal's area at 50 nautical miles range will be 2.6 mm × 0.32 mm = 0.83 mm^2. The tube area will be 1257 cm^2 or 125 700 mm^2. Thus the excitation time of ten times one microsecond is spread over a minute fraction of the screen and will only be renewed at each four second period – it's a wonder they can be seen at all!

Of course the viewer is aided by the tube phosphor's persistence and if the spot size was 0.5 mm, in the example above the ten samples would occupy 2.6 mm, so the spots would overlap. Thus integration would be present to enhance visibility. Nevertheless, such a real-time display has this obvious drawback of low refreshment rate. Another disadvantage follows from the need to put other data on the screen: alphanumeric characters, as in SSR, giving target identity and altitude. These have to be generated externally to the radar system proper. In the real-time example given above, a displayed range of 100 nautical miles will take a real-time of 100 × 12.36 = 1236 μs. The p.r.f. being 600 HZ gives an interpulse period of

1/600 s = 1666 μs. This leaves only 1666 − 1236 = 430 μs in which to exercise staggered p.r.f., to recover the display deflection system back to rest and to write any extra data on the tube. With writing speeds of the order of 25 μs per character it can be seen that not much extra data can be added. The way out of the dilemma is to use display techniques which operate in non-real time.

There are two main methods of achieving the aim of reducing the real-time necessary to display radar signals. One has already been discussed in parts 1 and 2 on primary and secondary radar – the technique of plot extraction in which target data (range, azimuth, coded data) are all extracted and stored as digital expressions. The other method, again a digital technique, preserves the real-time characteristics of the signal history but presents them at a faster rate after a radar period delay – this is the so-called 'radar video retiming' technique.

14.2 Plot extracted displays

14.2.1 General principles

The following text will describe the principles used in the technique. There are many variations to be found in numerous equipments but all are founded upon the same base: digital storage of data and its display in 'machine time'. Such techniques are possible because the data gathered on targets for a given antenna illumination will not be updated until the next antenna revolution. This data refreshment period is usually of the order of seconds. The user can normally tolerate data 'staleness' of up to a second or two, so the digital processing time between receipt of new data and its display is still much greater than in real-time systems.

The essence of the technique is found in fig. 14.1. Six target positions are shown. These will be indicated by bright spots on the tube at the correct location, sensed by the radar. The positional data enters as R, θ bits of the digital word expressed by the radar's plot extractor, which could be a great distance away from the display system. The digital words are converted into X–Y co-ordinates, usually (but not necessarily) taking the radar sensor location as the display centre's reference. The converted values are put as digital words into a store, the X and Y values for a given R, θ input being associated together.

Under control of a regular 'clock' signal which governs the display refreshment rate, the store is sequentially addressed. The X and Y values for a particular target position develop deflection voltages (or currents in the case of magnetic deflection systems). These are scaled to give direct linear proportionality to spot distance away from centre and target range, as in real-time displays. The X and Y values are held steady and await the arrival of a bright-up pulse initiated by the control circuits. When the bright-up pulse is applied to the CRT of the ppi, the beam will have already been deflected to

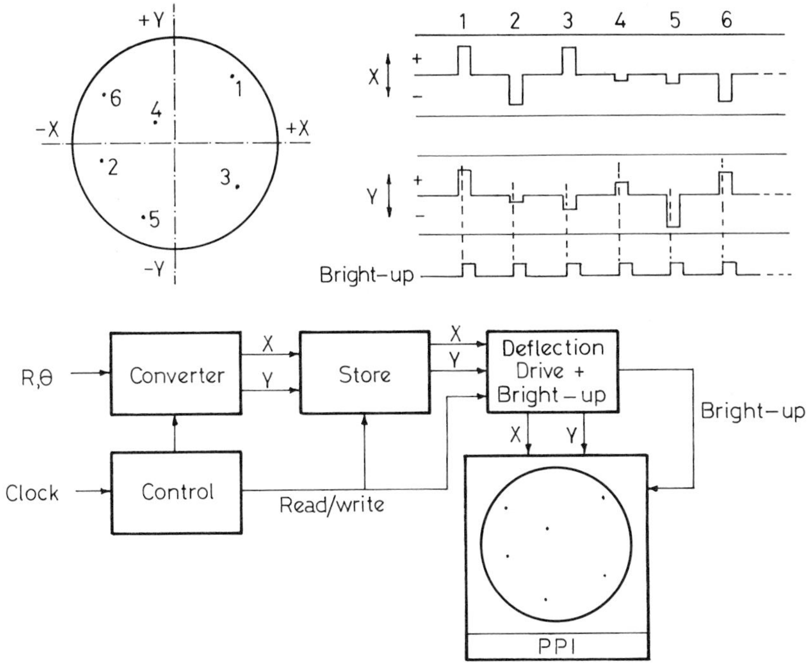

Fig. 14.1. By digital storage of X, Y position voltages, the crt spot can be moved to any nominated position. By repeating the 'read/write' cycle the bright spots persist at high intensity.

the appropriate position. The bright-up pulse allows the beam energy to increase and so excites the tube to display a spot. After the removal of the bright-up pulse, the store releases the next pair of X–Y values to be displayed. It will be realised that the 'read-write' cycle can be gone through many times per second before the R, θ value for a particular target needs updating. This allows the display to have high light output by repeated writing of the spot in one location.

The system display capacity now becomes a question of how fast the beam can be steered across the screen and the capacity of the store. From the waveforms shown in fig. 14.1 it is seen that the deflection waveforms are really a sequence of dc values of varying amplitude and polarity. By storing the past values of plot position it is possible to read them out in the same way as described above. The resultant series of dots form a track history enabling the user to estimate the position of future plots.

14.2.2 Writing symbols

Using the technique just described, the present and past positions of targets can be given. It is easy to visualise how it can be extended to write symbols associated with position. Suppose we wished the current plot position to be indicated by a square instead of a dot. One way of achieving this is to treat the position-determining waveforms as 'shelves' describing the symbol's centre of gravity and to superimpose small character-writing waveforms upon them. Figure 14.2 illustrates how this can be accomplished. The time frame has been expanded to show the action for one symbol more clearly. The order in which the sides of the symbol are written is shown together with its location on the screen. If a circle was required then the waveforms would take the form of two sine waves 90° displaced from each other. This is an example of the 'cursive' method of symbol writing as opposed to the 'mini-raster' or dot-matrix technique. Cursive formation is more difficult but leads to better character legibility.

In a practical system the *X* and *Y* deflection drives shown in fig. 14.2 would have these smaller waveforms added to their input. They would be called out from a pre-formed set, stored in a character generator, at the dictates of the

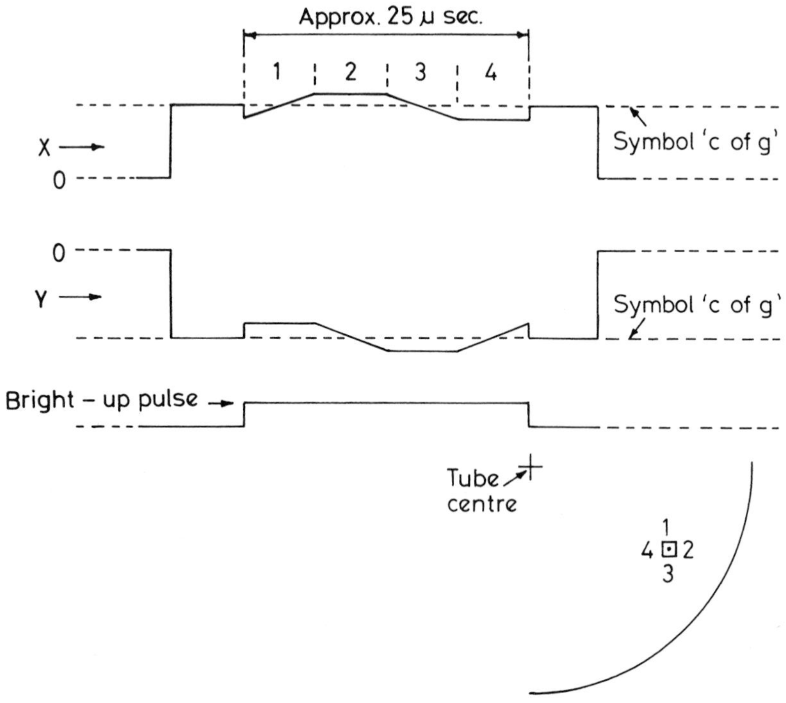

Fig. 14.2. Cursive method of symbol writing. The waveforms describing the square symbol are superimposed upon X and Y pedestals representing the symbol's centre of gravity. Waveforms for each side of the square (1 to 4) are indicated.

control circuits. Associated characters (squares, circles, diamonds, etc.) giving data on plots would be treated in the same way. Such data would be part of the digital word already containing the relevant R, θ bits and so complete association can be organised by suitable control signals. The method described above is sometimes called the 'synthetic display' technique. It operates in non-real or machine time. It is however possible to produce displays which preserve the real-time element and still give the advantages of the synthetic display – the 'mixed real-time and synthetic' system.

14.3 Mixed real-time and synthetic displays

14.3.1 Trace stealing

It takes time to display a large number of data characters. This leaves less time to devote to a real-time writing regime. How then to display real-time signals without sacrifice to data display time? Early attempts at this compromised the quality of the real-time display by 'stealing' trace time which should have displayed the radar signals and devoting it to symbol writing. Perhaps as many as one trace in twelve would be lost. It was claimed that if never the same traces in any antenna revolution were 'pinched' or 'stolen', then not much would be noticed by the viewer. It nevertheless constitutes a loss of target detectability, for which designers fight hard!

A solution has been found to the problem by use of a digital technique to produce a 'pseudo real-time' presentation. It is variously named as 'video retiming', 'video time compression', 'video digitising', etc. They all use the same principle which, with modern fast micro-chip devices, is delightfully simple.

14.3.2 Video retiming

The real-time display is characterised by its ability to preserve the time relationships between events from a starting reference. Take as example a row of dominoes stacked so that if the first fell it would knock its neighbour and so on to the end. If such a falling set was filmed and shown to an audience at rates different from the original, the essentials of the scene would still be there – the eleventh domino would not fall until struck by the tenth, the last would always be the last to fall.

In the real-time radar display much the same is happening; the speed of the electro-magnetic waves is much greater than that of the spot across the display screen. What has to be preserved is the *order* in which the events occur and to ensure their time proportionality. This is the core of the video retiming principle. It is best explained by reference to fig. 14.3(a) and (b). The first shows how the video input of signals and noise in time is graded, in

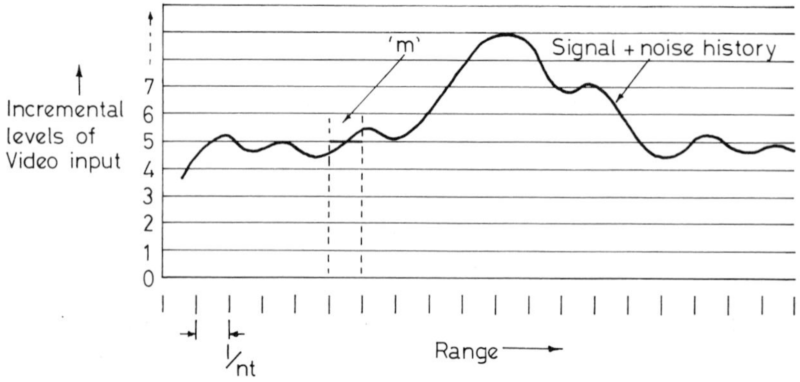

Fig. 14.3(a). Expressing an analogue signal in digital form. Range is divided into equal increments, typically one quarter of a pulse duration, t. In the example, the 'mth' increment has to be converted to a voltage level of 5 to equivalent binary value (10100 in a 6 bit system).

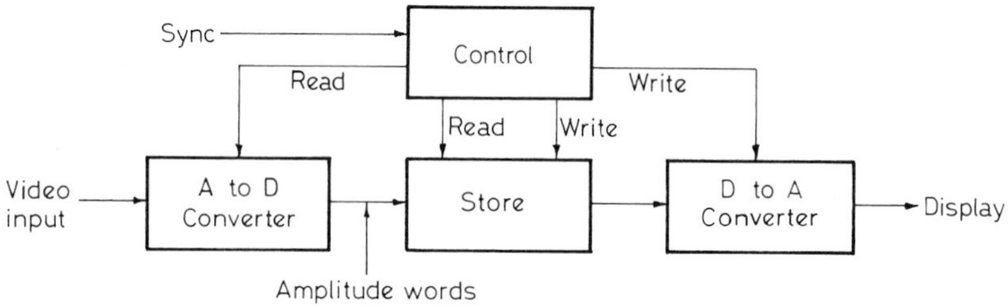

Fig. 14.3(b). Video re-timing system. Digital words expressing input amplitude in real time are stored. Read-out time and rate are under the designer's control. Output from digital store is re-converted to analogue form for display.

an analogue to digital converter, into amplitude values for each of successive time increments which are of duration $1/nt$ where n is greater than 2 and t is the received radar pulse width. Typically four samples per pulse are taken so in a 1 μs pulse width radar, $1/nt$ would be 0.25 μs. Under the control of signals synchronised to the radar system the digital expression of the input for each time sample is put into store as a parallel word. It would be usual to find the amplitude expressed in six bits. Thus the value put to store for the mth range increment exampled would be 5 = 101000, (1 + 4).

The sample rate has of course to be equal to the radar input rate and so a 100 nautical mile radar range would take 100 × 12.36 = 1236 μs to fill up. In the example above, the range would be expressed as a thirteen-bit binary word to encompass 1236 × 4 microsecond samples. When the radar input has been stored under the control of the 'reading' clock pulses, the digitised data in store is ready to be written out to the display. This can be done much faster than the reading speed, provided the time proportionality of the store contents is maintained. The store contents are written out to the display in range order at a constant clock rate to achieve this. The clock rate can be governed according to the amount of time available to display the stored video. The radar signals are displayed one radar period after their input to store thus the viewer is unaware of the process. In the example of a 100 nautical mile radar, requiring an interpulse period of 1666 μs, the time sharing between character and radar display writing could well be in the ratio of 3 to 1, giving 1250 μs to synthetic data writing and 416 μs for the 'pseudo real-time' regime.

The video signals written to the display have of course to be reconverted from parallel six-bit digital form to equivalent analogue form in a digital to analogue converter. Experiments have shown that the process of A to D, storage, and D to A conversion involves very little loss (about 0.5 dB). The technique has other advantages to the user. In normal real-time displays, a change of range scale from 100 to 20 nautical miles maximum displayed range involves a five-fold increase in the beam's writing speed. Even if the beam energy is altered to give the same threshold of excitation, the dwell time of the spot on a given small area changes by 5 to 1. Thus at the faster writing speed, legibility suffers and the display controls require adjustment. In the video retiming system the writing speed can be held constant if range scale adjustment is also made to change the rate at which the store is addressed. It is possible to use the allotted display time to show segments of the whole range scale; for example any 20 mile section of a 100 mile display can be selected by arranging a counter to start at the nth clock pulse from zero and to stop at the mth, the count in between representing 20 miles worth of 'writing' clock pulses.

It is necessary in using mixed display technique to have two means of generating deflection waveforms; that for the display of synthetic data will be of the type described in section 14.2, that for the video retimed data will be as described in section 13.3. A deflection switching system is employed to change automatically from one to the other as required.

14.4 Other azimuth telling methods

14.4.1 Digital devices

In section 13.3.2 an example of a simple azimuth telling sensor was given for the sake of clarity. Modern systems digitise the azimuth data at the very start of the process by using a variety of digitising sensors. Commonest among these is the 'digital disc' which outputs a pulse every time the antenna rotates by a small angle. Their discretion varies from ten to fourteen bits for a full 360°. Each increment of a ten-bit disc represents 360/1024 = 0.35°; that of a fourteen-bit disc, 360/16 384 = 0.022° (1.3 minutes of arc). In search radars, whose antenna rotation is constant and uni-directional these types are preferred since all that is required to produce an azimuth report is to count the successive pulses output from the sensor, after establishing (by a separate pulse output) where zero degrees begins.

The pulses expressing antenna rotation are usually called 'azimuth change pulses' (ACPs). They are generated by a circular disc of transparent material around whose edge a series of equispaced opaque radial bars is etched. The width of these bars, their number and spacing determines the number of increments into which the circle is divided. On one side of the disc is a small light source whose output is allowed to pass through the transparent spaces between the radial bars. The light passing through impinges upon a light-sensitive diode which conducts when exposed to light. The diode's changing resistance, as the light is switched on and off by the passing of bars and spaces, is used to generate electrical pulses forming the ACPs. The disc is mechanically coupled to the antenna turning shaft. Modern devices are actually mounted round the antenna's final driving torque tube. This obviates any back-lash in gear trains which otherwise must be used and can create azimuth error. Such devices are known as 'around the mast' digitisers.

The arrangement described above, generating ACPs and a north pulse is obviously a 'serial' system; that is, it only transmits changes in antenna position. It is obviously possible to make digital discs with multiple tracks of bars and gaps to express the antenna azimuth as a parallel digital word. In these, one track corresponds to each bit in the digital word. Using such devices it is possible to tell the antenna position in absolute terms without using a separate north or zero reference pulse. The absolute azimuth reporter is necessary if bi-directional rotation is used in the radar, e.g. for sector scanning systems.

In both serial and parallel cases, the digital words generated by the discs can be either used directly in plot extraction systems or put to digital resolvers, whose output consists of signals of amplitude proportional to the sine and cosine of the antenna azimuth position. Thus all display requirements can be catered for.

Optical discs have alternatives in various inductive or magnetic devices. All of these have the same design aim: to generate a pulse every time the antenna rotates a given small amount.

14.4.2 Analogue devices

Older equipments generally use analogue devices; one such was described in section 13.2.2. Most use the rotary transformer technique. Some called 'selsyns' are torque transmitting devices with enough mechanical output power in their receivers to be used to bring mechanical shafts remote from the antenna into azimuth alignment. In essence they are three-phase rotary transformers, coupled in parallel by remoting cables. The rotor of the transmitting selsyn is turned by the antenna rotation by mechanical coupling. It is energised by an alternating current source. As the rotor takes up various positions relative to the field coils of the transmitting selsyn, currents of varying magnitude are induced in them. These are reproduced in the field coils of the electrically coupled receiver selsyn at a remote point, whose rotor has the same electrical source as that of the transmitter. As a consequence of the magnetic reaction between rotor and field coils, the receiver's rotor takes up the same position as that of the transmitting selsyn. Sometimes selsyns are used in pairs coupled through a gear box to form a 'coarse' and 'fine' arrangement. Gear ratios of 30 or 36 to 1 are commonly used. This reduces azimuth correspondence error and increases the mechanical power available at the receiving end of long cables which naturally involve power loss.

Similar devices known as 'synchros' are sometimes used, again as coarse and fine pairs. These are small, which has certain mechanical advantages, but their small size makes them an unsuitable source of mechanical power. This is made good by a servo-amplifier system at the receiving end whose drive motor is used to force the receiving synchro's shaft positions into correspondence with the transmitting elements and has enough power in reserve for mechanical use in driving resolvers such as the 'sine/cosine' potentiometer already described. In synchro and selsyn systems it is possible to achieve azimuth accuracies of about 0.1° with careful design. The gear trains and transmission losses inevitably sets a limit to accuracy.

15
Limitations to performance

15.1 The analogue ppi display

15.1.1 Brightness

A number of factors combine to make the limited brightness of the ppi a serious drawback leading to variations of its design in an attempt to overcome it. As already pointed out, the brightness is a function of how much beam energy of the CRT can be converted to light. This in turn is dependent upon the chemistry of the tube's phosphor layer and the power incident upon it in terms of watt seconds per square metre – or in more practical units, milliwatt microseconds per square millimeter; for writing speeds and spot sizes are of this order of magnitude.

A technique which attempts to improve brightness is to use a television type method of display. In this, the radar data are stored in real-time at the normal refreshment rate of once per antenna revolution. The data are then repeatedly read out at a much faster rate. The most modern form of this is the so-called 'raster scan' display. Here, the radar input is quantised in time and amplitude and stored in digital form much as described in the video retiming process earlier. The stored data are then scanned as in the television technique, but having about 1000 lines per picture instead of the usually 625 or 819. Because the stored data are retained for as long as the user wishes and the play-out is continuous at a much faster rate than one antenna revolution, the picture brightness is of television standards. Early attempts at producing such a system used a special form of CRT with two electron beams, one for writing the real-time data upon a non-visible charge storing screen and the other, scanned as in television, to read the stored data into television viewing units. This 'scan conversion' technique never really found favour because of its tendency to produce flickering pictures on large display screens and to suffer from the too-coarse line structure of the final play-out. Modern raster-scan pictures have greater readability because of the finer 1000 line structure and the digital techniques used.

Another method is to use the direct viewing storage tube (DVST). As its name implies, this display has its own viewing screen, again converting the tube's beam energy into light, but with a profound difference. The DVST has two electron guns, both directed at the tube's phosphor-layered screen. Just behind the screen, a very fine mesh forms the means of storing the charge conveyed by the fine scanning beam. This is deflected and modulated as in the normal ppi. Thus any radar signals appear as small charged areas on the mesh. The charge will stay for very long times indeed. The second beam is very diffuse and not scanned at all. Its low velocity electrons continually 'flood' the mesh. Wherever the mesh has a charge upon its surface the low velocity electrons pass through and impinge upon the phosphor of the screen and their energy is converted into light. Because these low energy electrons are in a continuous stream they continually pass through the mesh's charged areas and so continually excite the same area of phosphor on the screen. The result is extremely high brightness – bright enough to be adequately viewed with sunlight directed onto the viewing surface. Arrangements are made deliberately to allow the charges stored upon the mesh to leak away in operator-controllable fashion, otherwise the screen becomes cluttered with literally hours of radar signal history. One such display, the distance from threshold indicator, still widely used in airfield control towers, is illustrated in fig. 15.1.

Fig. 15.1. A distance from threshold indicator (DFTI). It displays real-time radar signals in very high ambient light. Range rings are 5 nautical miles apart. Display centre is set so that zero range on the graticule is equivalent to runway touch-down point. Two aircraft on final approach are shown. (*Photo courtesy of Marconi Radar.*)

A serious limitation of the DVST in all its forms is the inability to write useable alphanumeric characters without undue technical complication. It also suffers from the effect of charge-spreading on the mesh which ultimately detracts from its resolution, because effectively the spot size grows with time. The screen can be 'wiped' clean by totally discharging the storage mesh with an electrical signal. Applying differential discharge (i.e. limiting it to controlled areas of the screen) is again extremely difficult, making any alphanumeric characters 'smear' as they move across the screen.

Another limitation associated with ppi brightness is present because of the CRT's constant spot size. Traditionally the spot is made to start its scan from a fixed point (often the tube centre) to represent the geographic location of the radar itself. The spot is scanned regularly from this point to the tube's edge and at each new radar repetition period the antenna has moved slightly in azimuth. If the radar p.r.f. was 600 Hz and the antenna rotation one per six seconds (10 r.p.m.), there would be $600 \times 6 = 3600$ radar scans per 360°, i.e. ten repetitions per degree. For a tube diameter of say 40 cm, its circumference is 40×3.142 cm $= 125.7$ cm so one degree occupies 0.35 cm which for a spot size of 0.25 mm is 14 spot sizes. A signal subtending an arc of 1° at the tube's edge is represented by ten bright spots, each of 0.25 mm width, separated by the equivalent of four spot sizes which have no signal excitation. The light output per square centimetre is therefore small. The ten spot sizes would lie exactly side by side with no gaps between at a place along the trace where ten spot sizes (2.5 mm) subtended an arc of 1°. This occurs at a distance from the scan start of about 14 cm. Out to this distance the ten spot sizes overlap, being totally overlaid at the scan start point.

A number of effects result from this, not the least of which is a tendency for the brightness per square centimetre to be greater at the tube's centre, diminishing towards the tube's edge. Brightness correction waveforms which, like 'swept gain', are range dependent are often used to compensate for this.

Another effect resulting from this tube/spot geometry is the variation of signal integration which takes place on the ppi screen itself. This was touched on in part 2, chapter 7 and produces a hidden integration loss factor described below.

15.1.2 'Collapsing' loss

W.M. Hall (reference 7), in his early paper on radar performance calculation, posited the loss factor which detracts from the full force of 'on the screen' integration. It is more readily understood diagrammatically as in fig. 15.2 which shows the video signals and noise which modulate the CRT beam energy. A tube spot size is also represented. Imagine the page is the screen and the spot, much magnified, is scanning left to right. As it moves the beam energy will be represented by the integrated values of the modulating video across the spot's area. The spot is shown straddling a 'noise only' and a 'signal

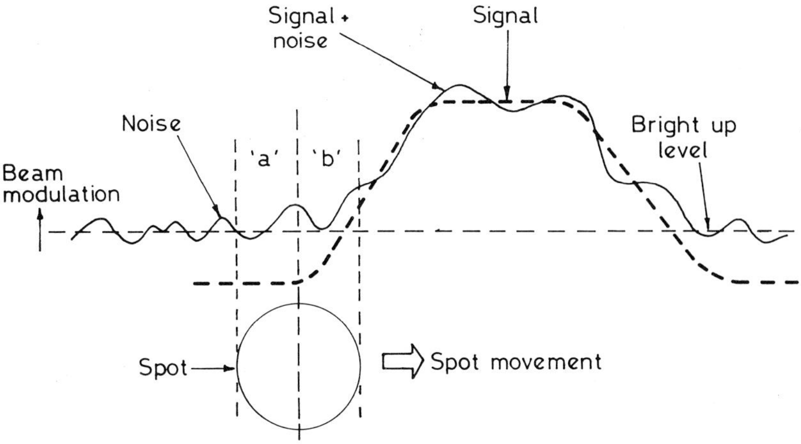

Fig. 15.2. Illustrating W.M. Hall's 'Collapsing Loss'. Dwell time of the spot on the screen can be long enough to integrate noise events across the spot which degrades the signal to noise ratio.

plus noise' region. Thus the 'noise only' energy in region a will add to that of 'signal plus noise' in region b. So the light energy from the spot at the indicated position will have extra and unwanted noise added to the already noise-polluted signal.

It can be readily appreciated that the loss introduced is a function of the tube writing speed and spot size. The first varies with changing range scales, the second is constant. W.M. Hall (reference 7) has formulated a method of calculating the loss which grows from negligible values at high writing speeds (short range scales where there are many spot sizes per pulse duration) to values of about 1 to 2 dB for long range scales wherein a spot size approaches or exceeds a pulse duration. The latter would be represented in a system having a 200 mm tube radius and a spot size of 0.5 mm where the range scale was 120 nautical miles per radius and the pulse duration 2 μs. In this case one spot size would encompass 1.5 pulse durations. It should be remembered that this 'collapsing' loss occurs at both the beginning and end of the wanted signal's history.

15.1.3 Dynamic range

The radar system itself has to preserve wide dynamic range in order to assist signal processing in clutter rejection. However, ppi displays are primarily designed to show the weakest signals with greatest effectiveness. This concentrates its performance at the noise level itself as indicated by the bright-up level in fig. 15.2. The display need not have very wide dynamic range since once the signal has crossed the bright-up threshold its presence will be manifest. This last statement is to some degree making a virtue of necessity, for the dynamic range of the phosphor on the screen is very limited, going from minimum visibility to maximum or saturation point with a signal amplitude change of about 10 dB or 3 to 1. Amplitude history when required, as in weather sensing radars, has to be inserted by other means such as special symbology. As an instance, areas of strong, medium and weak weather signals can be indicated by 'contouring' the boundaries between such levels using multiple threshold circuits and indicating areas between the contours by close or wide cross-hatching lines produced by a plot extraction technique and overlaid upon the real-time signals.

15.2 Synthetic ppi displays

15.2.1 Discontinuity

For many users, the traditional analogue ppi has the virtue of continuity in that it is such a perfect model of how the radar is operating. As the input signal changes, so does its displayed form. If, as is usual, the display is adjusted so that at least some radar noise events are visible, the viewer can make judgements regarding signal character and quality. With a little knowledge of factors affecting radar performance it is possible for the user to take account of input signal quality and strength and mentally put these into quite complex tracking algorithms which are virtually impossible to write into computer software. The following simple example is relevant.

Most users are familiar with the phenomemon of antenna beam shape and that its main beam has sidelobes. Targets large enough to create signals from sidelobes will appear in an analogue system as shown in fig. 15.3. The plane A, B, C, D represents the threshold above which signals will register on the display. The registration is depicted below the plane but deliberately expanded for clarity and is the plan view of the series of signal pulses expressed as the antenna sweeps across the target.

The experienced user will recognise the morse code letter R as caused by the main beam plus its two sidelobes. He is unlikely to be put off (as a plot extractor would be) by the elliptical shape of the main response caused by a target at a single range but generating individual signals varying slightly in range. Because the received signals are not of rectangular shape, having finite rise and fall times, the larger the signal, the earlier in time its amplitude will cross the threshold. This time variation evinces itself as apparent range variation. Automatic signal detectors based upon seeking range correlation from one repetition period to the next have to have special logic to account for this. Plot extractors also have to use counting logic which will ignore the (usually) small number of genuine signals created by the sidelobes.

The experienced user has these kind of rules already in his head and can attach 'quality' to such plots which in turn will modify his judgements about

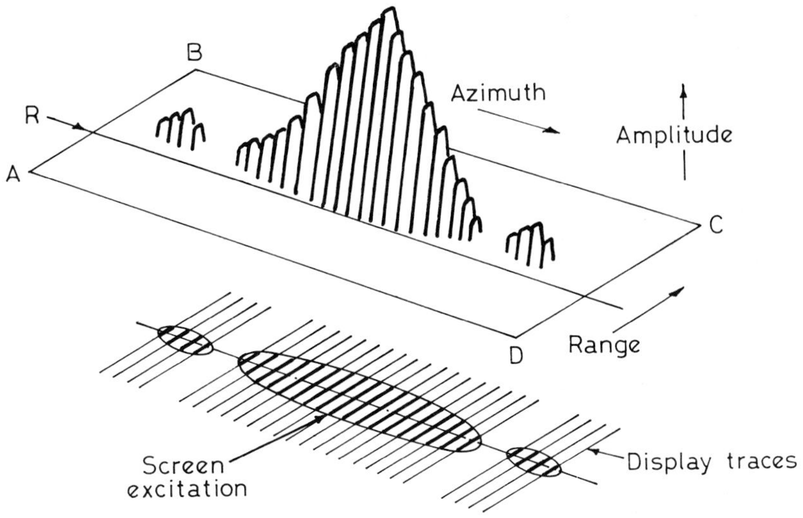

Fig. 15.3. Representation of a signal being registered upon the screen. Surface A, B, C, D is the threshold of visibility (bright-up level). As signals grow in strength across the antenna beam so their apparent range extent at threshold grows, imparting an elliptical shape to the resultant 'blip'.

their validity in tracks. Synthetic displays which replace the real-time history of the signals by a single symbol deny the user any kind of information about signal quality or at best can only express it crudely. Current moves towards introduction of track display are even further away from allowing the user real-time data. The track data are based upon plot association and by displaying only the latest plot plus vectors representing the sensed track it is hoped to avoid confusing the user with false plots. Let us hope that the designers do not use the power of tracking algorithms as an excuse for poor signal processing which can lead to false plots which the tracking function could hide.

15.2.2 Data clutter

There is a limit to the amount of data the user can assimilate when it is presented in alphanumeric form and symbols on the display. In modern equipments the storage capacity and writing speed are more than sufficient to cover the screen in non-flickering groups of symbols, numbers and letters in 'plaques', so much so that the plaques can merge. Arrangements can be made to correct this should it happen by either manual or automatic means. But there is a limit to the number of plaques of given character aperture (typically twenty-four per plaque) that can usefully be displayed. The larger the

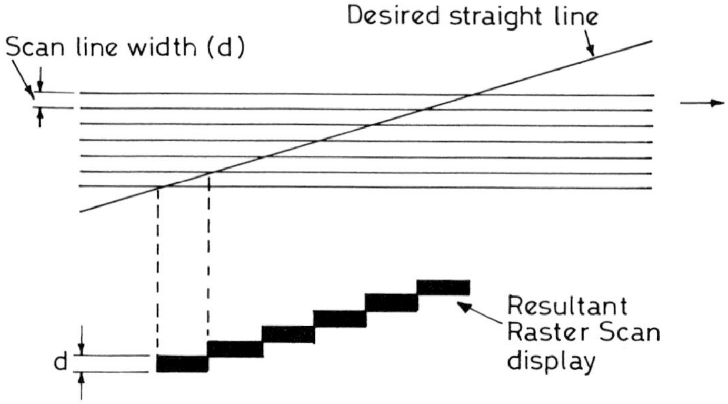

Fig. 15.4. Showing how lines can become ragged in a scan-converted display system. Smoothness is improved by increasing the number of scanning lines.

characters, the more readily they are understood. In turn the plaque size increases and their number reduces for a given display size.

15.2.3 Grating effects

In scan-conversion or raster-scanning display systems it is common to use straight lines to indicate geographic or navigational features. If these straight lines are scanned in television mode as a series of very closely spaced straight lines an unfortunate grating effect occurs which truncates the desired straight line into stepped strokes. The more parallel the desired line and scanning lines become the greater is the effect. An illustration is given in fig. 15.4.

Part 5

Into the 21st Century

16
Tracking

16.1 General

There are two classes of radar tracking. One is a technique wherein a radar with a narrow 'searchlight' beam or beams, symmetrical in shape in azimuth and elevation, acquires a target and uses the returned signals to steer the radar antenna mechanically and automatically so that its boresight always follows the target as it moves. The changes in antenna azimuth and elevation angles and the range are used to calculate the target's three-dimensional track. The technique is known as 'within beam tracking'.

The other class is one wherein a surveillance radar continually scans in azimuth. Plot positions of targets detected by it are stored in a computer. At each successive antenna revolution, plot position changes are sought and found. Speed and direction of targets are calculated from their displacement in the time taken for one antenna revolution. This is known as 'track-while-scan'. Both classes of radar tracking present fascinating problems for designers which will now be discussed.

16.2 Within beam tracking

This technique is widely used in defence radar systems for tracking enemy targets and guiding defensive missiles. Although many different engineering technologies and error measuring methods are used, they all use the same general principles of servo-mechanics. An example is as follows:

Figure 16.1 shows an arrangement of one antenna producing three beams – two, with receivers, symmetrically disposed about the antenna's mechanical boresight and one for transmission (an illuminator) central with it. In the diagram it should be noted that beam *shapes* are indicated, not their extent in range. The receivers are carefully controlled so that their gains are equal and so that their outputs for given inputs are carefully matched over a wide dynamic range (i.e., they 'track' together, as you will hear engineers say).

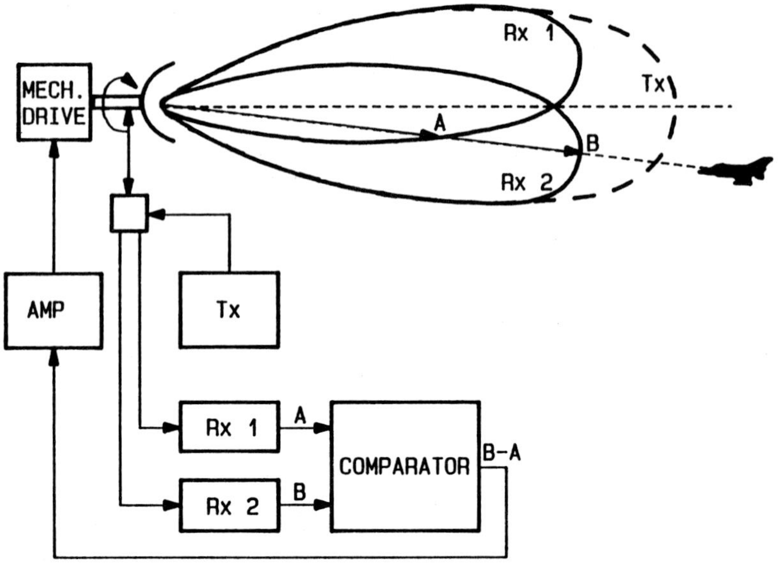

Fig. 16.1. Simple servo system. Two off-set beams produced from one antenna have cross-over points on the antenna's mechanical boresight. Each feeds two balanced receivers. Amplified signal comparison develops mechanical drive to move the boresight in line with the target. Another pair of orthogonal beams would give automatic target following in three dimensions.

Under these conditions, if the boresight is directly in line with a target, the signals from each receiver will be equal. If, as shown in fig. 16.1, the target is to the right of boresight, the signal from receiver 2 is greater than that from receiver 1. The comparator senses this difference and passes is to an amplifier which amplifies the difference to a level capable of driving the antenna mechanically in a direction which reduces the original difference. If, as is common, the boresight is driven past coincidence with the target bearing because of the momentum of the antenna (i.e., it 'overshoots'), the signal difference between receivers will once again grow. Again, the difference results in antenna drive power in the correct direction to restore coincidence between boresight and target bearing.

The art of servo-mechanical design resides in reducing this 'overshoot' characteristic without the need for 'neurotic' levels of overall system gain whilst producing the smallest residual error of correspondence between antenna mechanical boresight and target bearing.

The example above explains how the boresight is directed to a target in the horizontal plane. A practical system has another pair of beams orthogonal to these in the vertical plane with mechanical drive to match. Thus the whole system is capable of three-dimensional tracking targets. It is usual to operate this type of tracker in harness with a surveillance radar which directs the four beams to the correct azimuth area. An elevation search on this bearing suffices for the tracker radar to acquire the target in the vertical plane.

16.3 Track-while-scan

In essence, as described in section 16.1, this technique is simplicity itself – until we try to do it! As usual, errors of measurement cloud the issue, and once again they are largely probabilistic. Tracking is preceded by plot extraction; subsequent logic examines stored plot history, makes judgements on current plots and, commonly, makes complex predictions about the future expected positions.

16.3.1 Errors and their consequence

We have seen that position data from plot extraction are subject to range and azimuth errors. Both consist of two major components – bias errors and random errors. An example of range bias error would be the fixed small time difference between the system's true zero time reference (transmitter firing time) and that set into the processing and display system; typical values would be the equivalent of $^1/_{16}$th of a nautical mile in a 200 mile radar. In SSR another example follows from the I.C.A.O. specification of transponder delays. A fixed delay of 3 μs is specified between the finish of a valid interrogation and the start of the required reply. This is OK because the ground system can account for this in setting zero time references. However, a tolerance of ± 0.5 μs in the transponder is permitted to the 3 μs. Although this applies to the whole community of transponders, each individual unit will have its own fairly stable value within this range. That is, there can be two transponders in one interrogator beamwidth, one of which has a 2.5 μs reply delay and the other, 3.5 μs. Because of the stability of these delays over long periods, range bias errors of ± $^1/_{24}$th of a nautical mile result because the ground interrogator system would set its reference on the basis of a 3 μs delay.

Random range errors result from variation of target signal strength relative to a fixed system receiver noise level. From chapter 7.2 we note that, in primary radar, target 'glint' (or effective echoing area variations) over a number of successive antenna revolutions can produce massive variations in signal strength. Referring to fig. 15.3 we can see the effect of this. The plane ABCD represents the first threshold level in a plot extraction system. The weaker signals at the beam's edge cross the threshold at a slightly greater range than the stronger ones at the beam centre. This can be compensated for in the plot extractor by examining the signal history throughout the beam and relating this to the known beam shape. However, the same target at the next revolution of the antenna could exhibit a much smaller echoing area and all the signals from that target in the beam will be at a lower level. This introduces a rev-to-rev range error which is as random as the echoing area.

Equally, it has been shown that azimuth errors result from plot extraction because of the interaction between signal strength variations and the fixed logic in the extractor. Examples can be seen in fig. 11.9 but a better one is shown in fig. 16.2. Three major tracks are seen, the result of exposing the film for 20 successive antenna revolutions. The dots represent the plot extracted target position from the associated 'real time' target history. Variations of signal strengths from rev-to-rev are clearly visible; the number of pulses per beamwidth vary from about 10 to 20. The resultant plot position errors are equally clear and the job of a tracking routine is to recognise and smooth out the random errors.

Just by visual inspection of fig. 16.2 we can see the value of plot history. It is not difficult for us to perceive the average of past plots and imagine a smooth line passing through them closely resembling the true tracks. The relative speeds are also easy to judge because of the different lengths of track laid down in the same time. Indeed, since the range rings are at 10 nautical mile intervals, and the antenna rotated at 10 r.p.m. we can judge the 290° ~ 110° (NW to SE) track to

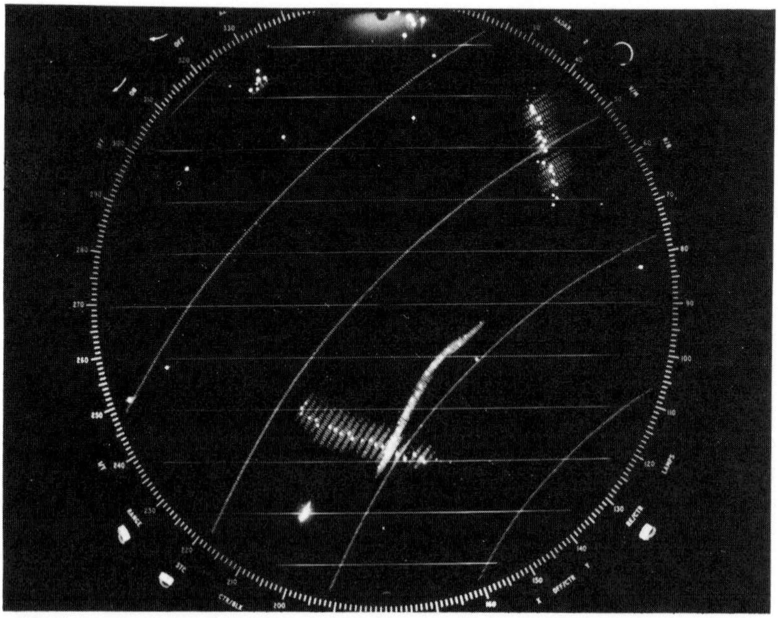

Fig. 16.2. Off-set and expanded ppi photograph exposed for 20 antenna revolutions. Three tracks are shown. Dots show the plot positions extracted from the corresponding radar returns (an average of about 15 pulses per beamwidth). Range rings are at 10 nautical mile intervals. (*Photo courtesy Marconi Radar.*)

have a speed of around 300 knots (20 plots in 10 nautical miles taking 20×6 seconds; 10 miles in 2 minutes = 300 in 1 hour).

This example has caused us to do exactly what a tracking routine does: deduce the speed and direction of targets. The radar has told us position data only; the rest we have obtained by studying plot history. But still this is a massive over-simplification. In studying the picture we have the benefit of seeing the real-time data from a succession of plots. Particularly for the north–south track (350–170°) we can see great variations in the quality of the basic radar data; some plots are based on very weak signals and thus less reliance should be placed on the plot detection for these. Many tracking routines are not given these data on the quality of the radar signals.

Our judgement on the direction of the craft is based on 20 revolutions of history. If we used the same mental processes on any five successive revolutions we might come to different conclusions and for any three successive revolutions our judgements could be in great error regarding both speed and direction.

A tracking routine emulates our own mental processes; i.e., it goes through averaging processes over a period of time.

If we represent the sequence of plots forming a track as a set of small arrows joining neighbouring plots, each arrow can be considered as a vector describing the target's movement at the time of the last measurement. These are commonly called 'state vectors' and it is really these that are the subject of our mental averaging and the source of a tracker's processing.

Uncertainty about true plot position means that each successive plot sample can introduce an 'across track' and 'along track' component of error. Averaging over a period effectively 'dampens' the sensitivity of the tracker to real changes of speed and direction – that is, the tracker has to experience a number of successive plots before a real trend shows in the continued averaging process. This can be appreciated from examining the tangential track (NE to SW) in fig. 16.2. There has obviously been a change in direction (one would judge by about 20°). But at what point in the track history would one be sure that this is true? Examining the 20 examples it looks as though it began at the seventh plot from its top end. By the tenth plot it looks certain. This means that even if we had seen this picture emerging in real time, we would have had to wait 24 seconds to make this judgement. In military radar systems this could become intolerable if the target was an enemy who would never announce his intention. For co-operative targets, the turn judgement could be made by a tracking routine (or by a human operator) if told to expect a manoeuvre.

16.3.2 Prediction

The most important decision to be made by the tracking programme is whether to include new reported plots into tracks – is this latest plot from the same target which made the others in a track? Or is it the first of a new track?

In fig. 16.2 there are about eight individual scattered and isolated plots. Are they the result of false alarms? Are they the first plots of new tracks? Programmes 'argue' in much the same way as you and I: on the basis of logic and probability. How would we begin to assess whether a new, isolated, plot was really part of a track or not? We would wait until the antenna came round again and see if another plot was registered in the same general area. If it did, and the possible target had a speed in the allowable bracket, then we would look for a third plot at roughly the same spacing as the first two and in the general direction of this possible virgin track. The word 'allowable' has been used above in the sense that targets have physical limits to their rate of movement, e.g., one would not expect a ship to travel at 200 knots or execute a 5 g turn.

The probability that two false plots would take us this far into the world of chance is pretty low but still not low enough for comfort; so we look for a third plot. But this time our search area would be much smaller than it was for the second plot, for we have made judgements about the target's speed and heading. If the third plot was indeed found in this area, the probability of its being a false alarm would now be very low indeed – low enough for our near-certainty that we have a new track. As tracks progress, this search area can be adjusted in size, dependent upon its history and making assumptions about the target's ability to change speed and direction (and for aircraft, height). Most tracking programmes do all these calculations all the time and much more accurately, faster and tirelessly than we. But the machines encounter the same logical doubts as we. For instance, suppose a fresh plot turned up in the expected 'track association' area, but it had a companion plot in the same area. Which would be the right one to choose? Should we consider both are real and that a new track should be looked for? Many computer trackers use 'branching' or 'forking' routines to resolve this situation; again using the same sort of arguments that you and I would, including taking account of the higher probability of false plots in clutter areas.

A special case is often encountered. This is when two estabished tracks cross each other. Given the high manoeuvrability of modern fighters it would be entirely possible for them to be wrongly identified when the crossing point had been passed. This is where IFF and SSR come into their own, for we have very few doubts on target identity if both carry different known codes and have different known altitudes.

16.4 Tracking algorithms

Although there are a number of ways to do this, the most commonly used is the Kalman filter, named after the inventor who first proposed it in 1960. The Kalman filter is really a mathematical process which operates as follows:

Suppose we have an already established track. The process will have had track history to guide it. From this the speed and heading of the target will have been estimated by calculation from earlier state vectors and so the next expected

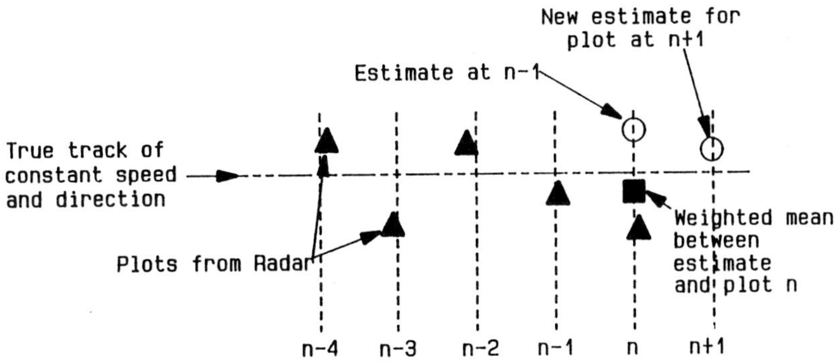

Fig. 16.3. Kalman filter action. Successive plot positions are used to estimate where the next plot has the highest probability of being. Difference between this and the actual reported plot position is used to form a 'weighted mean' position. This is used as output.

plot position can also be estimated. When the radar obtains this next position (by measurement) it is compared to the estimate. Almost certainly there will be a difference. From this, a mean value is calculated, together with a *value for the error in estimation*. This latter value is used as a 'weighting' factor in setting the next estimate. The process is illustrated in fig. 16.3.

In some tracking systems (the so-called 'alpha–beta' trackers) the weighting factor is fixed at a value calculated from a knowledge of the radar's own errors and how they are distributed. In others the weighting factor is varied automatically, the variation (up or down) being based upon the success of the estimating process which is judged by the magnitude of correspondence errors with time. This weighting factor has two components: A (alpha), which is concerned with the position error; and B (beta), which is concerned with the velocity error. The advantages of automatic adjustment to these factors are obvious when one considers the great variability of plot measuring accuracy in the radar (the radar's measuring errors, variable signal strength, targets in clutter, the vertical polar diagram's lobes and gaps, wide range of target sizes, etc.).

16.5 False plot filtering

In the tracking process there always has to be an initial plot. To get maximum detectability, the false alarm rate should be set to as high a value as possible. This makes the system generate a large number of plot declarations which do not form tracks; these would be intolerable if displayed to the user – he'd literally have 'spots before the eyes'! Most modern trackers and plot extraction systems have a false plot filter to prevent these rogue plots from reaching the display or the tracker. It is relatively simple in operation. Every plot generated (true or false) is stored and is the subject of a potential track. Only after the track is established with very high probability are the plots released to the tracker or to the display. Commonly this is at the third associating plot, the first two having been used to establish the potential track. Thus, each new track will have the first two plot reports suppressed, but commonly retained in store for possible later track associations. Obviously, plots which don't form tracks will not be displayed and the 'spots' will have disappeared. Using this kind of tracking and false plot filtering allows the radar system to be operated at very high sensitivity with false alarm rates as low as 'one in a hundred' instead of the common 'one in a million' (ten thousand times greater).

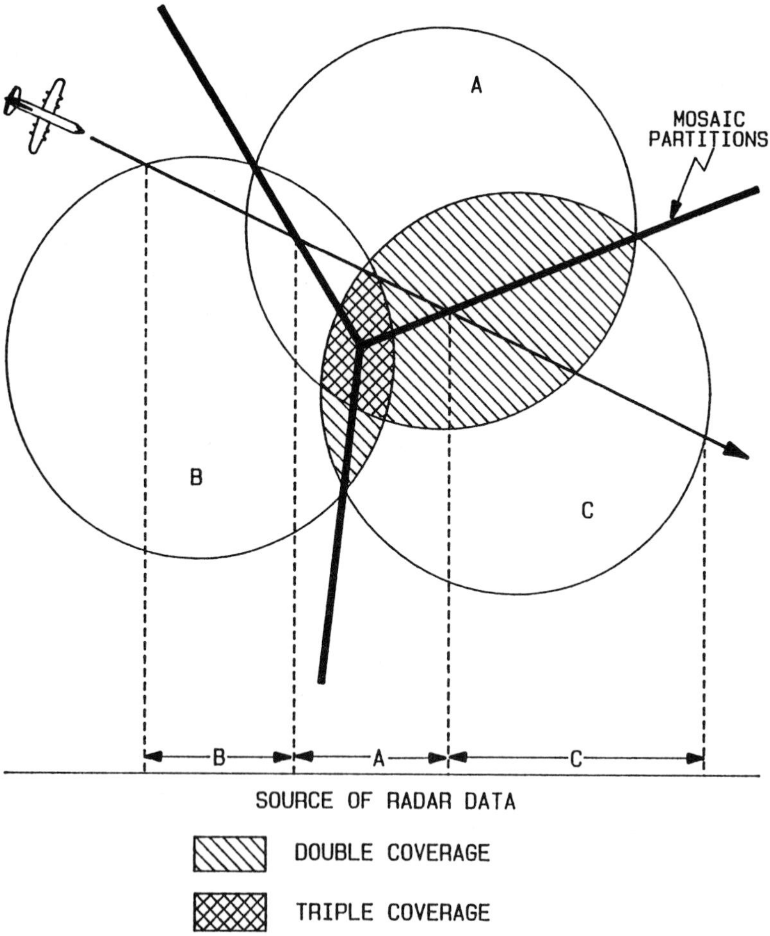

Fig. 16.4. Simple mosaicing. Joint coverage areas are partitioned electronically. In the exampled track above, the source of signal is shown changing from radar B to A to C. Other choices can be made on the basis of signal quality from each radar on given targets in joint coverage.

16.6 Multi-head tracking (MHT)

Figure 16.4 shows three radars, A, B and C, with overlapping coverage. The target illustrated, as it flies its track, will generate position reports first from radar B then from A and B, on to A, B and C, through A and C and finally from C alone. In the MHT technique, outputs from all radars are combined to make best use of positional data on given targets. There are three main methods used to effect this: mosaicing, true multi-head tracking (TMHT) and distributed tracking (or local track averaging – LTA).

16.6.1 Mosaicing

In this, all radar outputs are fed to a joint user, and enter a processor which divides the combined coverage area into pre-set regions. Within each region, even if it has overlapping cover, target position reports from only one nominated radar are selected for display. Obviously that with the better quality of output is chosen.

A variation on this rigid selection method is sometimes used. In this, the radar outputs are used in a coincidence gate circuit. In regions of overlapping coverage, targets will generate contemporary outputs from two or three radars. The processor seeks the one with the highest signal-to-noise ratio and uses it as final output. In cases where the radar outputs are plot extracted, the same principle would apply but now based upon a 'plot quality' factor in the output plot message from each radar: that with the highest quality would be output.

In the example shown in fig. 16.4, the value of this technique can be appreciated from the following imaginary case. Suppose, in the triple cover area, the target was over a patch of high ground. It is unlikely that radar A would be relying heavily upon its mti system's sub-clutter visibility to produce a useable output from the aircraft. It is also likely that this high ground would not be illuminated by radars B and C because of the greater range and earth curvature.

Because of the target height it is also very likely that both radars B and C would produce target signals – free from mti processing losses – which would be stronger than that from radar A and therefore would be more useful. The mosaic system would use the best of B and C for output. There is, however, a distinct weakness in mosaicing in that it totally discards useful information in favour of the single source the system has chosen.

16.6.2 True multi-head tracking (TMHT)

In this technique the system seeks to improve the final track quality by using every available plot output from the whole group of radars. Every radar has position reporting errors, as we have seen earlier. The random azimuth errors will show as 'across track' displacement for radial tracks and 'along track' (speed) errors for tangential tracks. Most positional errors will be a combination of both 'along' and 'across track' components. The effects of azimuth error will be exagerrated as range increases (0.5° azimuth error represents 0.5 nautical miles displacement error at 60 nautical miles range and 2 nautical miles at 240 nautical miles range). These will be expressed by the Kalman algorithm, as described above.

In the TMHT system a central processor uses a more complex form of Kalman filter which now has to associate a number of plots from a given target

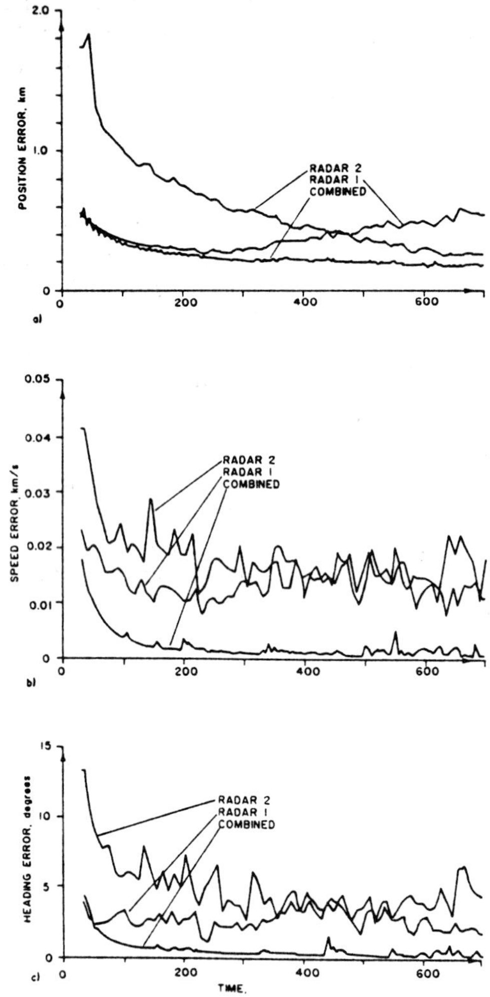

Fig. 16.5. The effect of true multi-head tracking (TMHT). A ten minute flight on an aircraft's known track through overlapping cover of two radars. The improvement in track accuracy with TMHT is self-evident (data by courtesy of GEC-Marconi Research).

into a single track and give a new estimate based upon the combined concurrent plots, which have 'time of detection' included in their report data. Each plot in a concurrent group will have errors as described above.

The tracking routine will then have to consider how much importance to give to each plot report on the same target. The 'weighting factor' coming from the Kalman process determines this 'importance'. Obviously, when making the new position estimate, the plots with the highest weighting factor will be taken as of greater value than the others. But this does not mean that other plot reports with a lower weighting factor are ignored – the data from these are also valuable, but have significance proportional to their weighting factor.

The efficacy of TMHT can be seen from the example in fig. 16.5 which shows the improvement in target data accuracy achieved relative to that from individual radars in the TMHT net.

16.6.3 Local track averaging (LTA)

Here, a number of sensors each have a local tracker and give track data instead of plot position as output to a central remote track merger. The average of these is taken and form system tracks for individual targets. LTA is commonly found as an addition to groups of sensors which provide track data to control centres (such as airfields) close to the sensors themselves. The addition of this central LTA facility now allows wide area control from a central establishment provided the sensors give contiguous or overlapping cover.

A distinct drawback results from what is essentially 'double-averaging' or duplicate error smoothing. The averaging or smoothing process at the local tracker delays detection of manoeuvres (turns and changes of speed). The extra averaging taking place in the track merging process delays this even further.

Readers with a mathematical bent are commended to reference 22, the current 'tracker's bible'.

17
Phased arrays

17.1 Principles

Once again the engineering fraternity has been lax in its use of terms; in the literature you will see the title 'phased arrays' when what is truly meant is 'phase-steered arrays'. *All* radar beams are generated by careful control over the phase of energy across the antenna's aperture and thus are 'phased arrays'. The following text will attempt to explain how beams can be steered in a controlled fashion by varying the phase of energy.

Although phase-steered arrays are not yet generally in use, their advantages are such that they will quickly enter the wider market, and should become ever more worthy of description during the lifetime of this book. What are they, what do they do and how do they do it? Again, the principle of operation is simple. In the earlier part of this book the manner in which radar beams are generated has been described. Figures 3.1 and 3.2 of chapter 3 show how a linear array combines the signals received by its elements to form an output which is the vector sum of all contributions. The connections between each element and the summing point are arranged to have equal electrical length so that the phase of the signal at each element undergoes the appropriate phase shift before the summation is done. It doesn't much matter what the shift is, provided it results in equality of phase for all signals at the summing point.

As a consequence a signal source which is at a great many wavelengths' distance from the array, and at right-angles to it produces equal phase of the signals received by each array element; the beam shape is at its maximum. The consequences of shifting the signal source to left or right of the array's boresight are shown in fig. 3.2 – the beam's shape is expressed.

In phased arrays, this principle is used, as it were, back-to-front. Phase shifting devices are introduced into each connection between the combiner and each antenna element, as shown in fig. 17.1.

If all these were set to the same value, the antenna's beam would produce its maximum orthogonal (i.e., at right angles) to the line of the elements (fig. 17.2(a)). Suppose the line of elements was to be mechanically fixed at this same

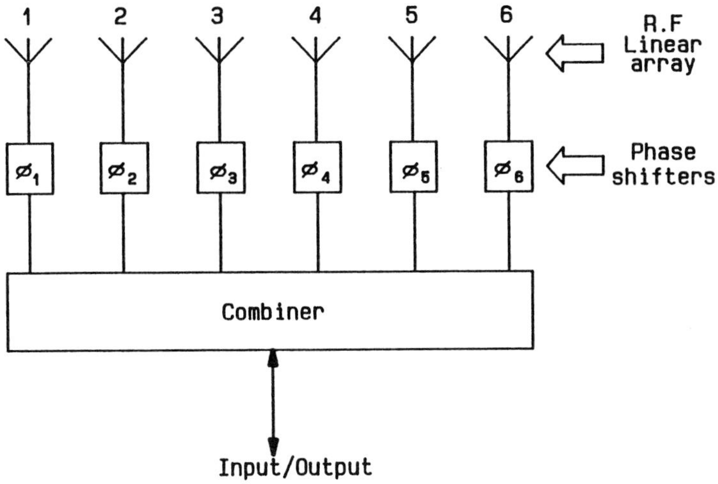

Fig. 17.1. Linear array with phase shifters interposed at r.f.

position, its beam maximum pointing at, for example, North. Now we introduce phase shifts into each element's connection before the combiner, each n° greater than its neighbour, as shown in fig. 17.2(b).

The physical line representing signals of equal phase from all the elements has now effectively rotated by an angle (Θ) which is a function of the phase shifts (n°) introduced across the array.

17.2 Practice

In the example of the 6-element array given, it is easy to see how the steering angle (Θ) is related to the phase shifts in each element by the following reasoning:

Suppose the elements were spaced, as is common, a half wavelength apart and we wished to steer the beam 30° away from its natural position, as shown in fig. 17.3. The whole array now has to produce an equi-phase line 30° from A–A. To produce the negative steering angle required, energy from element number one has to be phase delayed most and that from number six not at all. So across the whole array we need to produce x° of phase shift. In the example, $x° = \text{Sin } 30° \times 2.5$ wavelengths (2.5λ being the sum of the element spacings). Sin $30° = 0.5$; thus $x° = 1.25\lambda$ and since one wavelength is the equivalent of 360°, $x° = 450°$. This phase delay has to be progressively and equally shared among the array's elements; so each must have $\frac{450°}{5}$ phase shift more than its neighbour. Reflecting this back into fig. 17.2(b), we find that if $\Theta = 30°$, then $n° = 90°$.

We can now consider the array as an antenna whose beam can be 'steered'

Fig. 17.2(a). Phase shifters set to equal values produces a beam maximum orthogonal to the array.

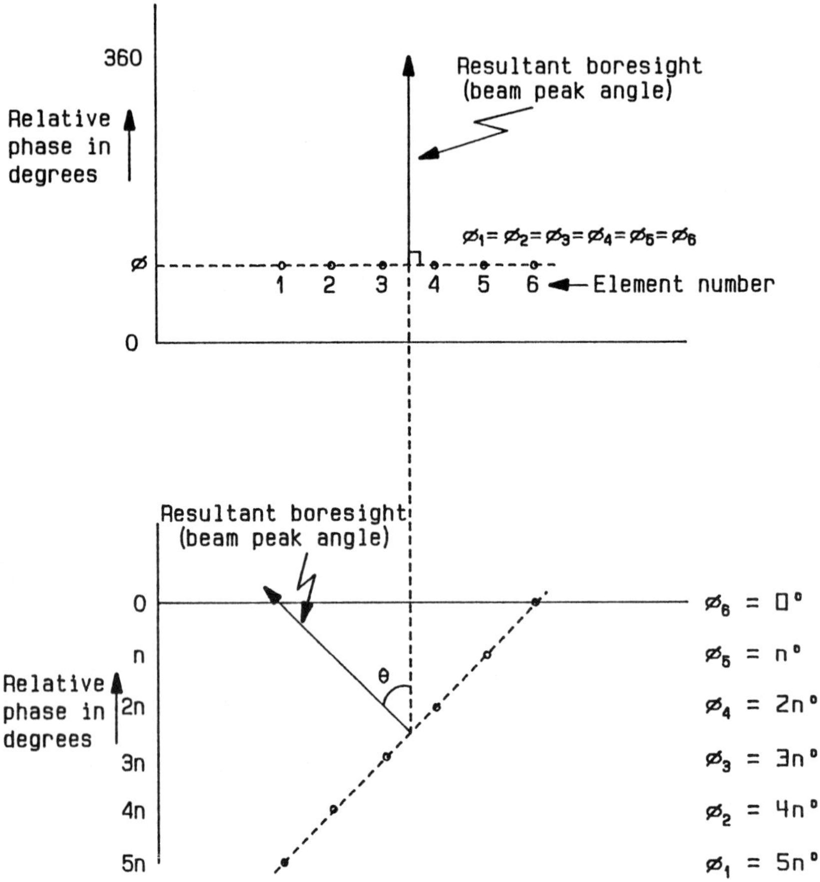

Fig. 17.2(b). Progressive equal phase shifts move the beam peak away from the orthogonal.

either side of its natural boresight in a controllable fashion by the added variable phase shifting elements. Because these can be introduced and switched by very fast microwave switches, extremely fast beam steering is possible (in microseconds). This is of immense value in modern adaptive radars in dealing with targets of known position and track: instead of waiting for a mechanically rotated antenna to come round to the target bearing, an electronically steered beam can be rapidly directed to targets and dwell on them for controlled periods.

As usual, in using this technique there are problems to be overcome. The first

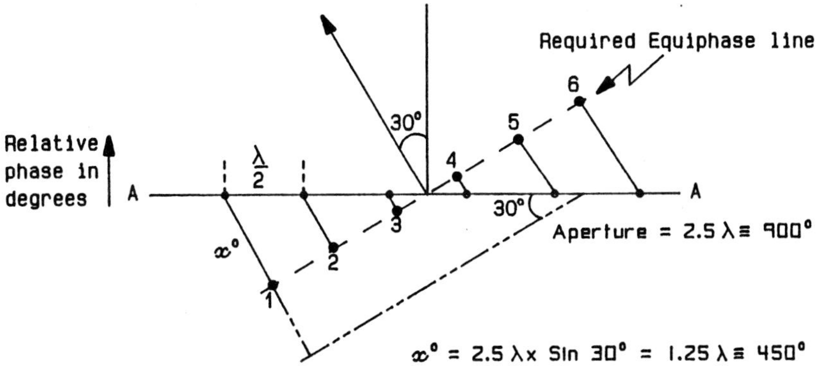

Fig. 17.3. Progressive phase shift of 90° between elements steers the beam by 30°.

is fairly readily seen. From the earlier description of antenna performance in chapter 3 it will be remembered that antenna gain is related to its physical size (in terms of how many wavelengths it measures). If we look at the linear array in our example, the gain is a direct function of the distance between elements number 1 and 6 (i.e., its physical aperture. A target on its natural boresight bears the full force of the whole aperture.

A target at 45° to the boresight would only 'see' 0.7071 (aperture × cos 45°) of the full aperture. Thus, if we steered the beam by the phased array technique, the gain of the antenna would effectively reduce as the scan angle increased away from its natural boresight. In the limit, a scan angle (Θ) of 90° would produce zero gain – the target would be looking at the antenna end on!

We also remember that the beam shape broadens as the physical aperture of the antenna reduces; thus angular resolution becomes worse as the phased array scan angle increases. At 45°, the beamwidth of the linear array would become

$$\frac{1}{0.7071} = 1.414 \text{ times wider.}$$

These problems have a number of different solutions, dependent on the radar's role. In surveillance radars, the phased array radar can take varying forms. One of the earliest is of monumental proportions – the 'B MEWS' radar, illustrated in fig. 17.4. Another form, which overcomes the difficulty of beam shape changing with scan angle, is to make the array into a cylindrical structure. A good example of this is a design for an electronically scanned array (ESA) for SSR (Mode 'S') use. It is illustrated in fig. 17.5. A right cylinder of identical and equi-spaced radiators is formed. A continuous group of radiators is chosen – the mid-point of the group representing the centre of the aperture of the antenna formed by the group. Phase shifters of differing value are introduced

Fig. 17.4. The massive BMEWS (Ballistic Missile Early Warning) radar at Thule in Greenland. (*Photo courtesy of Raytheon*).

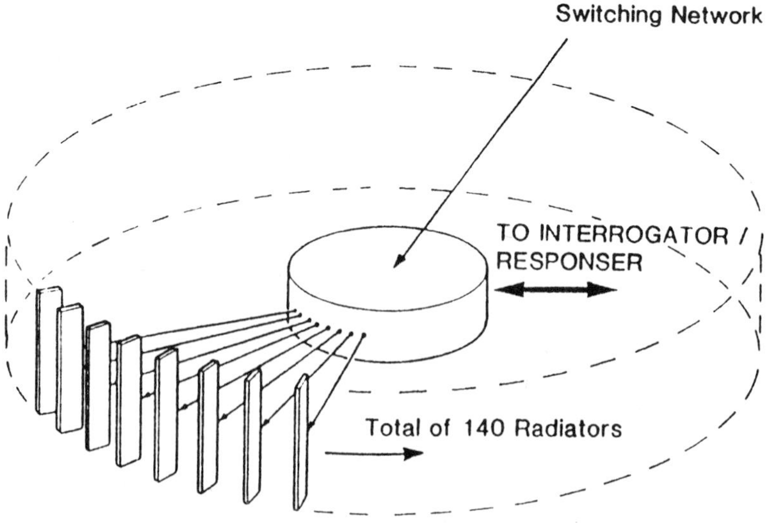

Fig. 17.5. A possible electronically scanned array (ESA) configuration. Contiguous groups of elements are selected by the switching network and excited as an array whose boresight is on the line joining the array's middle element and the cylinder's centre.

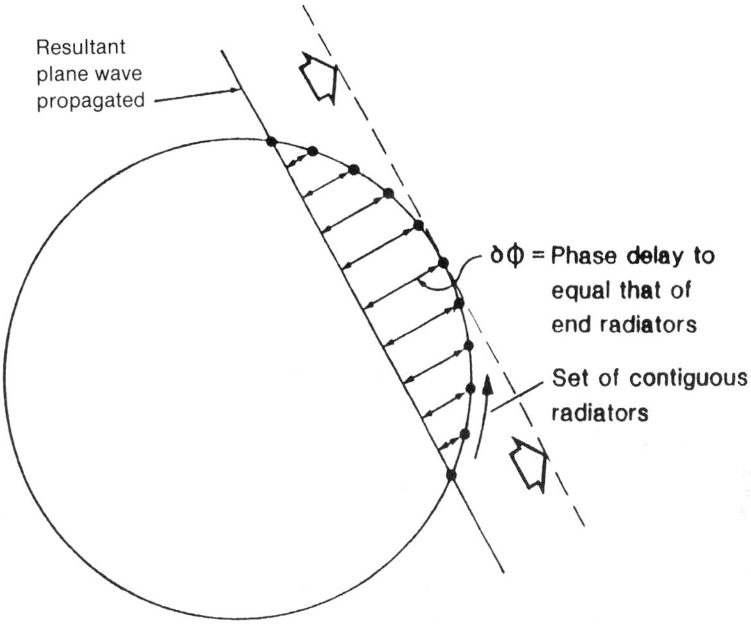

Fig. 17.6. Illustrating the need for phase correction in ESAs to produce effective plane wave propagation.

to each element in the group so that energy radiated from them all have the same phase across the tangent line of the cylinder at the group's centre point. The principle is shown in fig. 17.6. In practice, the amplitude of energy to each radiator is also varied, as well as the phase, so that 'tapered illumination' of the horizontal aperture is achieved (as in normal antenna design – see chapter 3). The radiators themselves are each of the form which make up the LVA antennae described in section 11.4.4. Thus the whole array can produce consistent beam shapes and gains throughout the 360° azimuth domain but in discrete azimuth increments; so if the cylinder was made of 180 equi-spaced elements, the azimuth increment would be $\frac{360^\circ}{180} = 2^\circ$.

17.3 I.F. Beam forming

The techniques described above are all implemented at the radiated frequency (r.f.). There are other ways of producing phased array beam formation, but at the much less sensitive and lower intermediate frequencies (i.f.) of the system; typically $60 \approx 100$ MHz.

This relies upon the principle (detailed in the Appendix) that, if two r.f.

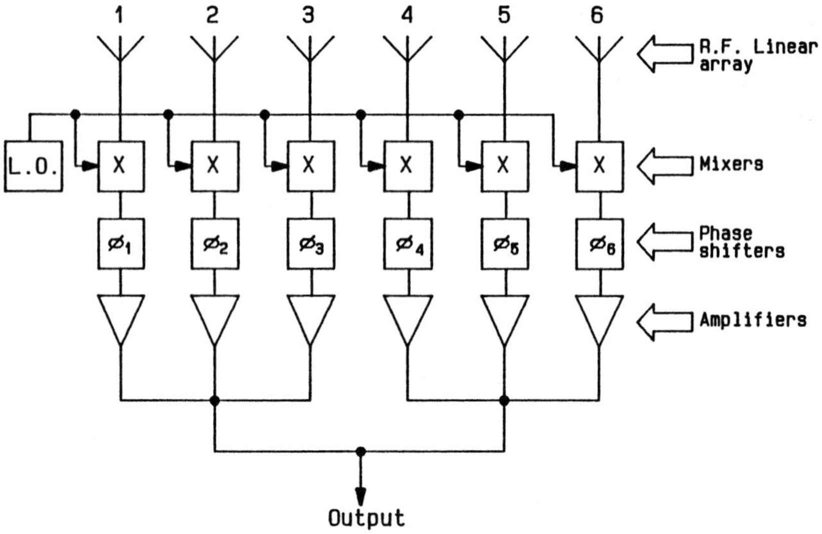

Fig. 17.7. Beam forming can be achieved by introducing phase shifters for each element's contribution at the i.f. stages. They must share a common Local Oscillation (LO) to maintain phase referencing.

signals of different phase are down-converted to i.f., then, provided the down-conversion of the r.f. signals share the same local oscillator, the i.f. versions of the two r.f. signals will retain their original r.f. phase difference. Thus, using the example of our 6-element linear array the summing of the signals from all the elements is performed at the i.f. after down-conversion, as shown in fig. 17.7.

The same principle of producing electronic scanning as was described above for the r.f. technique can be applied at i.f. That is, if controlled phase shift is introduced into the i.f. signals before the summation point, the same beam steering relative to the natural boresight of the array can be produced.

Engineering problems of maintaining constancy of the phase relationships are now compounded. Temperature variations in the r.f. components create phase drift – on top of this, comes the problem of stabilising the i.f. amplifier gains and phase shifts. This is commonly achieved by regularly circulating a test signal of very stable characteristics through the group of i.f. stages and governing the performance of each stage by a servo-system which increases or decreases the gain as dictated by the test signal's level at each amplifier's output.

There is an important difference between r.f. and i.f. beam forming which may not be immediately obvious to those outside engineering disciplines. The r.f. technique doesn't involve any frequency translation and if no amplifiers are used, the process is what engineers call 'reciprocal', i.e., it can be applied to transmission and reception without any active change – the antenna beam shape is obtained from a single input/output socket. An amplifier is not a 'reciprocal' device. If it has a gain which magnifies a 1 mV input to a 1 V output, applying one 1 V to the output socket will not produce 1 mV at its input socket! The i.f. technique is based upon frequency conversion and must use mixers and amplifiers, so it doesn't have 'reciprocity'. The example given in fig. 17.7 shows how the beam can be formed on reception by down-conversion. If one wanted to use the same technique for transmission then a separate set of different electronics would be required – one wherein the desired set of signals, each with its own phase and amplitude carefully calculated to match its neighbours, was developed by i.f. synthesis techniques, then up-converted (by amplification and

Fig. 17.8. A 3–D 'Martello' surveillance radar. Each of 40 radiators produce a beam narrow in azimuth. Phased array techniques are used to produce a single 'cosecant-squared' transmission pattern in elevation and eight separate beams for reception, each narrow in elevation and at fixed and known elevation angles. The array is mechanically rotated in azimuth. (*Photo courtesy of Marconi Radar.*)

mixing) to its r.f. version and fed to individual elements in the antenna array. Some systems compromise by using the r.f. technique for transmission and the i.f. technique for reception.

17.4 A practical case

A number of types of surveillance radars capable of three-dimensional position fixing have now seen many years of field operation. They are almost exclusively used in military defence roles. One such is illustrated in fig. 17.8 and like its companion types it uses phased array techniques for producing a multiple set of beams: a rotating 'bill board' consisting of 40 identical radiating horizontal 'planks', each of which has 62 identical r.f. transmission and reception modules. Thus, the whole array is made from 2480 individual elements, each of which can have the phase and amplitude of its input and output controlled. For transmission, this control is exercised to produce a cosecant-squared pattern in elevation with a narrow search beam in azimuth. On reception, the i.f. beam forming technique is used to create eight pencil-shaped beams each fixed at different elevation angles and slightly overlapping their neighbour beams. All the beams have the same narrow azimuth shape. Thus, on transmission, all targets are illuminated by the single cosecant-squared radiated pattern. However, on reception, targets will appear in one or more beams. If as is usually the case, the target is captured by two neighbouring beams, their signals will be recognised as emanating from a single target because they share the same azimuth and range. Their amplitude difference will give a measure of the target elevation angle because the shapes of the eight beams are known by measurement and the range is known. Readers may remember reference to this amplitude-sensing method of monopulse direction finding in chapter 11. In practice, heights with an accuracy of about 1000 ft at 250 nautical miles are achievable, which – although vital information in defence scenarios – is not nearly good enough for air traffic control whose SSR systems in any case, can give accuracies of ± 100 ft.

18
New radar techniques

18.1 Synthetic aperture radar (SAR)

18.1.1 Introduction to synthetic aperture radar technique

For once the scientific fraternity has given an accurate name to describe what is being done: to make a radar whose antenna has an aperture which is not 'real' but made by the synthesis of relatable parts. Once again, the principle is simple in essence, and again it is helpful to refer to earlier parts of this work. In chapter 3, the notion of producing directional beams by the linear array has been described. The relationship between the number of elements in the array, their spacing and the resultant beamwidth has shown that the greater the number of elements, the narrower becomes the beam shape produced. In chapter 9 the technique of digitising continuous (analogue) processes has received explanation. The SAR technique relies upon combining both these techniques to produce a radar having incredible resolving power at massive distances.

18.1.2 Fundamental ideas

Suppose a 24-element linear array was to be mounted on the fuselage of an aircraft and its beam directed downwards to the earth's surface. If the elements were half a wavelength apart, the beam in azimuth would be about 4° in width. This would have very poor azimuth resolution (about 4 nautical miles at a range of 60 nautical miles; a convoy of ships seen broadside at the same range would appear as one long echo).

Essentially, this poor resolution is the result of summing signals from only 24 elements. If this summation was to be made over a larger number then the effective beamwidth would become narrower and the azimuth resolution improved. The SAR technique does exactly this.

18.1.3 How SAR works

The 24-element array in our examples has a single summation point. A single transmission from it would yield signals reflected within its 4° wide beam. After

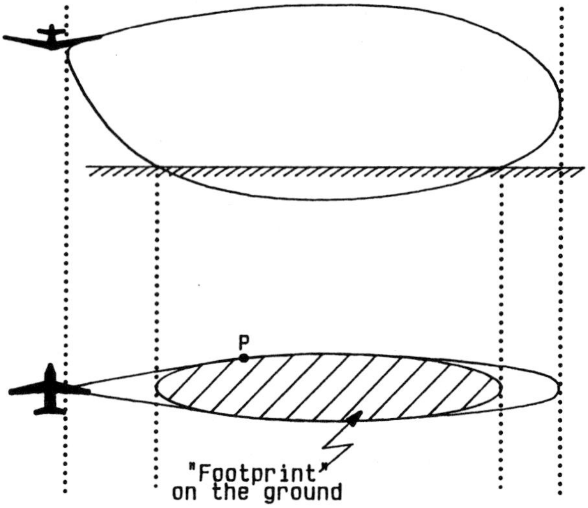

Fig. 18.1. An airborne synthetic aperture radar (SAR) illuminates the ground. Its 'footprint' is very large and the natural resolution cells are extremely big. Fine resolution is restored by repeated examination of given points (P) within the beam during the illumination period, which can be many thousands of radar repetition intervals.

detection these can be lodged in a range-ordered store. At the next transmission the aircraft will have moved along its track and the process can be repeated. If we were to govern matters such that the time between transmissions was equal to that taken for the aircraft to move about one half-wavelength, we could regard the second sample as being that from a second element of a linear array. By repeating this process and then combining the results from a large number of successive samples (N), we would in effect have constructed a linear array with an aperture of (N) half-wavelengths. With modern digital technology the vast storage required presents little difficulty and integration over thousands of samples is taken.

Take now a practical example. Imagine the 4° beam emanating from the side of the aircraft exampled above. In elevation it would be shaped to illuminate the ground below. It would therefore create a 'footprint' on the ground as shown in fig. 18.1. As the aircraft moves along its track, the footprint moves over the ground. At a range of 100 km the minor axis of the footprint's ellipse is 7 km. Imagine a point P just coming within the beam. It will stay illuminated by the radar beam for the time it takes the craft to move 7 km. At 250 knots (465 km/h), this takes 54 s. If the SAR repetition rate was 500 per second, the point P would experience 27000 radar pulses. As in normal radar processing, the range domain is broken into a string of contiguous, small, range memory cells. The received signals, from each transmission, after coherent detection, are lodged within the string of cells. The results of many successive detections are combined, range cell by corresponding range cell, in each string as illustrated in fig. 18.2. Thus for each range cell the combined output is the result of integrating over 'n' transmissions. At each successive transmission the aircraft will have moved along its track. The combined output for each cell will thus progress from summations 1 to n, 2 to (n + 1), 3 to (n + 2), etc.

Each new input to a given cell will confer a small but detectable difference upon the resultant integrated output. Suppose the integration was performed over **5000 successive transmissions, i.e., an integration time** of $\frac{5000}{500} = 10$ s. Because the actual antenna on the craft combines the outputs of its 24 elements at one summing point, the subsequent integration can be likened to erecting an antenna of 5000 elements – but over the time of 10 s as opposed to the 24 elements which are simultaneously and instantaneously present. The beamwidth has now synthetically been narrowed by a factor of $\frac{5000}{24} = 208$; the 4° natural beamwidth of the antenna has effectively become $\frac{4}{208} = 0.019°$. Thus the azimuth resolution at 100 km has improved from 7 km to 34 m.

It is easy to appreciate a number of significant points from this. The greater the number of successive repetitions integrated, the longer is the integration time and the finer is the resolving power. It is also obvious that the integration period must be no greater than the time it takes for the 'footprint' to pass a given point. As the integration period grows, the more likely it is that certain features, such as aircraft, fast vehicles, etc., will have moved from one resolution cell to another or one string to another. This upsets the system's 'visibility' and thus

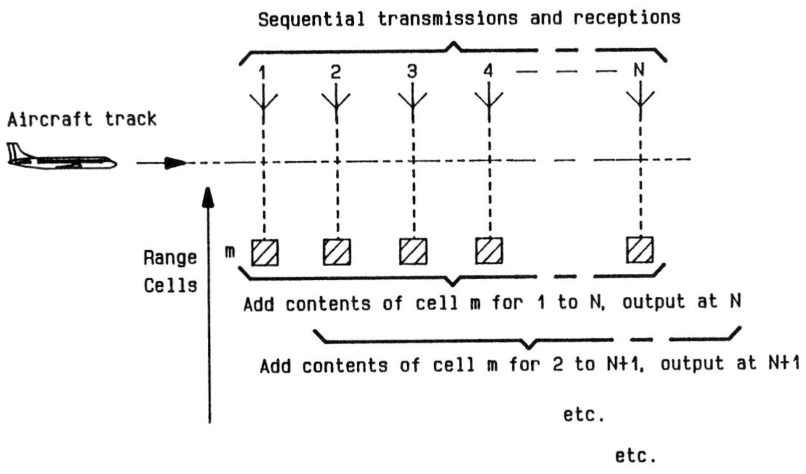

Fig. 18.2. SAR processing principle. All points in the beam's 'footprint' are examined thousands of times (N) in its passage across them. Integration over N repetitions is performed for each individual range cell, sometimes made very small by pulse compression technique.

Fig. 18.3. SAR image from the ERS-1 Satellite of the Dartford area of Kent in the UK. This 25 km^2 area is taken from a 100 km^2 master, whose data gathering time was 15 seconds. The new bridge across the river (top centre) is clearly visible. Resolution cell is 15 m^2 but capable of being made even smaller. (*Photo courtesy Matra Marconi Space UK.*)

SARs find their greatest usefulness in mapping stationary features. The effect is analogous to taking a photograph with an extremely small camera aperture, requiring a very long exposure time; a person moving quickly through the scene will be a mere blur. Satellites at great height produce very long 'footprint' dwell times and fantastic resolution of photographic clarity is obtained. A graphic example of the SAR technique in action is given in fig. 18.3.

Not so readily appreciated is an apparent paradox. As pointed out in chapter 3, the natural beamwidth of an antenna has azimuth resolution which improves as the beam is narrowed – in the SAR technique the opposite is true! As the real beam is widened it permits longer integration times and it is this that governs the resolving power – the longer the integration time the finer is the resolution. In fact some SARs use only a single radiating element!

For the system to be of equally fine resolution in the range domain it is usual to employ the pulse compression technique described in chapter 8. To reproduce the extremely narrow pulses (of the order of tens of nanoseconds, e.g., 0.02 μs) the system requires very large bandwidth, e.g., 50 MHz for a 0.02 μs pulse. This in turn means that the frequency sweep across the long transmitted pulse has also to be of this value and engineering problems are eased if the radiated frequency is made large, e.g., 3000 MHz (10 cm wavelength).

18.1.4 Problems

The SAR relies heavily upon the radar's coherence – the maintenance of phase relationships between transmitter and receiver. The problem is made more acute in SAR because the short-term stability (of concern in mti systems) must now be present over much longer periods, i.e., the total integration time of many seconds. The aircraft, as the SAR 'platform' exampled above, will have yaw, pitch and roll during its flight. These could upset the SAR performance if they introduced across-track errors which disturbed phase relationships. Thus the antenna is commonly stabilised; any deviations from a straight line flight must be corrected by introducing the appropriate phase compensation to the received signals before processing. Failure to compensate the errors correctly defocuses the image. The signal processing itself is usually carried out in very fast digital processors which now have the necessary massive storage and speed supplanting the older, time consuming, techiques of optical processing which involved photographic methods and equipment. Thus modern SARs can produce their finished output in virtually real time.

18.1.5 Refining the principle

The principle of the SAR, as outlined above, is a very simple one: A large aperture linear array is produced by scanning a smaller one over a period of time using a coherent detection system. The larger aperture is a section of the SAR's straight-line track in space. It will be remembered, from chapter 3, that to form a beam efficiently in a linear array the energy it intercepts must be essentially a plane wave; that is, on boresight, all contributions from the individual elements must be of equal phase at the summing point. In practice, this is done by arranging that the electrical lengths (in terms of wavelengths) of the connections between each element and the summation point are equal.

In the SAR, the same principle applies but has to be implemented differently, because the elements of the array are not simultaneously present; they appear at different points in sequence as the SAR moves along its track.

Consider the circumstance in fig. 18.4. It shows a point (A) which is on the boresight of the narrow beam we wish to synthesise. The boresight will be at right angles to the SAR's track and pass through A at the midpoint of the integration period of N samples; that is where the 'N/2th' element is exercised. The range to A at this point will be R. We have seen that for the array to produce a properly formed beam, with its maximum in line with A, contributions by all elements 1 to N from a source at A must all be of the same phase. This is patently not the case here. The first sample will have been taken earlier and hence the range from element No. 1 to A (R + δr) will be greater than R. This difference (δr) will reduce when sample number two is taken, and finally fall to zero at sample number N/2 and grow once again to (δr) at sample number N. Each of these small range differences produces proportional phase differences. If they were not removed, their effect would be tantamount to spoiling the beam shape through defocusing. Modern SARs are 'focused' by compensating for the phase errors created by these small range differences by

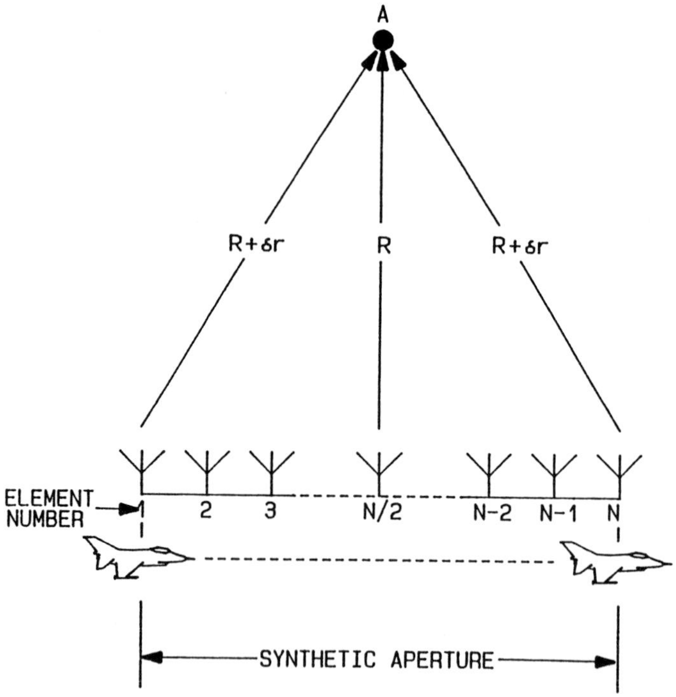

Fig. 18.4. Illustrating the need to apply phase corrections to samples taken during beam formation in SAR. Small range differences (δr) across the effective aperture (1 to N) produce phase errors which, if not corrected, will 'de-focus' the resultant narrow beam.

introducing phase correction (φ_1 for element 1, φ_2 for element 2, etc.) during the final integrating process.

18.1.6 Inverse SAR (ISAR)

The SAR technique is based upon using a radar on a *moving* platform, whose trajectory is known. ISAR uses a *fixed* radar beam and lets the target fly through it. The same techniques as in SAR are used, combining the contents of corresponding storage cells in a long integration period. But now rate of change of range and doppler frequencies of targets are sought and processed. From these, target motion can be calculated. In a practical system it is usual for the host radar to have its beam directed to the general target area and then to be held stationary, allowing the target to move through or within the fixed beam.

Because the ISAR is a coherent system and integration times of the order of seconds are possible (dependent upon beamwidth and target speed), extremely fine azimuth resolutions of about 1 to 2 metres at 10 km can be obtained at a wavelength of 10 cm. This means that complex targets such as a large jet aircraft can be examined for what might be called their target signature resulting from large scatter areas such as their engines and tail planes which have known size and physical relationship. Some targets can move a considerable distance in the observation time. This means that although the fine resolution allows us to know *what* the target is, we are not so sure about *where* it is.

ISAR techniques can produce images of fine detail. Bernard D. Steinberg (reference 23) describes in very clear prose, a 32 element array used in a 'microwave camera' in 1988 to view aircraft flying into Philadelphia International Airport. The fuselage and wings, tail plane and engines are clear enough for the aircraft type to be recognised.

18.2 Over-the-horizon (OTH) radar

18.2.1 Introduction

Most radars, indeed all discussed in this book so far, use microwave frequencies. Wavelengths range from 50 cm to 3 mm. Their propagation has been discussed in chapter 4 and the notion of the 'radar horizon' (a greater distance than the visible horizon) was introduced, together with the idea of the '$^4/_3$ Earth'.

The basic mechanisms and effects which make it possible for the radar to operate beyond the 'radar horizon' have long been known from early work in communications and primitive radar systems which operated at much lower frequencies (3 MHz to 30 MHz).

The advent of the semi-conductor, microchip, very fast and compact signal processing and computing engines have led to the re-exploitation of these early effects and principles in OTH radars. They utilise two modes of propagation – ground wave and sky wave.

18.2.2 Ground wave propagation

At frequencies in the range 2 MHz to 30 MHz, propagation over sea water exhibits an unusual characteristic. The electro-magnetic energy penetrates the salt water and propagates across its surface with very little loss and at a slilghtly slower speed than in free space. The penetration into the sea 'couples' the two media, water and air. As a consequence, the wave is, in effect, captured by the surface. At h.f. wavelengths the sea's surface is electrically smooth and also highly conductive, so the vertically polarised energy's electric field remains at right angles to the surface. Because the energy is directly coupled into the surface the wave travels beyond the radio or radar horizon. Thus any objects on or near the sea surface will be illuminated and reflect some of the impinging energy, which at hundreds of kilometres distance is in large enough quantities to be detected with good signal-to-noise ratios. A comparison between normal microwave radar coverage and OTH h.f. radar in low elevation regions is illustrated in fig. 18.5.

18.2.3 Sky wave propagation

The technique of achieving long distance communication by sky wave propagation has been in use for decades. It relies upon predictable behaviour of layers of ionisation in the upper reaches of the earth's atmosphere. Their height, thickness and electron density all vary with time. The electron density can become so high that the layer acts as a reflector of radio waves in the h.f. band.

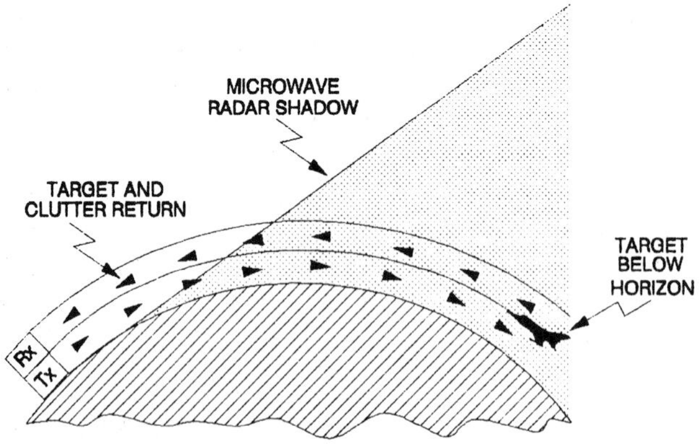

Fig. 18.5. OTH radar ground wave principle. High frequency radio energy launched across a sea surface carries far beyond the radar horizon of a microwave radar at the same site.

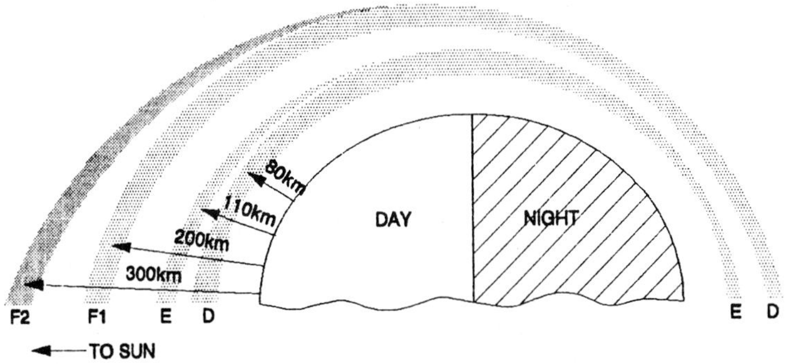

Fig. 18.6. Radar reflective layers above the earth and their general behaviour under the sun's influence.

The reflectivity is itself a function of frequency, and successful operation depends upon keeping track of what frequencies in the band are best to use as the layers change throughout the day and night – we shall see how later. Figure 18.6 shows the general day and night changes. The layers are not as sharply defined as in the diagram, their density growing and fading gradually with height.

The reflective property of the layers comes from free electrons which have been released from the gases in the atmosphere by radiation from the sun – hence the day/night variations. The extent of the electron density variations with time and height can be judged from the sample given in fig. 18.7. See references 24 and 25 for more information.

The reflective property of the layers allows radio energy from the ground to be 'bounced' off the layer and reflected back to the ground, thus creating a radio path between two points which can be as much as 900 to 3000 km apart. In communications systems this allows transmitter/receiver stations to establish reliable links by making frequency changes throughout the day and night to match the changing nature of the ionosphere.

In radar terms, radio energy from a transmitter, reflected back to earth at these vast distances, will be reflected by the earth's surface itself and also from objects on or near it. Receivers located near the transmitter, or at a point sharing a similar propitious link path, can receive some of the reflected energy and by special doppler processing, ships and aircraft can be detected.

18.2.4 OTH radar equipment and its working – A typical ground wave system

A simplified diagram of the different ways OTH radars can be organised is given in fig. 18.8. It shows combinations of ground wave and sky wave elements in monostatic and bistatic systems. It will be remembered that 'monostatic'

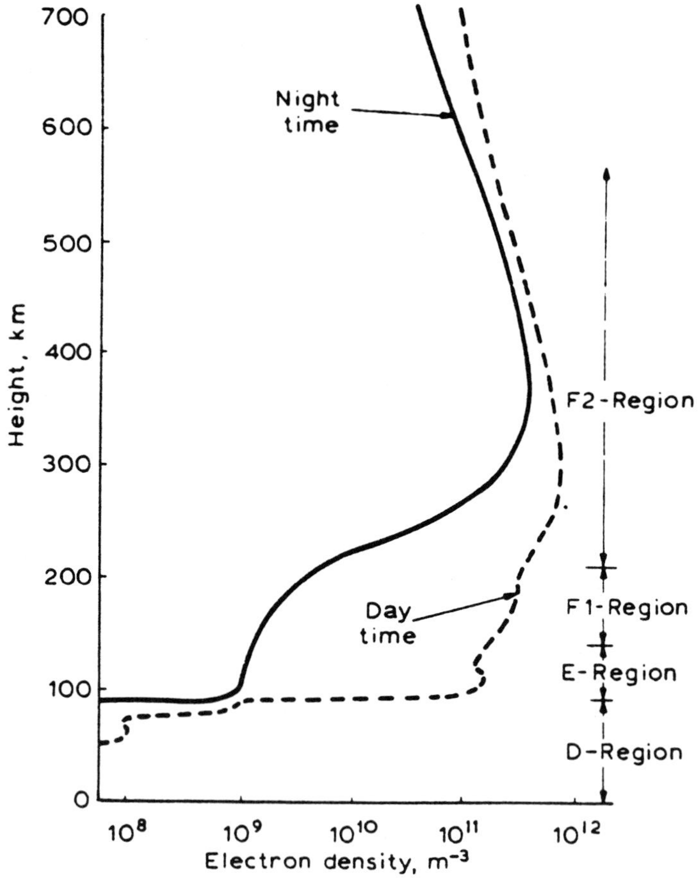

Fig. 18.7. OTH radar-sky wave operation. Reflective layers in the ionosphere are formed by free electrons. Their density varies with time. The greater the density, the higher is the frequency which can be reflected. (Reproduced from ref. 24)

implies one location for transmitter and receiver, and in 'bistatic' mode the transmitter and receiver are at different places.

First find a large tract of ground, up to 1 km in length and no more than about 100 m from a shoreline. Now mount a number of h.f. radiators, vertically polarised, along the shore to form a linear array at 10 MHz (30 m wavelength). An array of 8 elements, 15 m apart, would serve to produce a fan beam of about 15° in azimuth and 20°–30° in elevation. This would be used to carry transmitted pulses of about 50 kw peak power and 200–400 μs duration. The array's closeness to the sea would allow the pulse energy to radiate within the beam and across the sea's surface in the manner described above. Using the phased array techniques described in section 18.1, the beam can be steered

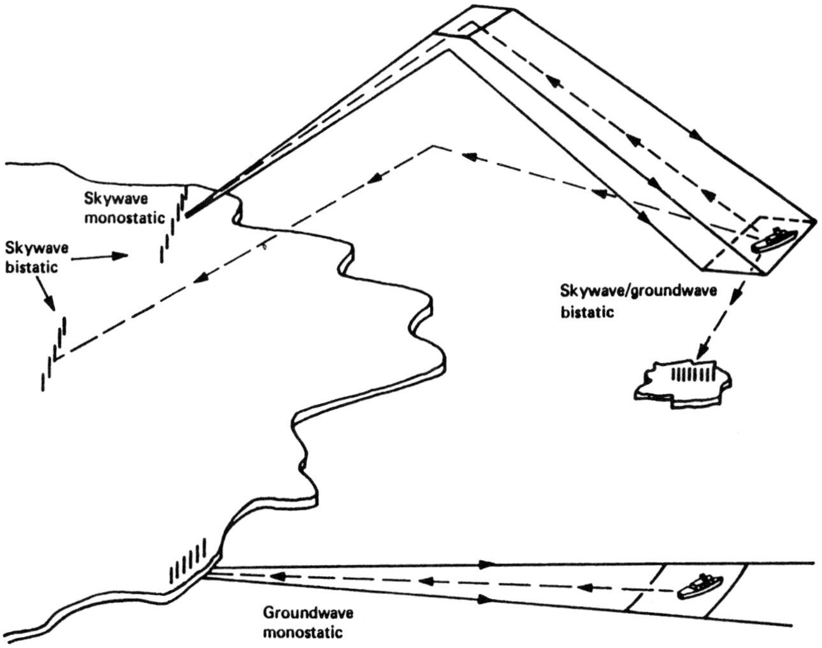

Fig. 18.8. Various configurations of OTH radars are possible. (Reproduced from ref. 25)

either side of its natural boresight to cover a large sector, typically 90° wide. Now erect another similar linear array alongside it and make it have a much smaller (e.g. 2.5°) azimuth beamwidth by increasing the number of elements. Again, by use of phased array techniques, the array can be made to create a cluster of contiguous narrower beams which are simultaneously present and steered to be coincident with the transmitted beam, by synchronising the phase switching of the transmitter beam and receiver beams. The transmitter and receiver beams can then be directed to cover the same 15° arc in the quadrant and successively switched to cover the full 90° in six steps. The energy reflected from transmissions will be received in one or more of the narrower receiver beams. The principle is illustrated in fig. 18.9. A functional block diagram is shown in fig. 18.10. The whole system must be coherent and so the transmitter, receiver and processing all share a common accurate frequency reference source.

The transmitter and receiver beams are formed and slewed by the phased array techniques described in chapter 17. In the transmit mode the peak powers of about 50 kw (shared among the antenna's radiators) are amplitude modulated by a pulse which is usually 'rounded rectangular', i.e., its sharp edges deliberately blunted to reduce high frequency spectral lines. It will be

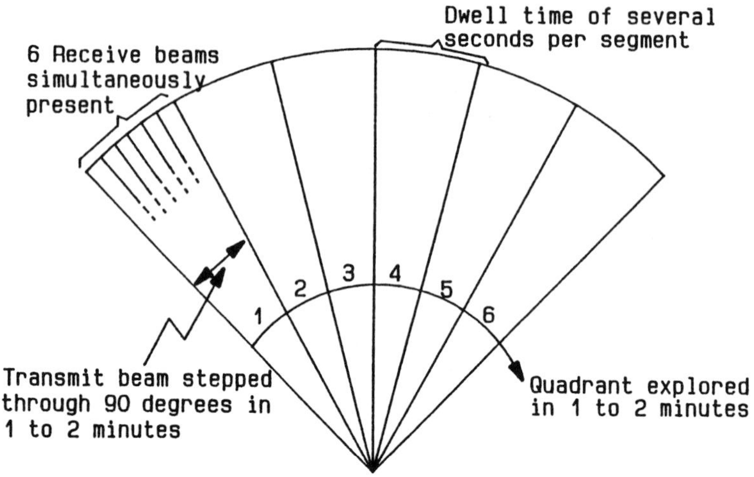

Fig. 18.9. The beam structure and management of a ground wave OTH radar.

remembered from chapter 1 that a rectangular pulse has a very wide spectrum and the lines far out from the centre represent 'out of band' radiation which causes interference to other users. Thus, typically, a waveform such as 'cosine squared on a pedestal' would be employed; this effectively rounds off the top corners of a truly rectangular pulse and slows down the speed of its rise and fall, reducing the higher frequency components in its spectrum.

The receiving antenna in our example is organised to produce six narrow beams side by side, each 2.5° wide. This implies a total array of at least six times the number of elements used to produce the 15° transmitter beam – hence the requirement for a large 'real estate' for the station. If the element spacing was half a wavelength (15 m at 10 MHz) then the receiver arrays would be $6 \times 8 \times 15$ m = 720 m wide, and separation from the transmitter array of some tens of metres would be necessary to prevent crosstalk from transmissions damaging the receiver inputs.

Beam forming for both antennae would employ the phased array techniques described in chapter 17 and their directions governed by control from a systems management function of a central computer. The pulse duration is, by normal radar standards, very long and so intrinsic range resolution without pulse compression is very coarse (about 16 nautical miles, 30 km). This can be improved, however, by comparing the amplitude of signals from a given target in the range cells immediately before and after the one being examined. By this means an average range accuracy of about ± 2 nautical miles can be achieved, because the pulse's transmitted amplitude history is known.

A more important process is carried out on received signals; this process performs a detailed analysis of the doppler components in the received signals.

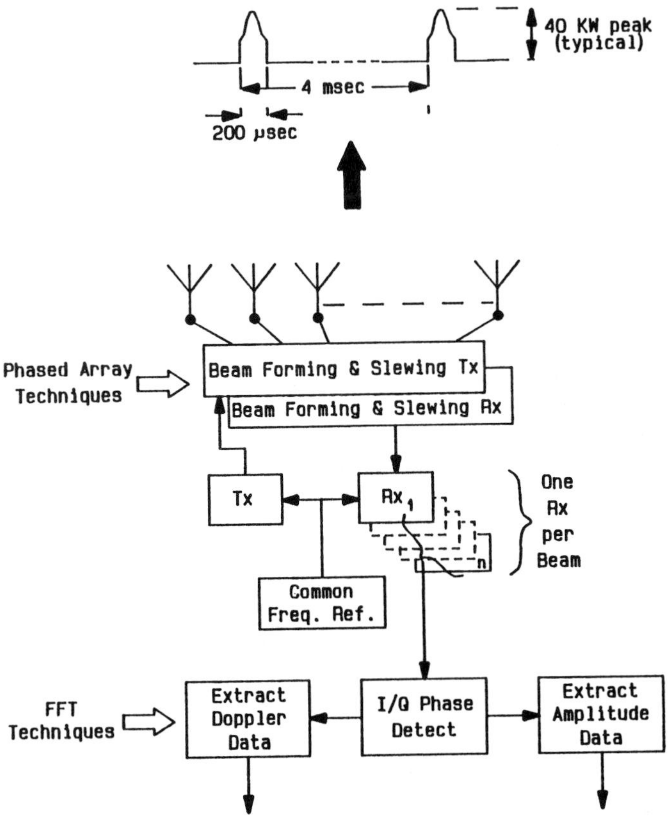

Fig. 18.10. Functional block diagram of a typical ground wave OTH radar. Beam forming and slewing is by phased array technique and use of a common frequency reference. Resulting coherence permits Doppler frequency analysis and display.

In each resolution cell there could be signals caused by the waves of the sea, ships, buoys, low flying aircraft, etc. These all create their own contribution to the compound signal in the cell. Because the system is coherent, these contributions will contain doppler frequencies, determined by the reflecting object's radial velocity relative to the radar. We have already seen in chapter 9, which deals with signal processing, that in a coherent system it is possible to construct a bank of filters each tuned to a specific doppler frequency (see fig. 9.18 (b)). These are, in an mtd radar, relatively few in number – typically 8. Thus the resolution in the doppler domain is quite coarse. In the OTH radar of our example, this principle is used but put into practice by very fast and powerful computer-based processors to create a large number of narrow band filters effectively acting as a complete spectral analyser of the doppler components.

The technique goes by the name of Fast Fourier Transform (FFT). In signal

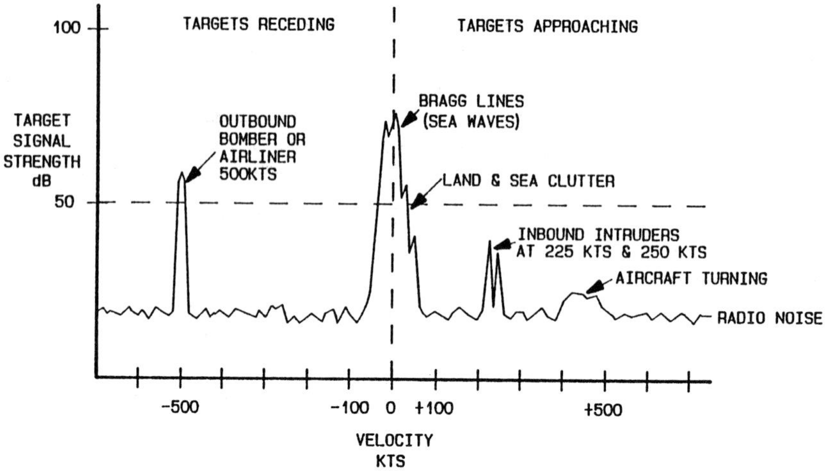

Fig. 18.11. Typical signals from a ground wave OTH radar from its doppler processor. These greatly improve tracking performance by supplementing the radar's coarse position – fixing data. (*Data courtesy of Marconi Radar.*)

processing parlance the 8-filter bank cited above would be called an '8 point' FFT. This gives each filter quite wide bandwidth (many tens of knots radial velocity). The computer-based FFT can execute many millions of operations per second and can be made to perform many thousands of FFTs. In our OTH radar example one would expect to find a 4096 point FFR to be embodied. An example of the power of such detailed analysis can be seen in fig. 18.11 obtained in trials a few years ago. It shows the doppler contents of one resolution cell which in the trials system was about 20 km square. The discrete velocity data are of great value in tracking routines and are a powerful supplement to the relatively coarse range data.

The analysis of each resolution cell can only be this fine if its contents are examined for the equivalent of 4096 successive samples. At a pulse repetition frequency of 250/s the beam dwell time must therefore be $\frac{4096}{250} = 16.4$ s. Thus it can be seen that finer range resolution cells would complicate matters, for aircraft would easily and often move across range cells in this time.

As exampled in fig. 18.9, the quadrant contents take a long time to explore. Other methods of beam formation and their exercise are possible, such as 'floodlighting' the area and making more received beams. This requires more transmitter power and a larger array of receivers and processors, but produces a faster data rate.

18.2.6 A typical sky wave OTH radar

The ground wave system described above is relatively simple to manage – its propagation is across the land–sea interface and over the sea and does not depend upon ionospheric effects which, as we have seen, are very frequency dependent. An essential element of sky wave systems is a frequency management sub-system which will show the frequencies which will best achieve reflection down to the earth's surface at pre-determinable places. Figure 18.12 shows a hypothetical OTH radar station in Italy covering a large quadrant of the earth's surface. As in the ground wave case, beams are formed to be relatively narrow in azimuth. Radiation through such a beam, wide in elevation, is reflected by the ionosphere and will be directed downwards, forming a 'footprint' on the earth's surface. The radar system has to manage its transmission frequencies so that the footprint moves in range at a given azimuth in an overlapping fashion; its azimuth is governed by use of the phased array technique described for the ground wave system. As each azimuth is selected, frequency management steps the footprint in the range domain. When the full range has been covered, a new azimuth is selected and the process repeated.

Commonly, an 'ionospheric sounder' is employed to discover the frequency to use for given footprint positions. There is a worldwide network to provide these data which, among other things, give the virtual height of reflection layer as a function of frequency together with the elevation angle which provides the propagation path. The elevation pattern of the antenna systems is usually very broad, with little directional power, and the often-represented 'beam' in

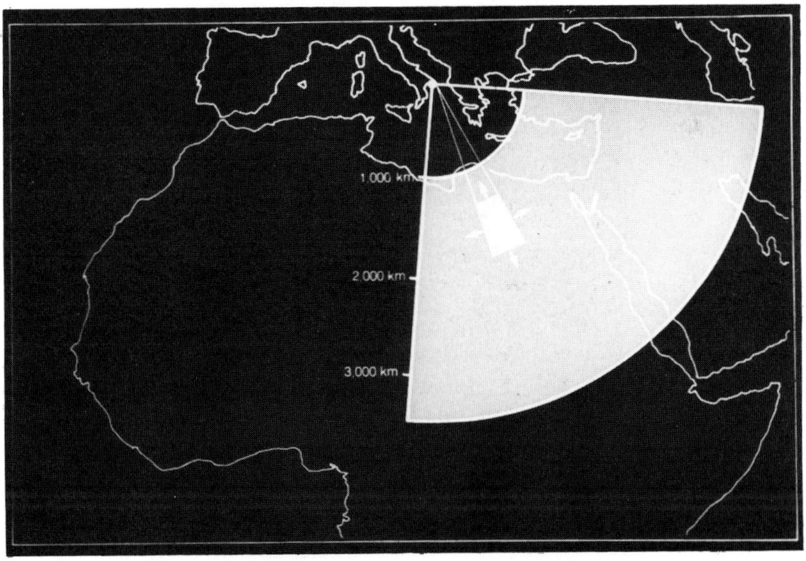

Fig. 18.12. A hypothetical Sky Wave OTH radar in Italy showing the extent of its coverage. The 'footprint' is idealised. It is stepped in range by frequency management and in azimuth by phased array technique.

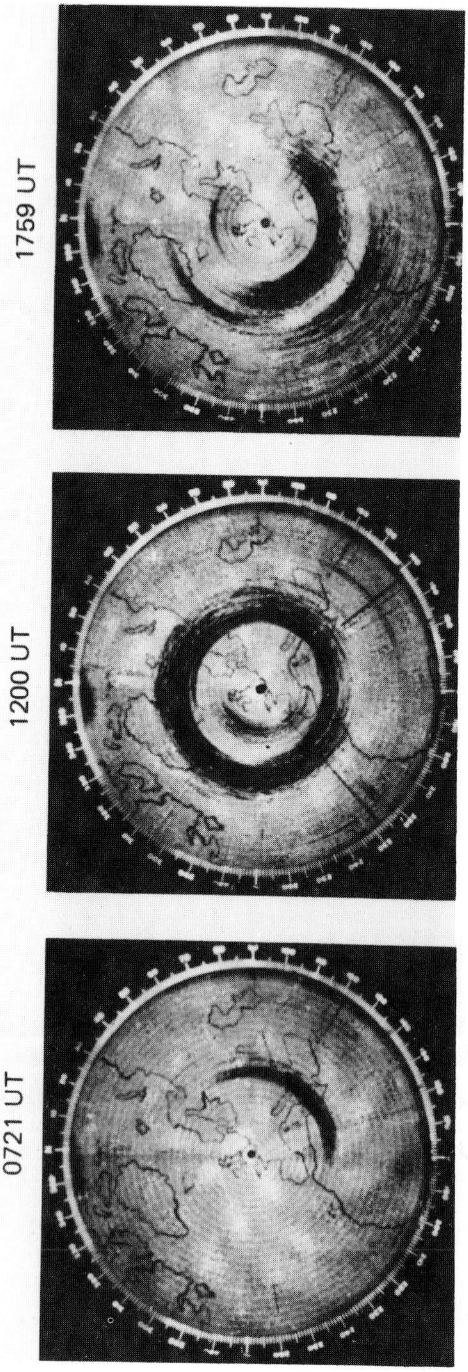

Fig. 18.13. Ionospheric backscatter soundings from a station at Slough in the UK at 17 MHz. The black areas are 'reflectors'. It shows F_2 opening up to Eastern Europe at dawn, complete F_2 azimuth coverage possible at noon, and wide cover in the SW quadrant at dusk. (Reproduced from ref. 25)

elevation is really an expression of the 'path of greatest transmissibility' or least loss which varies both with time and frequency.

A knowledge of these data permits the stepping out of footprints to desired areas according to a calculable frequency sequence. The ionospheric soundings produce tabulated data for the station's management but a good idea of their contents can be seen from fig. 18.13, taken from reference 25. The figure shows, in global terms, those areas of the ionosphere which can be used in a sky wave system at particular times. However, the areas in which such data are reliably useable are not very large and it is common to find sky wave systems including a 'sounder' fairly close to the radar itself.

Figure 18.14 shows a typical sky wave OTH radar system which bears a great likeness to the ground wave system, including doppler analysis, except for the all-important frequency management sub-system. Beam management in terms of position of footprint and dwell time is called for, and the same type of processing is used. A major difference occurs in the type of transmission. The

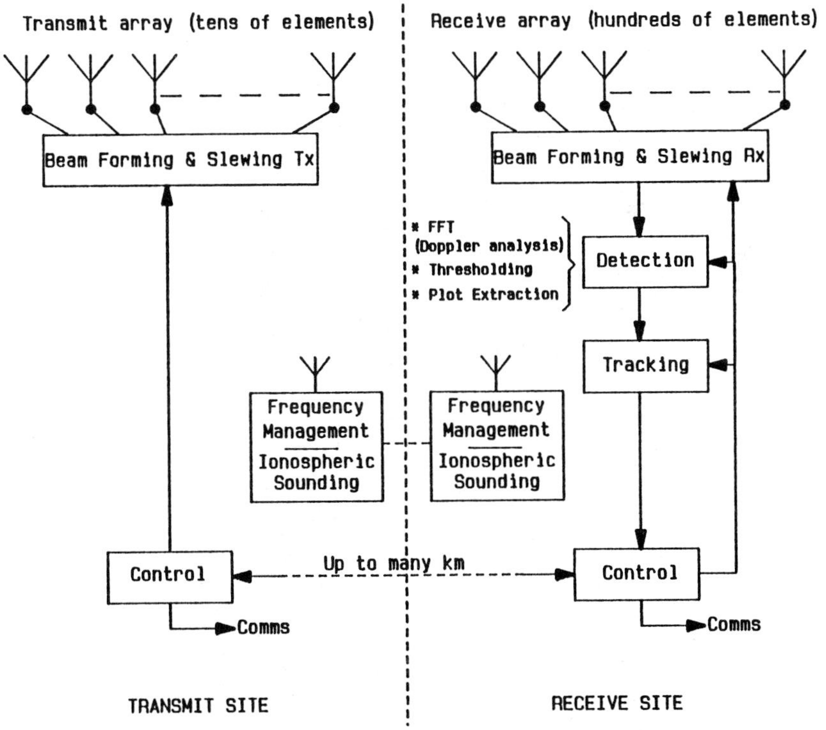

Fig. 18.14. Functional block diagram of a typical sky wave OTH radar. The receiving station can be a great distance from the transmitter. Frequency management follows from data gathered by the ionospheric sounding system and also provides common frequency source for coherent detection.

carrier wave frequency has to follow a sequence dictated by the frequency management system to step the footprint across the available range extent (900–3000 km). To get unambiguous range information at 3000 km requires a low p.r.f. of 50 Hz (20 ms periods). This is not an efficient way to use the available transmitter power if pulses of only 200 μs are used – it represents a duty cycle of only 1% from tubes capable of 100%.

An alternative frequency modulated carrier wave (FMCW) technique can be adopted. In this, a carrier wave's frequency is modulated so that it gradually increases linearly with time (much as in a pulse compressed radar) at a known rate. Echoes from its radiation which are received many milliseconds after launch will bear the same frequency as at their launch time. When the signals reach the receiver at this later time, their frequency difference from that being now transmitted will be different and because the transmitter modulation is linear, this difference is directly proportional to range. For instance, if the total modulation sweep was 2 KHz for a 500 km segment of range, a signal received from a target at the end of this segment would be 2 KHz different from that of the transmitter at its time of arrival. At 250 km the difference would be 1 KHz. By erecting a series of filters through which the received signals pass, they can be assigned ranges by filter number. If the filters were 100 Hz wide, 20 of them would elicit the range to 25 km accuracy.

When one realises the vast areas which can be covered by these techniques it will be appreciated that their complication is very worthwhile. Such systems have very many peacetime uses giving early warning of tidal waves and incidence of violent storms, their passage, growth and decay.

18.3 Multi-static radar & multilateration

18.3.1 Introduction

To many, the terms 'multi-static' and 'multilateration' themselves create a barrier to understanding. Fortunately, they describe quite simple principles, used long ago in radar's history in systems which did not have the benefit of narrow and rotating beams. The principles have already been introduced in earlier sections for specific systems but here the topics will be treated as generalities with wide application. The term 'multi-static radar' – as we have seen in section 18.2, OTH radar – means that transmitters and receivers at multiple and different locations are used. If there are two locations the system is bi-static (although classical scholars will be more content with 'di-static'!) If there was one transmitter and two receivers it would be 'tri-static', etc.

Multilateration ('multilateral' means, literally, 'many sided') is a technique where location of target position is obtained by measuring its distance (the length of the sides) from a number of known points. If the only data available are those of distance then three points must be used to avoid ambiguity. If an angle between target and one known point is known then only two distances are necessary.

The earliest UK radar (the 'Chain Home' (CH) system) was sometimes used in a multilateration mode. A number of stations around the coast produced wide area static beams about 60° wide in azimuth. Because of the beam's width, given aircraft (or a group of planes) produced signals at a number of CH receiver stations. Their range and crude bearing was reported to a control centre where lines of proportional length were drawn out on a map in the appropriate directions from the reporting stations. Their meeting point gave the slant range and a much more accurate position in space of the aircraft, because range could be much more accurately measured than bearing.

18.3.2 A modern example of multi-static radar

In this, a rotating beam, very narrow in azimuth, transmits pulses in the normal radar fashion. A receiving station about 200 km distant detects their reflections from aircraft. The positions of transmitter and receiver are known and the receiver is told the transmission time and the transmitter antenna's bearing. From these data the target position is calculated.

The method is based upon the geometry of the ellipse. For those unfamiliar with this, the following model may be helpful. The main property of an ellipse is that the sum of distances in a straight line from its foci to a common point on the ellipse is a constant. This can be illustrated 'inside out', as it were, by the following method. Make a loop of string and lay it on a flat surface. Now put two pins at different points within the loop. Put a pencil point in the loop and move it outwards to stretch the loop into three straight sections, one of which forms a straight line between the pins. Now trace the pencil point on the flat surface so that the three sections are always straight – the loop always taut. The pencil will have drawn an oval shape perfectly symmetrical about the two pins. It is an ellipse with the two pins as its foci. Because the loop has been kept taut and its total length as well as the distance between the pins were both constant, the sum of the other two varying sections muist also have been constant. The model above is one of 'plane geometry', i.e., the two-dimensional surface is flat. It also holds true for solid geometry where the loop is allowed freedom of movement in a third dimension. In this case the point would move over an elliptical spheroid which again would be an egg shape, symmetrical about the foci.

Translating this to the bi-static radar exampled above, we have two focal points, the transmitter radar and the receiving station, at known locations 200 km apart. The plan position of the situation is represented in fig. 18.15. The rotating transmitter beam is about 0.5° wide in azimuth and its bearing is reported continuously to the receiver station to a 12-bit accuracy, that is, in increments of $\frac{360}{4096} = 0.088°$ (5.27 min of arc). The receiver has 20 equi-spaced beams, generated by a fixed electronic static array using the i.f. beam forming

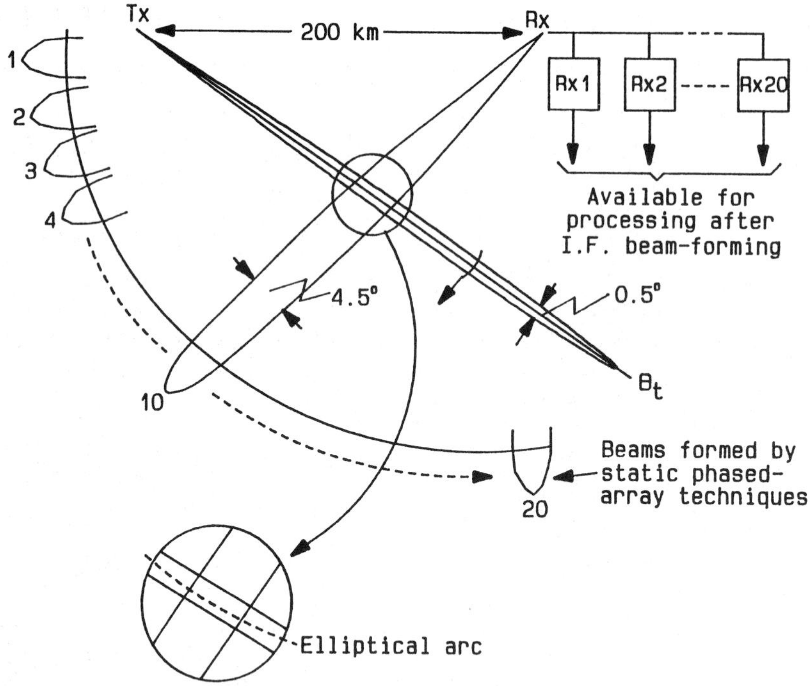

Fig. 18.15. A multi-static radar operated in the UK. A narrow rotating beam transmitter illuminates targets. Signals are received and processed by a static electronic scanned array 200 km distant. The two stations are very accurately time referenced. Detected targets lie on an elliptical arc within the two beams' intersection point.

technique. They are simultaneously present, covering a 90° sector. Imagine an aircraft within the crossover area of the transmitting beam and receiver beam number 10. By very accurate time referencing shared by both receiving and transmitting stations, and a common synchronising pulse, the receiver station knows the precise moment that each pulse is launched into space. Energy reaching the target will be scattered in many directions and some will be gathered by the receiver beam in line with the target (number 10 in the example). The received signal's time of arrival can be measured very accurately.

Because the velocity of propagation is constant and the energy is travelling in straight lines we can now see how the elliptical model comes in: The distance between the foci is known, the total time taken for energy to reach the target and onwards to the receiver is known and thus the total distance travelled is known – the sum of these is the total length of the 'loop'. Thus, somewhere in the beam crossover area, is an arc of an ellipse, calculable from the gathered and available data, on which the target lies. Successive transmissions illuminating the same target at slightly different azimuths will yield further measurements, including

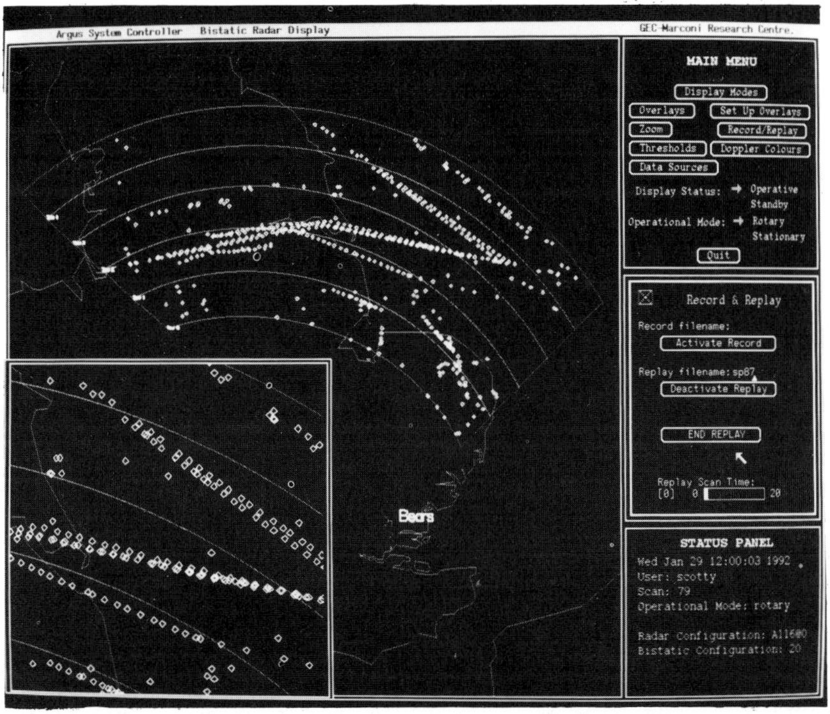

Fig. 18.16. A working example of multi-stasis. This time-lapse photo is of the plot extracted output of the BEARS (Bi-static Electronic Array Radar System) at Great Baddow, Essex, UK. No tracking routines are active, so many false plots are present. The bottom left 'window' is a blow-up of the top-centre of the picture. Range rings are at approximately 30 km intervals. (*Photo courtesy of GEC-Marconi Research Centre.*)

velocity calculated from the doppler components introduced by the target's movement. From these a more accurate target position is calculated.

Because the receiver beams cover a wide area, all targets of sufficient size within the area will be detected, and plots representing them derived in a plot extractor. Over successive transmitter beam revolutions, tracks can be formed. Because the target positions can be predicted from the tracks, and the times of transmission are known to both transmitter and receiver, expected times and directions of arrival of signals at the receiver can be predicted. This allows the small number of processors to be switched at a very high speed among a large number of beams at times when signals are expected within these. This process is sometimes referred to in the literature as 'pulse chasing', which is a bit confusing since what is really meant is 'pulse catching'.

An example of the method in action is given in fig. 18.16 taken from an experimental system, funded by the U.K. Ministry of Defence, using a transmitter at Great Malvern, near Worcester (Royal Signals and Radar

Establishment, RSRE) and a static electronically scanned array at Great Baddow, Essex (GEC–Marconi Research Centre). The area covered by the 20 receiver beams is shown in the 'playing area' graticule with plots of aircraft detected.

18.3.3 Multilateration

The principle is illustrated in fig. 18.17. Consider, first, the simple case of a plane surface. Three known points are shown. Any point on the surface is defined by its straight-line distances to each of the three known reference points. If only points 1 and 3 are used, the position of point P becomes ambiguous – it could equally be at P or F. However, if the angle Θ is known, the ambiguity is resolved. In a three-dimensional hemispheric space such as is occupied by vehicles on or above the earth's surface, the above still holds true except that for the two-point reference case, one extra angle is required (the elevation angle of one distance line).

This principle is used in a number of ways, the most modern being position fixing by geo-stationary satellite clusters. In this, satellite signal receivers at point P measure the time of arrival of identical signals sent simultaneously from the satellites whose position is fixed relative to the earth and at known positions. Because the velocity of propagation is known and constant, the signal's arrival times are directly proportional to their distance travelled. Point P is calculable from a knowledge of three distances. The time reference to synchronise the sending of the signals is extremely accurate (to about 1 second in 36000 years!)

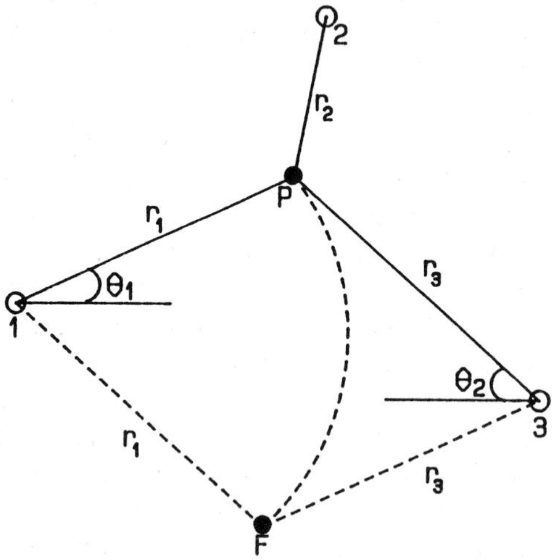

Fig. 18.17. Sensors are at known points (1, 2 and 3). If the range from only two of these was given (e.g., 1 and 3) ambiguity of position results (P or F?). Either ranges from each sensor, or from 2 and one 'angle of arrival' are necessary to resolve the ambiguity.

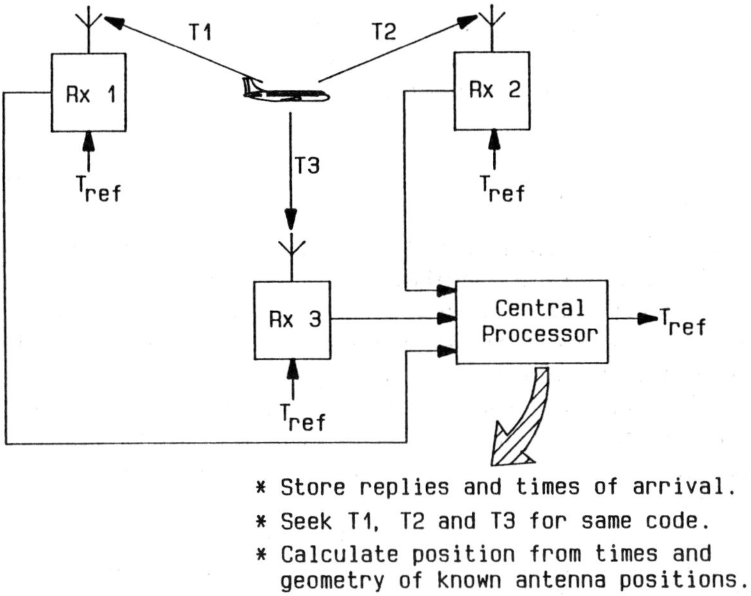
T1
T2
Rx 1
Rx 2
Tref
Tref
T3
Rx 3
Central Processor
Tref
Tref
* Store replies and times of arrival.
* Seek T1, T2 and T3 for same code.
* Calculate position from times and geometry of known antenna positions.

Fig. 18.18. Multi-lateration in SSR. Three receivers at known points with a common time reference are the source of range and code data on given aircraft, because range is proportional to time. Code correlation is sought and from the corresponding times of arrival, the target position is derived.

and permits calculation of receiver position to fine limits (about 10 m using the system's full signal coding). A succinct description of the NAVSTAR system will be found in reference 26.

This is not, however, a true radar system. A good illustration of the use of multilateration in radar is given by a possible application in SSR. From chapters 10 and 11 on SSR it will be remembered that vehicles (usually aircraft) issue pulse coded replies from a transponder in response to pulse coded interrogations. Usually interrogations and their replies flow through the interrogation station's beam which is regularly rotating and narrow in azimuth; the transponder's antenna beam is, in essence, omnidirectional.

Imagine now that three independent SSR receiver stations with omnidirectional antennae are erected and form a triangle, as shown in fig. 18.18. Any replies from the aircraft will be received by each station and their times of arrival accurately associated with them. These are passed to a central processor which looks for the same code emanating from each station. These may be in response to requests for the aircraft identity or its altitude code. In any case either of these reply pulse trains would appear at each station. The processor would, through code association, take their different times of arrival and calculate the target position from a knowledge of the station geometry.

Identity and altitude interrogations are usually issued in an alternating sequence (modes A, C, A, C, etc.). If the aircraft responds to both identity *and* altitude interrogations the accuracy of the time referenece would be good enough to calculate, from the three range values and a knowledge of each station's height, how *high* the target is. This is now introduced into the processor and turned into an SSR altitude code for comparison with the next set of codes to be received from the aircraft. By this means, altitude and identity codes from the same aircraft can be separated.

In the above example it will be noted that no knowledge of the *times* at which the interrogations are issued is required at the receivers. Indeed, the system could work without an interrogator because SSR transponders are required to issue replies stimulated, not by received interrogations, but by their own internal receiver noise in a random fashion at a rate of between 1 and 5 per second (the so-called 'squitter' rate laid down in the I.C.A.O. Annex 10 specifications, reference 10).

Practical forms similar to the above system are in use in the USA using Mr George Litchford's patented ideas (G. Litchford–Litchstreet Company, Cherry Lawn Lane, Northport, New York 11768, U.S.A.). Other studies have shown that the technique can be used to measure SSR equipped aircraft to about $\pm$ 100 ft at an altitude of 25 000 ft using stations about 15 nautical miles apart.

19
SSR – mode 'S' in action

19.1 The Aeronautical Telecommunications Network (ATN)

Since this book was first published in 1985, a great deal of progress with Mode 'S' has been made. Of the greatest significance is its internationally agreed and recognised place in the ATN, whose grand and noble aim is to provide the means of ground–air–ground communications in all parts of the global airspace.

19.1.1 ATN Architecture

Its structure is seen in fig. 19.1. The two way link between ground organisations and aircraft is to be provided by any one of three data 'carriers':

- Satellites.
- Mode 'S'.
- VHF.

Satellites will be used over vast oceanic areas where the other two carriers are either not able to operate or are impractical to use. Mode 'S' services will be provided in the volume of the large number of the world's busy airspaces to give radar surveillance as described in chapter 12. The massive data capacity of Mode 'S' (5×10^{33} which is 5000 million, million, million, million, million messages in both the up and down links) will allow all kinds of automatic digital data interchanges between ground and air of great service to a.t.c., airlines, meteorological and other users. The VHF carrier is already in use as a data link, providing automatic a.t.c. clearances beyond the Atlantic coast of Canada.

19.1.2 Standard protocols

To enable the large number of sub-systems and equipment elements to work together end-to-end, the data communication is intended to follow the 'Open

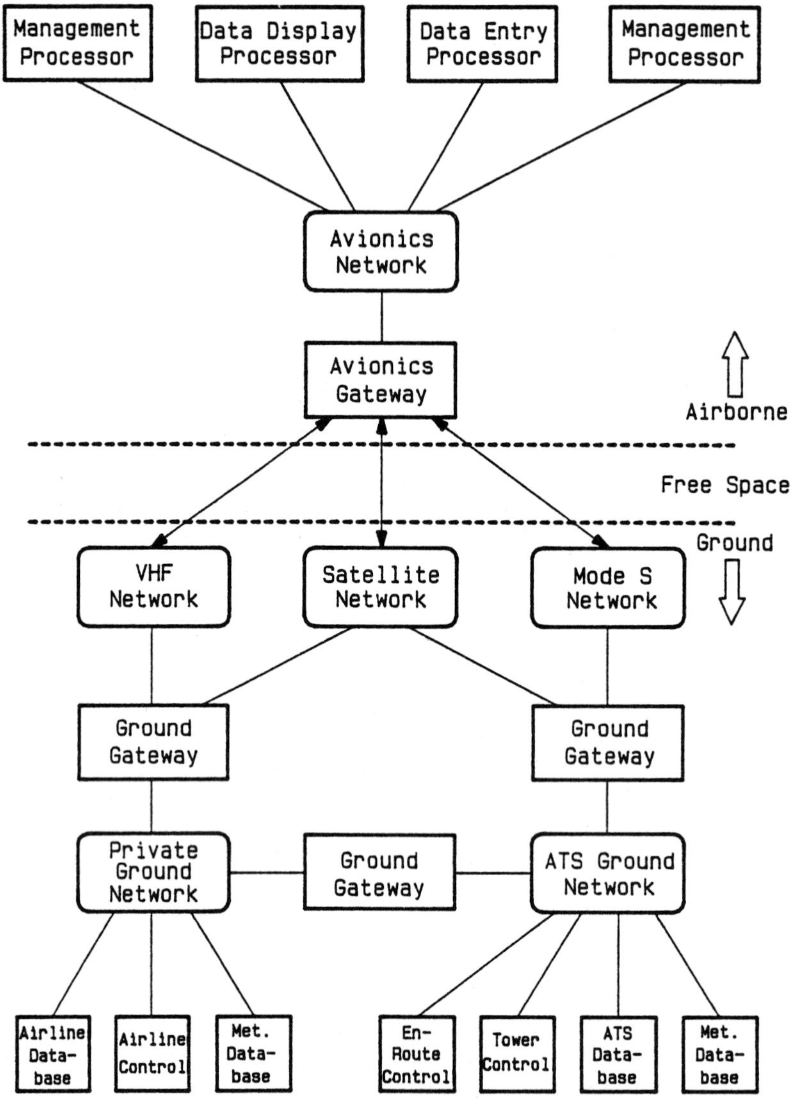

Fig. 19.1. The Aeronautical Telecommunications Network (ATN) Structure. VHF, satellite and SSR-Mode 'S' links provide ground-air-ground communications. The ATN aims at giving service for all craft in global airspace by subscription to International Standards.

Systems Interconnection' (OSI) reference model agreed by the ISO. This defines seven 'layers' of functions and message structures at both ends of the data link. An idea of these functions is given in fig. 19.2. The intention is to

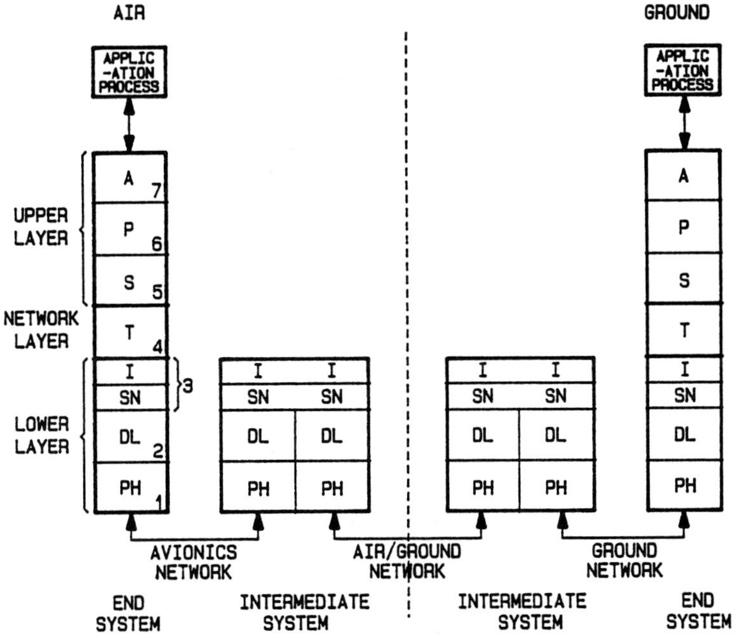

1. Physical layer (wire, fibre optic, radio)
2. Data link layer (to manage the physical layer operations)
3. Inter-network and sub-network sub-layers (decides data routes)
4. Transport layer (to establish reliable end-to-end service)
5. Session layer (to establish a relationship between two end users)
6. Presentation layers (controls format and appearance of data)
7. Application layer (controls access to comms system)

Fig. 19.2. The OSI model – a layered set of interfaces each requiring a defined message structure.

allow any user at either end to define an addressee, construct a message and send it without the need to specify the route; this will be automatically selected on the basis of what carrier is available, the tolerable message delay, its length etc.

Of the seven layers, only the three lower levels are currently defined (1991); many data systems are already using them successfully in, for instance, message switching telecommunications. The remaining four are still being discussed and designed. Their design is particularly difficult. This can be appreciated when one realises the huge imbalance between the message content itself and the large number of identification 'bits' which must be attached to the messages to define parameters such as their route, the connection time required, the need for acknowledgements, and error checking, etc.

19.1.3 Mode 'S' and the ATN

We have seen that Mode 'S' is to be a vital component in the ATN. How will it be constructed and how used? Figure 19.3 is a block diagram showing the major components and their functions.

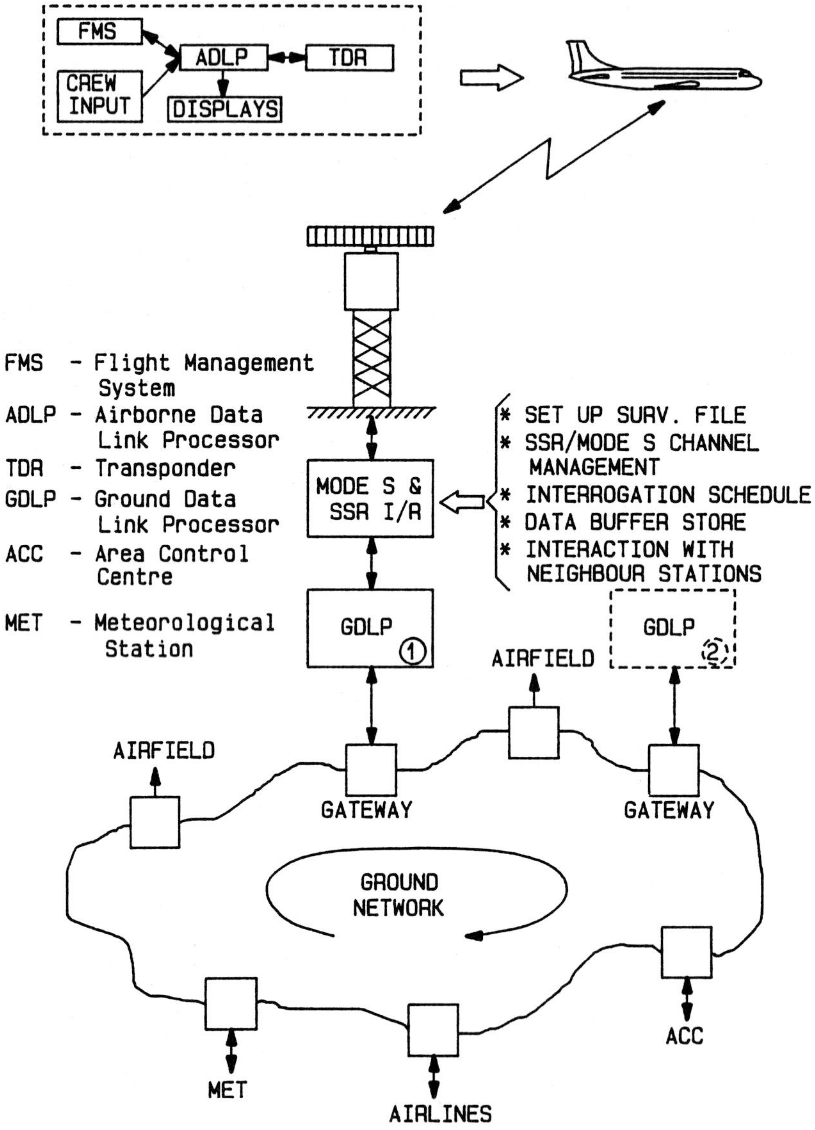

Fig. 19.3. Block diagram showing major components of Mode 'S' in the ATN, and their functions.

Essentially, we see a number of potential users gaining access to the Mode 'S' interrogator/responser sensor station via a Ground Network and a Ground Data Link Processor (GDLP). A group of sensors can serve a wide volume of airspace, producing over-lapping coverage through which aircraft can be continuously tracked. A second GDLP is shown as an example of this.

As the sensors explore the airspace volume they will issue Mode 'S' all-call interrogations at the rate of one or two per beamwidth dwell time. A Mode 'S' equipped aircraft not already being tracked will respond to these, giving its unique address. From this knowledge of its position, subsequent interrogations with this address can be issued in the next antenna revolution at the appropriate bearing and the aircraft's track can subsequently be built up and 'logged' in the ground station's Surveillance File. These surveillance data are made available to users via the GDLP and the ground data network.

Suppose, now, that an air traffic controller wants to request a change of altitude to an aircraft in his charge. He types in the aircraft call-sign and the message content expressing the request. When this is released into the ground network it will be automatically routed to the correct sensor by the system's knowledge of which station is covering the aircraft in question (from surveillance files). The channel management function of the interrogator/responser will assemble the request in the correct digital form for passing to the aircraft as modulation of P_6 in the Mode 'S' format shown in chapter 12 (fig. 12.1(b)). When the sensor's antenna approaches the aircraft's bearing, this

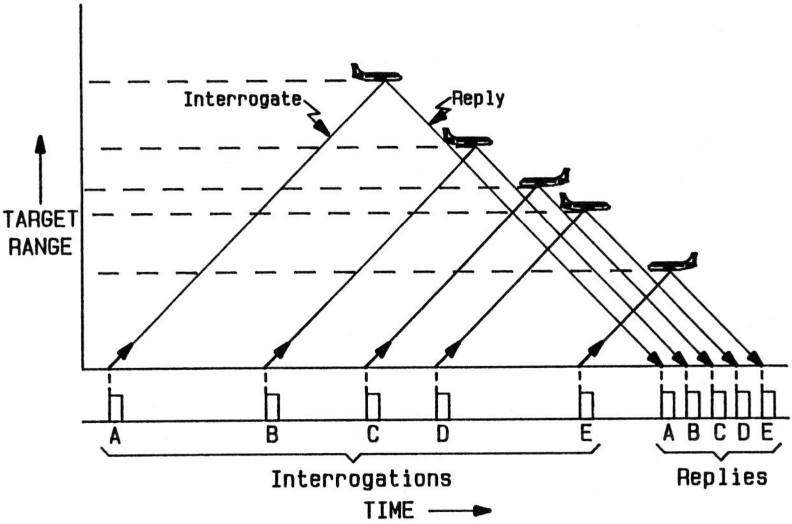

Fig. 19.4. Mode 'S' interrogation time-scheduling. Once aircraft range is known (from surveillance files) a number of craft in one beamwidth can be served in time order so that their replies do not overlap and create garble.

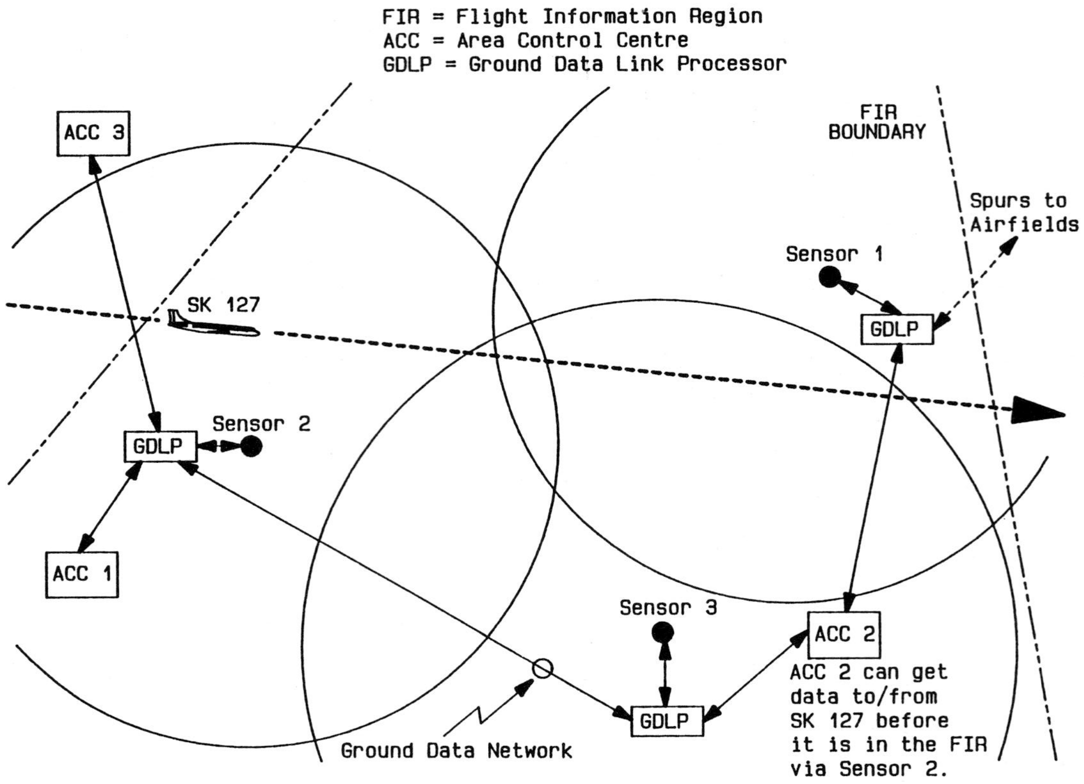

Fig. 19.5. A network of mode 'S' sensors provides communications to and from aircraft to ranges beyond that of the local sensor, using the access to GDLPs via a wide ground network.

interrogation will be made during the beam's dwell-time on the aircraft. Trials have shown that a single interrogation and its reply have very high data integrity because of the error sensing and correcting methods used. Should uncorrectable errors arise, the cycle of interrogation and reply will be repeated.

A major aim of Mode 'S' is to obviate the circumstance of 'garbling' (the stimulation of replies from aircraft close in range at the same azimuth). This is achieved by time scheduling interrogations to specific aircraft among a number in one beamwidth, using their discrete address codes, so that their individual replies arrive singly in time order with no overlap. This is illustrated in fig. 19.4. An example of two aircraft at 'garbling' range (C and D) is shown. From a knowledge of the required message lengths it is possible to calculate the appropriate time (to microseconds) to release interrogations C and D so that their replies arrive separately. 'What happens', you ask, 'if two aircraft are at the same range and *both* are required to respond to an "all-call" interrogation? This is handled by a technique called 'stochastic acquisition' (stochastic = determined by random processes). Effectively, the interrogation includes data bits which cause only the aircraft with the highest transponder receiver sensitivity to reply. Once the reply has been received, accepted and 'logged' in the Surveillance File, that aircraft no longer replies to all-call interrogations; thus the second aircraft's subsequent all-call reply will be in isolation. There are many other cunning techniques in the use of Mode 'S' and those wishing for more detail are referred to the Guidance Material on Mode 'S' associated with reference 10.

19.1.4 Mode 'S' – another OTH radar

These days, people concerned with improving the use of radar and communications technique in all kinds of traffic control are commonly addressing problems of 'space management'. In general they are interested in *who* is going to be *where* at *what time*. A supreme example of this is the current concern for the optimum management of airspace for civil aviation. Mode 'S' is a prime example of a system designed to give the masses of data required in a short space of time.

The system's ability to use a network of intercommunicating Mode 'S' stations will be of great value to airspace managers. As shown in fig. 19.5 a group of Mode 'S' stations with overlapping coverage can encompass vast areas. By exercising the ground network, air traffic managers, aircrew and air traffic controllers can get and give data from and to aircraft at great distances far beyond the reach of a local sensor – and thus far beyond the local radar horizon. The advantage of the 'lead time' for planning and control is obvious.

Expanding this notion to the whole ATM system, it will be possible to enter dialogue with a craft leaving Tokyo and govern the route, time and altitude of its passage to its destination – London.

20
Raster scan displays

20.1 Background

The display units described in chapters 13 to 15 use mainly cursive writing techniques. By this is meant that the beam of the display unit's cathode ray tube is made to 'write' characters and symbols on the screen in a continuous smooth line; they are written just as you and I write with a pen on paper. This technique produces symbols and characters of great clarity but because it is essentially an analogue method – the beam's deflection waveforms changing smoothly with time (see fig. 14.2) – it doesn't easily match the digital systems that often precede it.

The raster scan technique overcomes these difficulties by truly digital methods and in modern systems, produces a clarity of display indistinguishable by eye from those given by the cursive writing method. It uses essentially a 'television' type of technology and multicolour displays are therefore much in evidence.

Realising this, one can be excused for asking the question 'if it is a colour TV technique, why have we waited so long to see such displays brought into radar use – we've had colour TV for decades now?' The answer lies in the long time it has taken to produce colour TV tubes with sufficient resolving power to match that obtainable with cursive TV tubes.

20.2 Concerning resolution

Raster scan displays, indeed, displays of any kind, increase their usefulness if various colours can be used. But this usefulness in turn depends upon the ability to produce the required resolution.

Let us look first at the principle upon which colour TV tubes are based. There are two main types; the colour dot matrix tube and the 'Trinitron' tube.

20.2.1 The dot matrix tube

In this, the screen consists of a mass of minute circular dots each made of a phosphor radiating a specific primary colour (red, green or blue) when excited by an electron beam. The dots are in groups of three ('triads') and are extremely accurately aligned with a 'shadow mask' which is perforated with circular holes, one hole for each triad. Three electron beams are generated by separate electron guns, one for each primary colour, set in a delta configuration mounted in the neck of the tube. The shadow mask is meticulously positioned so that the three beams converge at the holes, and is so spaced from the phosphor screen that each beam can only illuminate its corresponding dots, the mask creating a 'shadow' for the other two dots in the triads. The arrangement is illustrated in fig. 20.1(a). In high quality 'monitor standard' tubes such as used in many raster scan display systems, the triads have a spacing (pitch) of about 0.3 mm; the pitch is relaxed to about 0.7 mm in domestic colour TV sets. From the magnitude of the tube design and manufacturing problems, one can begin to realise why we have had to wait so long for such tubes to be a viable and reliable proposition – and also why they are, relatively, so expensive. The physical

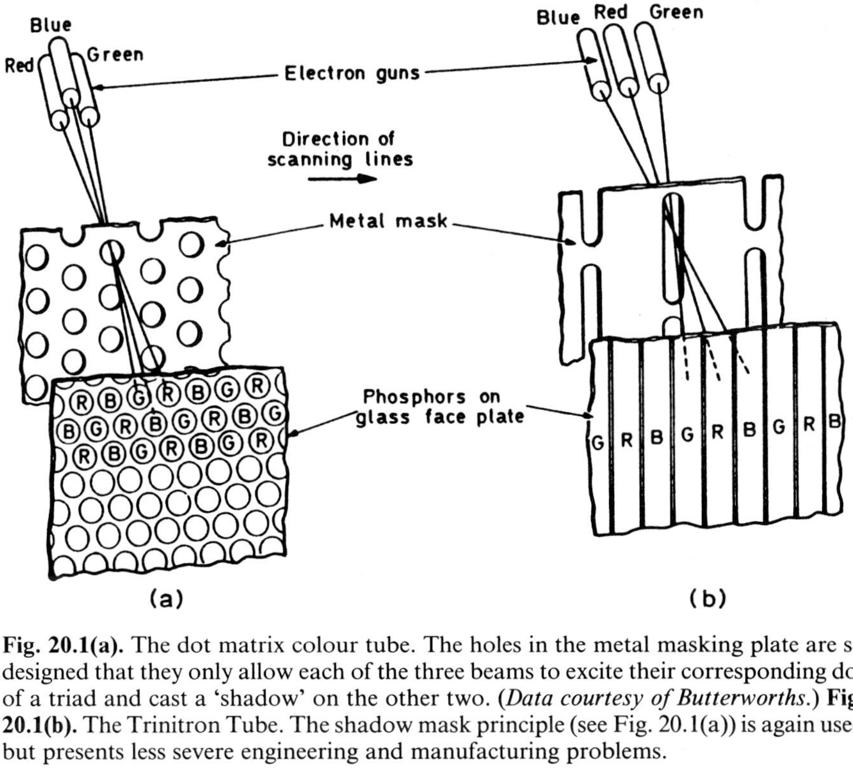

Fig. 20.1(a). The dot matrix colour tube. The holes in the metal masking plate are so designed that they only allow each of the three beams to excite their corresponding dot of a triad and cast a 'shadow' on the other two. (*Data courtesy of Butterworths.*) **Fig. 20.1(b).** The Trinitron Tube. The shadow mask principle (see Fig. 20.1(a)) is again used but presents less severe engineering and manufacturing problems.

accuracies, quality control standards, chemical purity, and mechanical strengths needed in manufacturing a 50 × 50 cm tube are so stringent, it is to me a miracle that they ever work – but they do! In such tubes the screen would contain 2.7 million triads at 0.3 mm pitch.

The three independent beams are scanned together across the shadow mask by deflection waveforms. Their intensity is varied in their passage, as dictated by the colour signals generated by the scene being televised – the colour camera tube separates the scene's colours into its primary colour constituents which are transmitted to the receiver in three separate channels. The whole spectrum of colours is reproduced by different mixes of the three primary colours. For example, various intensities of red and green will produce various shades of brown; total mixing of red, green and blue produces white, and different intensities of the three, various shades of yellow. The mixing, according to classical theory, is performed by the eye's retina provided the sources of the individual colours are not discernible by one's own visual resolving power.

Because the triads are so closely spaced their individuality is not seen by the viewer; normal human visual resolution has a value of about 2 minutes of arc. That represents about 0.3 mm at 0.5 m range so that at a greater viewing distance the individual triads with 0.3 mm pitch would not be perceived by the naked eye and an unexcited screen would appear as a plain grey or black surface.

The resolving power of raster scan displays is usually expressed in terms of the 'number of lines' in the picture. Figures of 1000, 1500 and 2000 lines are common. Putting this into perspective with practical tube structures we see that a 50 × 50 cm tube can easily support a 1500 line resolution with triads at 0.3 mm pitch (50 cm contains 1666 triads). A 2000 line resolution demands triads of 0.25 mm pitch. A 19 inch diagonal tube with a 5:4 aspect ratio has a 15 × 12 inch screen (380 × 305 mm). With triads at 0.3 mm pitch the screen can be exercised by only a 1000 line system without reaching the limit of normal human acuity. Thus we see a relationship between triad structure, tube size and resolution limits; the most important of these is the viewer's own resolving power – for to exceed this is uselessly expensive. Equally, by giving the electronic system driving the tube a higher resolution than the tube itself, the tube remains the resolution-limiting factor and makes best use of its capabilities. Hence, a 1600 line resolving tube is exercised by a 2000 line electronics back-up.

20.2.2 The 'Trinitron'

Trinitron tubes use the same principle as the 'dot matrix' tube, but instead of triads of dots, the tube's screen has triple strips of minute thickness running from top to bottom of the screen's face. Each strip has a phosphor emitting red, green or blue light, as in the dot matrix tube, when excited by an electron beam. Again a 'shadow mask' is employed to prevent beams from impinging upon the wrong colour strips. In this case the circular holes of the dot matrix mask become thin vertical slots and the three electron guns of the tube are now mounted in a horizontal line, as shown in fig. 20.1(b).

Because the colour strips are continuous in vertical lines the problem of accurate mechanical registration between the mask and the strips, although difficult, is easier of solution than its equivalent in the dot matrix tube case and the Trinitron tube is now more widely used in raster scan displays.

20.3 Reprise

Readers may recall that in chapter 1 the 'scientific method' was discussed, drawing attention to the need to modify theory if valid experimental results are ever found to disprove the theory. The theory of colour vision is a particularly good example of this and I introduce it now because it is highly relevant. The 'classical' theory is that we perceive colour by the mix of the primary colours emanating or reflected from the object viewed, receptors in our visual system being stimulated by red, green or blue wavelengths of the received light in various intensities. This can be, and has many times been, verified by scientifically controlled measurement.

However, this theory doesn't explain a time-honoured effect, noted for centuries – that shadows cast by incident and single colour-filtered light have a hue that is only possible if another colour is present. For example, a shadow cast by a blue light has a yellow hue. How can this be? Because the theory says you can only perceive yellow if the blue is accompanied by green and red.

Dr Edwin Land, the inventor of the 'Polaroid' camera, carried out other experiments producing similar startling effects, wherein a coloured scene was recorded by two black and white slides, one photographed via a red and one by a green filter. When projected via a red light for the 'red' slide and a white light for the other, and collimated to produce one image, it has full colour reproduction, including blues and yellows. Another effect, which many may readily see, is that where a chequer-board of white and black squares is seen on TV, the black squares are blacker than the tube face itself. Subsequent theory to explain this and other anomalies in the 'classical' theory indicate that the brain constructs the perceived reality by processing the eye's inputs in different parts of the visual cortex and is much more dependent upon integration of a whole picture than upon the physical reaction to one particular small area of the scene. This doesn't mean that current theory is wrong, only that it's incomplete.

20.4 Raster scan technique

20.4.1 Pixels

The TV technique explained earlier employs a raster scan method. The group of three electron beams is scanned from the top left hand corner of the tube across its face in a horizontal line. The group is then deflected slightly down so that the next line is scanned again from left to right just below the first. The process is repeated until the whole screen has been scanned to the bottom. This takes about one twenty-fifth of a second. The picture refresh rate is fast enough to be imperceptible to the naked eye.

As the name implies, the raster scan system uses the same technique to deflect the group of three beams across the screen. A major difference, however is found in the method by which the final image is achieved. At source, the picture content is described by digital words which relate to the number of picture elements – 'pixels', as they are called – in the tube's screen. To keep resolution equal in both horizontal and vertical planes, triad geometry in both is made the same. Thus a 1000-line system would use a tube which contained at least 1000 triads in both horizontal and vertical dimensions. The screen would be made of 1000 × 1000 pixels. We can therefore liken the screen to a large wall made of tiny square bricks each of which can be addressed from a digital store (i.e., line 528, row 17). As in building, it is possible to create 'windows' in the wall through which one may 'see' all kinds of different scenes. In some raster scan display

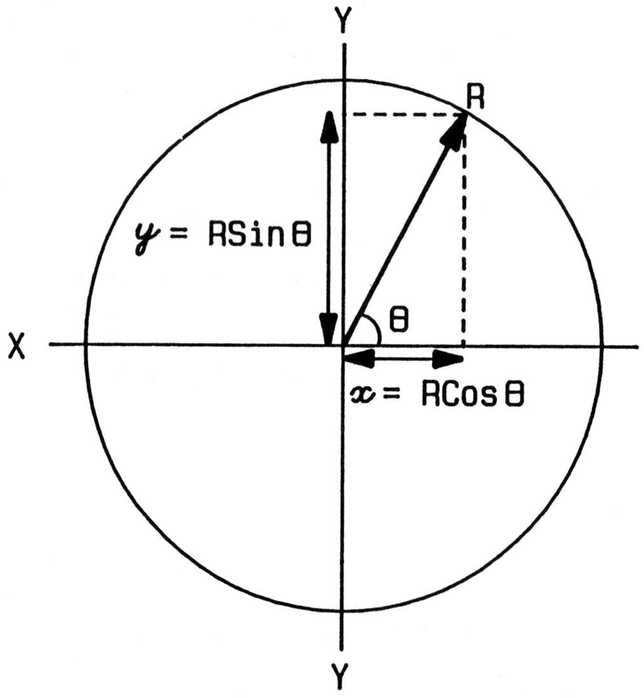

Fig. 20.2. Co-ordinate conversion. The vector RΘ can be expressed as cartesian co-ordinates (x and y).

systems, these windows are in fixed positions (nominated by the user at the design stage): more complex systems allow full freedom for the size and position of windows to be varied at will by operators. Immediately the usefulness of this is apparent because it becomes possible for many kinds of data related to the operator's task to be given on one display, allowing freedom to change its contents, or its position, or for it to be shown or inhibited, as the user wants.

20.4.2 Building the radar picture

Surveillance radars generally report target range using the radar antenna position as reference and target bearing relative to north. It is therefore natural for the reports to be in these terms (RΘ). The ppi display reconstructs the radar picture knowing these references. Many military systems use a grid reference system based upon geography – a 'Georef' system. Plotting positions in a georeference grid is more conveniently done if target reports are in Cartesian co-ordinates – X and Y axis distances from a reference point. Many radar systems are designed to serve georeferenced operations and their radar position data are converted from RΘ to X–Y terms. This is often performed at the radar site itself and in plot extracted systems the digital data words describing target position undergo this conversion before output to the display. In these days of microprocessors this is no longer the problem it used to be. In essence the process of co-ordinate conversion follows from the diagram in fig. 20.2. The radar senses the range and bearing of a target; the conversion to X and Y co-ordinates is by simple trigonometric calculation performed in a digital processor.

It will be realised that raster scan displays are very like a georeferenced system in that their technology is based upon similar principles, using rows and columns equivalent to X and Y dimensions.

Co-ordinate conversion can be carried out at the remote display site; indeed, this is a common requirement because large numbers of radar systems use the RΘ method of reporting detected plots. However, a further translation is necessary so that the radar data input to the display can fit in with the 'pixel' organisation of the raster scan picture.

20.5 Organising the picture

There are different architectures used in raster scan display systems but all embody the functionality represented in fig. 20.3. It illustrates a typical radar display showing airways, target position (triangles), past history of position (dots equivalent to target track) and three different sized and overlapping windows; there could be many more if desired and one would almost certainly contain a 'menu of functions' such as: range scale, height and other filters, brightness, picture off-setting, etc. The operator uses 'man-machine interface'

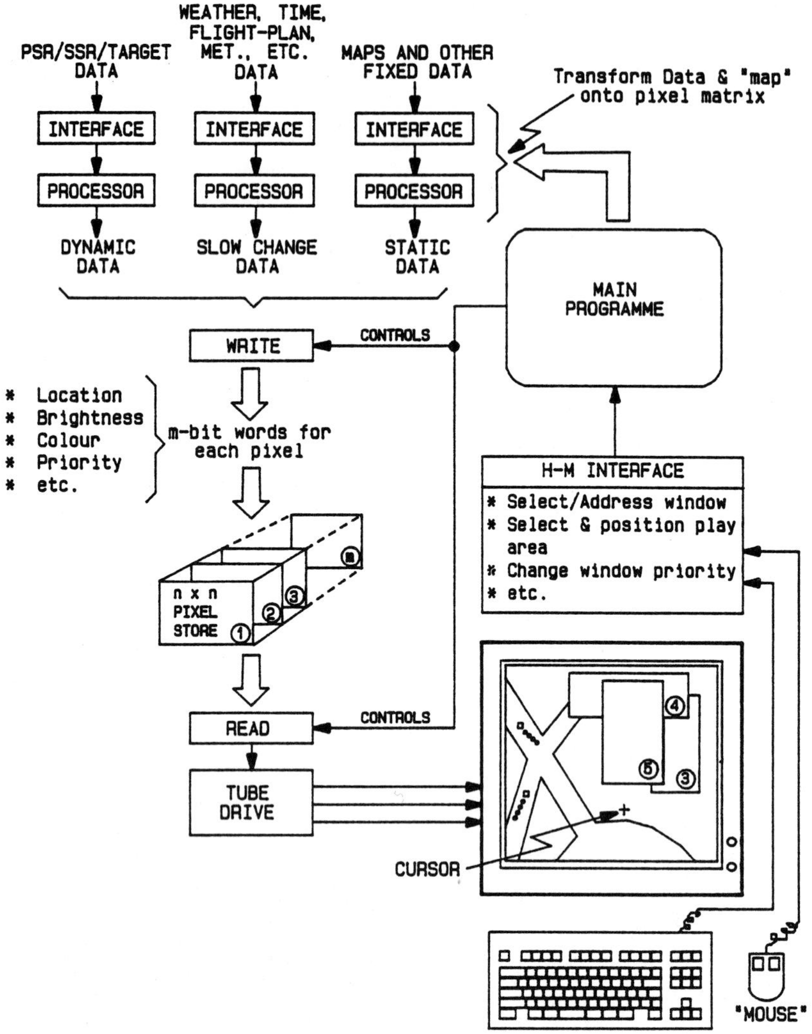

Fig. 20.3. Typical raster scan display – functional diagram. External data is translated, scaled and formatted to match the display system's pixel structure. The operator can prioritise position and enter data by use of a cursor, keyboard, mouse or other HMI devices.

(MMI) – or, in these hypersensitive days, human-machine interface (HMI) – tools to move the cursor about the screen. In the example shown, the HMI device is a 'mouse' – so called because the first examples of this were indeed mouse-like in size and shape. As the mouse is physically moved over a surface (usually a flat plate) the cursor on the screen moves in sympathy; the flat surface becomes a model of the screen area – mouse up, cursor up, mouse left, cursor left.

Data are normally classified into 'static' and 'dynamic' groups. The static group includes items such as various maps, whose content would seldom change. The dynamic group would be target data such as plot position, and any other associated information liable to change (e.g., height of an aircraft). Weather data also come in this category although their rate of change will be slower than those for the relatively fast moving targets. These various sources of data are input to processors which effectively scale in digital fashion the overall input scene to the pixel playing area. Suppose it was a 1000 line system. They are then put to a 1000 × 1000 bit store at the correct location. That is, each input source is 'mapped' from the real world onto the pixel matrix and thus each of the 1 million elements of the input has a pixel location number. The digitised input for each pixel is expressed typically as an 8-bit word. In some systems this is extended to 12 or even 16 to permit higher data capacity and picture refresh rates. These 8-bit words describe, for each data source, such things as the hue, brightness and priority required by the source to excite each pixel making up its complete picture. Take, as an example, the aircraft in the top quadrant of fig. 20.3. This is dynamic radar data. The radar input to the store would, in effect, say: '. . . make pixels number 75 to 87 in line 351 yellow and give them priority number two'. This would represent the top of the square symbol describing the plot (e.g., 'primary radar only'). The rest of the square

Fig. 20.4. A mobile small-scale raster scan work station. Clusters of displays can be built into complex work stations and operated from one set of HMI devices.

symbol would be drawn by similar digital words in the store. By giving the plot symbol 'priority level two' and the map data priority level one, we ensure that the symbol will be visible in pixels containing parts of both; the higher the number the higher the priority.

In fig. 20.3 various 'windows' are shown. These have been set up by the operator to show different types of available input data. As illustrated, the data in window 5 have been given highest priority and thus will be entirely visible. This has been achieved by the 'priority' bits in the digital words stored for all pixels in this window. Although data from other sources are available for display in this area, those with the highest priority 'over-write' the others until the operator chooses otherwise. If he reversed the priorities 4 and 5, the horizontal oblong of 4 would pop to the top and window 5 would be sandwiched between it and window 3. If target data were required in that area the operator would be able either to move the windows, or temporarily suppress them by use of the HMI.

The picture is made by scanning the store of 1000 × 1000 locations, i.e., pixel by pixel, as in normal TV practice. The digital words are converted to analogue form by the picture drive circuits, stimulating the tube electron guns in accordance with the data stored for each pixel in turn.

The flexibility of the system will be obvious. The example used above is in the context of air traffic control. Little imagination is needed to translate its use in other areas such as transport control systems, or emergency services (fire, ambulance, recovery etc.). All of these need representation of the position of vehicles, their identity, means of communication, routes that are accessible for their movement, mission status, journey times, etc. The raster scan display technique is ideal for such systems.

References

There are numerous standard works on radar theory. In the author's opinion, the most accessible to users is that in reference 1 below, itself a source of deeper reference. Another book with a wealth of easily understood descriptions of techniques, electronic devices and their working is the *Electronic Engineer's Reference Book* (reference 4 below). Those users wishing to broaden their knowledge of such things are recommended to consult it.

1. Skolnik M.I. (1962) *Introduction to Radar Systems.* International Student Edition. McGraw-Hill, New York.
2. 'Atlas of radio wave propagation curves for frequencies between 30 and 10,000MHz, (1955). Japan: Prepared by the Radio Research Labs, Ministry of Postal Services.
3. Bean, B.R., Cahoon, B.A., Samson, C.A., Thayer, G.O. (1966) *ESSA Monograph 1.* Washington: U.S. Government Printing Office.
4. Turner, L.W. *ed.* (1976) *Electronic Engineer's Reference Book.* 4th edn. 5th edn ed. by F.F. Mazda. Newnes-Butterworth, London.
5. Ridenour, L.N. *ed.* (1947) *Radar Systems Engineering.* Radiation Laboratory Series. McGraw-Hill, New York and London.
6. Skolnik, M.I. *ed.* (1970) *The Radar Handbook.* Naval Research Laboratory. McGraw-Hill, New York.
7. Hall, W.M. (1956) 'Prediction of pulse radar performance'. *I.R.E. Journal* **44**, 224.
8. Skolnik, M.I. (1962) *Introduction to Radar Systems.* McGraw-Hill.
9. Cole, H.W. 'The choice of technology in atc radars'. *The Controller* 3/1982, 3/1983, 4/1983.
10. Annex 10 to the 'Convention on international civil aviation' (1968) *International Standards and Recommended Practice.* Aeronautical Communications. London: H.M.S.O.
11. Blake, L.V. (1961) 'Recent advances in basic radar range calculation technique' *I.R.E. Transactions MIL* **5**.
12. Cole, H.W. (1967) 'SECAR – A modern SSR ground interrogator and decoding equipment'. *Journal of I.E.R.E.* **33** (1).

13. Marchand, N. (1974) 'Evaluation of lateral displacement of SLS antennas'. *I.E.E.E. Transactions on Antennas and Propagation* **22** (4).
14. Cole, H.W. (1980) 'The suppression of reflected interrogation in secondary surveillance radar'. *Marconi Review* 2nd Quarter. Marconi Research Laboratories, Chelmsford.
15. Lord Rothschild (1978) 'Risk' The Richard Dimbleby Lecture 1978. British Broadcasting Corporation, London.
16. Wyndham, B.A. (1978) *Reflection Suppression in SSR.* London: H.M.S.O.
17. Cole, H.W. (September 1980) 'The future for SSR'. *ICAO Bulletin.*
18. Ullyatt, C. (1969) 'Sensors for the ATC environment with special reference to SSR'. U.K. symposium on electronics for civil aviation paper no. 1, cat. 1, section C. I.E.E., Savoy Place, London.
19. Kirkness, R.H. (1963) 'A discussion of SSR systems modified to facilitate ground to air data transmission'. Proceedings of conference on electronics research and development for civil aviation. London: I.E.E.
20. I.C.A.O. Circular reference 174-AN/110 (1983) 'Secondary surveillance radar mode 'S' advisory circular'. Montreal: International Civil Aviation Organisation.
21. Shinn, Dr D. (1964) 'The Effect of Atmospheric Refraction on Height Finding'. *Marconi Review* vol. 27, no. 155.
22. Blackman, Samuel S. (1986) *Multiple-Target Tracking With Radar Applications.* ARTECH HOUSE CO.
23. Steinberg, Bernard D. FIEEE (Dec. 1988) 'Microwave Imaging of Aircraft', *Proceedings of the IEEE* **76** (12).
24. Mazda, F. (1983) *Electronics Engineer's Reference Book*, 5th edn. Chapter 11. 'The Ionosphere and the Troposphere'. Butterworth & Co, London.
25. Scanlan, M.J.B. (Ed.) (1987) *Modern Radar Techniques.* Chapter 5. 'Over-the-Horizon Radar'. BSP Professional Books, Oxford.
26. Kendall, Brian (1992) *Manual of Avionics*, 3rd edn. Blackwell Scientific Publications, Oxford.

Appendix

Showing: (a) That the phase relationship between a transmitted pulse and consequent received signal is preserved when translated to a lower frequency by means of a stable local oscillator (stalo)

and (b) That the output of the mti phase sensitive detector (p.s.d.), as an additive mixer, has a Cosinusoidal form.

(a) Let the transmitted pulse be expressed as:

$$A_0 \sin (w_1 t + \alpha)$$

where α represents the phase angle for any particular pulse,
and $w = 2\pi \times$ frequency of r.f. energy in the pulse.

A returned signal will be expressed as:

$$A_1 \sin (w_1 t + \alpha + \beta)$$

in which β is the phase change due to path length (the factor $n(\lambda/2) + \delta(\lambda/2)$ in chapter 9, equation 9.1).

If the stalo output is $A_2 \sin w_2 t$, multiplicative mixing with the signal gives:

$$A_1 A_2 \sin(w_1 t+\alpha+\beta) \sin w_2 t = \frac{A_1 A_2}{2} [\cos \{(w_1 - w_2)t+\alpha+\beta)\} - \cos \{(w_1+w_2)t+\alpha+\beta\}] \quad \text{(A1)}$$

The difference frequency is passed on by the i.f. circuits to form:

$$A_3 \cos [(w_1 - w_2)t+\alpha+\beta] \quad \text{(A2)}$$

Mixing a sample of the transmitter pulse with the stalo output gives:

$$A_0 A_2 \sin (w_1 t+\alpha) \sin w_2 t$$

$$= \frac{A_0 A_2}{2} [\cos \{(w_1 - w_2)t+\alpha\} - \cos \{(w_1+w_2)t+\alpha\}] \quad \text{(A3)}$$

This sample forms the coherent oscillator's phase-locking pulse of the form:

$$A_4 \cos [(w_1 - w_2)t + \alpha] \qquad \text{(A4)}$$

In the 'coho-stalo' mti system the coho oscillation is preserved in this phase until the next repetition period and so the locking pulse can be said to establish the coherent reference signal applied to one arm of a phase detector. The other arm accepts the i.f. signal of equation A2.

Note that if A_3 and A_4 are made equal, the difference between equations A2 and A4 is β, the wanted original phase difference between transmitted and received signals.

(b) Inputs to the phase sensitive detector are as follows:

$$\text{Signal} = A_3 \cos [(w_1 - w_2)\, t + \alpha + \beta] \qquad \text{(A5)}$$
$$\text{Coho} = A_4 \cos [(w_1 - w_2)\, t + \alpha] \qquad \text{(A6)}$$

The p.s.d. is an additive mixer and its output is given by A5 plus A6, i.e.

$$A_3 \cos [(w_1 - w_2)t + \alpha + \beta] + A_4 \cos [(w_1 - w_2)t + \alpha] \qquad \text{(A7)}$$

If we call $(w_1 - w_2)$ the i.f., w_3 we have:

$$\text{Output} = A_3 \cos [w_3 t + \alpha + \beta] + A_4 \cos [w_3 t + \alpha] \qquad \text{(A8)}$$

Thus

$$A_3.A_4.2\cos \left(\frac{[(w_3 t + \alpha + \beta) + (w_3 t + \alpha)]}{2}\right) \cos \left(\frac{[(w_3 t + \alpha + \beta) - (w_3 t + \alpha)]}{2}\right)$$
$$= A_3.A_4.2\cos \left[(w_3 t + \alpha) + \frac{\beta}{2}\right] . \cos \frac{\beta}{2} \qquad \text{(A9)}$$

The term $\cos (w_3 t + \alpha + \beta/2)$ is filtered from the output circuit of the p.s.d., leaving $A_3.A_4.2\cos \beta/2$. The term $[A_3.A_4.2]$ will be operated upon by various amplifications and attenuations to form a resultant $\bar{A}$. Thus the phase detector output has the form:

$$\bar{A} = K \cos \frac{\beta}{2} \qquad \text{(A10)}$$

Index